The Einstein Hoax

The Disastrous Intellectual War on Common Sense

by

H. E. Retic

© 1997, 2001 by H. E. Retic All rights reserved.
No part of this book may be reproduced, restored in a retrieval system, or transmitted by means, electronic, mechanical, photocopying, recording, or otherwise, without written consent from the author.

ISBN: 1-58820-235-6

The H. E. Retic Co.
West Caldwell, NJ

Initial Copyright:- July 1997
First Revision Copyright:- December 1997
Second Revision Copyright - June 2000

This book is printed on acid free paper.

1stBooks - rev. 08/01/01

Summary of "The Einstein Hoax"

Much of this book was written a long time ago when the author first became aware of the concepts of Special and General Relativity. At the time he assumed that the subject was well thought out and understood by people more knowledgeable and more intelligent than himself. His goal was to learn from them and to understand the phenomena that was being described. He could not, however, accept the prevalent idea that the workings of Nature were beyond the ability of mere mortals to understand at the "common sense" level. As he studied the subject, he learned that this dictum was an intellectual "con game" by men who were acting in the manner of a priesthood defending the "true faith" rather than scientists and who, when challenged on points which a bright physics or engineering student would easily understand, showed a rather limited ability to reason and covered their limitations by asserting that it was the challenger's limitations which prevented understanding.

The dictum is not true, the mechanisms by which Nature operates are, for the most part, easily understood by reasonably bright (and interested) high school students, and the necessary mathematics and arguments to evaluate their validity are within knowledge of sophomore students of engineering and/or physics. (A knowledge of Integral Calculus, Elementary Physics, and Dimensional Analysis is all that is required for reader to check the material presented for himself.) It is the purpose of this book to make these mechanisms understandable to the general reader. This book contains three sections. The main section, "The Einstein Hoax", starts with a description of the early history of Special and General Relativity and proceeds to show that the earlier published Lorentz Transformation-Aether Theory is, except for philosophical interpretation, identical to Einstein's Special Theory of Relativity and, in fact, is a special case solution of that theory. As such, the Aether Theory must be a legitimate option if Special Relativity is to be considered valid. The Aether Theory, however, is easily understood in terms of "common sense" while an examination of the conclusions of

Special Relativity yield logical absurdities which no rational mind should be willing to accept.

But, fair is fair. The proof of a theory or its interpretation is, and should be, in the experimental evidence. Fortunately, experiments in the field of Quantum Physics have demonstrated that photons emitted as pairs are coupled by their "quantum numbers" (in this case their direction of polarization) and changing the plane of polarization of one of the photons changes the plane of polarization of the other. The interesting conclusion of these experiments is that the velocity of that coupling is at least four times the velocity of light and may even be infinite. A minor modification of the experimental setup would allow the effect to be used to determine the absolute velocity of the laboratory through space to an accuracy of better than 400 miles per second. Since the interpretation associated with Special Relativity cannot tolerate the idea of an absolute velocity, we must revert to the Lorentz Transformation-Aether Theory because Dr. Einstein's interpretation of Space-Time falls on its face.

General Relativity is based upon the Principles of Relativity and of Equivalence, but a glance at its conclusions (the gravitational equivalents of the Lorentz Transformations for velocity effects), shows that it erroneously yields results inconsistent with those principles. It yields a gravitational transformation for time which is not multiplicatively commutative (the property which allows Special Relativity to work) and yields a transformation for length of unity (the equivalent Lorentz Transformation for Length is the reciprocal of the Lorentz Transformation for Time). These deficiencies, which resulted from an error in mathematics of a type which would not be tolerated if made by a freshman studying Calculus, led Dr. Einstein on a merry chase for about 18 months until he adapted the incomplete geometry of Riemann and concluded that space was curved. Unfortunately the resultant theory was relativistic and yielded results which appeared to work because the Sun's gravitational field is so weak (about a million times weaker than the field around a neutron star) and Relativistic Science was off and running, foolishly bringing us "Black Holes", "Singularities", "Wormholes", etc. At the same time, while claiming to explain gravitation, General Relativity failed to explain the most significant gravitational effect, the source of gravitational energy and the force which holds you to your chair.

When one applies a few "thought experiments" using readily tested and understood principles to the phenomena of gravitation, it is easy to derive gravitational transformations which actually are consistent with the Principles of Equivalence and Relativity. The result is surprisingly rich, a simple application (done after they had been derived and not before) of these transformations reveals the source of gravitational energy and its consistency with the absolute validity of the Law of Conservation of Energy in terms consistent with "flat space", the reason why gravitational collapse is limited, and the inevitability of a Universe similar to our own being created within every object which approaches the 'black hole" radius. From inside of that object, it would appear, late in its history, to be an almost empty Universe expanding from a single point which "inflated" into existence. (The author was completely surprised by these results.)

For those who wish to dig deeper, the author provides a rigorous derivation of the nature of the gravitational field in the appendix "Gravity" (copyrighted 1987). Unlike the "The Einstein Hoax", it does not require the reader to take the author's word for anything. Every step is clearly spelled out both in words and in simple equations so that the reader can verify the conclusions for himself. Also included is a second appendix entitled "Corrections to Residual Errors in Special Relativity". This appendix derives the Lorentz Transformation for Transverse Force correctly, defines mass correctly so that Special Relativity can be used for accelerated reference frames, and shows the location of kinetic energy, which in turn, reveals the nature of inertial mass.

I hope that you will read this book with an open mind and judge for yourself the validity of the conclusions. I do promise that it is closer to reality than current concepts and, if any reader can find a significant flaw, it will either be corrected or the book will be withdrawn. **It is time to end the "Einstein Hoax".**

Table of Contents - "The Einstein Hoax"

Chapter	Title	Page
1	Introduction	1
2	Historical Background	7
3	The Nature of the Einstein Hoax	17
4	Does the Aether Exist?	21
5	The Resurrection of Absolute Velocity By Quantum Experiments	41
6	The Nature of Reality	51
7	Applying the Lorentz Transformations Properly	77
8	Generating the Gravity Transformations	87
9	Dr. Einstein's Error and the Introduction of Curved Space	109
10	Gravitational Contraction and Collapse	121
11	Gravitational Collapse and the Creation of a Universe	147

Index - "The Einstein Hoax" - Continued

Chapter	Title	Page
12	The Space Time Continuum	157
13	The Nature of Particles	167
14	Adding the Quantum Effects to Our Understanding	185
15	Changing the Paradigms	203
16	What Can We Conclude?	219

Appendix 1 - "Gravity"
(Copyright 1987) 225

Appendix 2 - "Corrections of Residual Errors in
Special Relativity 399

Chapter 1 - Introduction

1.1- As a young man, the author had sufficient arrogance to believe that, while a lack of time prevented an individual from understanding all of Nature, there was no aspect of Nature that could not be understood at the intuitive level by a reasonably intelligent and adequately motivated individual. As a result the statements which appeared in the texts of the time that the effects defined by the Special and General Theories of Relativity were beyond such an understanding and could only be treated by mathematical manipulations were a challenge. Inherently, mathematics is a science of 'how much' and not of 'how come', and it was the 'how come' that the author wished to understand. To meet that challenge, the author proceeded to study the concepts involved with the good faith belief that they had been well thought out and well verified by men far better trained and wiser than himself. However, the deeper the author probed, the more disillusioned he became. It became more and more apparent that the effects represented were quite easily understood at the intuitive, or common sense, level and that the reason that they appeared mysterious was that those who purported to be experts did not actually understand the subject matter and that their assertions of its incomprehensibility were rationalizations to cover their own limitations. It turns out that the subject matter is readily understood by anyone with a good ability to visualize physical reality particularly if he is familiar with Physics and Calculus at the college freshman level.

1.2- The author's initial confusion resulted from the fact that, while Special Relativity was presented as the epitome of physical wisdom, initially it was impossible for him to find a meaningful distinction between it and the Lorentz Transformation-Aether Theory which had preceded it by two years. It finally dawned on the author that the Special Theory of Relativity was actually the Lorentz Transformation-Aether Theory without the constraint imposed by the requirement of an absolute velocity reference (the Aether). Then, contrary to the rules of evidence which would be employed in a court of law, the academic community forced the acceptance of the idea that, since both theories had demonstrated that our absolute velocity through space could not be observed, the Aether had no significance and **was not to be used** as the basis for a physical theory. This position was taken even though Dr. Einstein had maintained a belief in absolute time (equivalent to a belief in the existence of the Aether) for about 25 years after Special Relativity had been published. He had also warned that the non-existence of the Aether had not been proven, what had been proven was that its use was not necessary in mathematical analyses of physical processes.

1.3- Both Special Relativity and the Lorentz Transformation-Aether Theory demonstrated that the observed velocity of light was independent of the velocity of its source. Therein lies the rub. Such a result is obvious if light is a wave propagating through a medium (the Aether). By abolishing the Aether, Special Relativity lost the velocity reference the Aether represented and asserted instead that light propagated as ballistic particles (photons) through empty space. Advocates of Special Relativity provided no explanation as to how Nature performed such a remarkable feat of speed control without using the Aether as a reference and instead, they strongly asserted that any doubts a questioner had resulted from his limited intellectual capacity rather than from a legitimate concern. That attitude has run into some trouble in recent years. Observations of the radiation background of space have shown that the Earth has a velocity of 300 kilometers per second with respect to that background, observations by the COBE project have found the initial point from which the Universe started to expand, and experiments by quantum physicists have demonstrated that our absolute velocity through space can be measured in the laboratory. As we shall see, these observations make an overwhelmingly strong argument for the validity of the Lorentz Transformation-Aether Theory and the artificiality of the Special Theory of Relativity.

1.4- The author's disillusionment became deeper when he began to study General Relativity. That theory supposedly explained gravity as a phenomena resulting from a curving of space caused by the presence of matter. However, General Relativity blithely fails to discuss the most significant characteristic of gravity, the force which presses you to your chair. Compared to the need to explain that force and the energy it represents, the corrections General Relativity provides to the Newtonian orbits of planets, the path of starlight, or the rate of passage of time are rather trivial. Incredibly, not only does General Relativity fail to explain the source of that force (and the energy it implies), many texts on the subject actually deny that the force exists. When the derivation of General Relativity is examined carefully, it is found to contain a fundamental error of a type which would not have been tolerated if it had been made by a college freshman who was studying Calculus. Even with that error, the truth of the basic premises of General Relativity (the Principles of Relativity and of Equivalence) insured the error would not be revealed by observations made in the weak gravitational field of the Sun or by observations of a distant binary star system. In order to partially compensate for the effects of his mathematical error, Dr. Einstein introduced the artificiality of curved space. This concept has had the unfortunate effect of leading a large number of highly trained astronomers and cosmologists down the garden path and led to such absurdities as Black Holes, Wormholes, and Singularities. Dr. Einstein may have recognized the existence of a defect(s) in General Relativity since he is reported to have been uneasy about its extension into regions of intense gravitational fields.

1.5- The author's disillusionment with the job performance of those from whom he had hoped to learn led him to study the subject matter by starting from basics using an approach which was suitable for the analysis of relativistic phenomena (i.e.- velocity effects and gravity effects). This capability is not possessed by the Tensor Calculus commonly used by physicists and which allows them to by-pass the need to understand the phenomenon they are investigating. The author recognized, as does a surveyor who routinely corrects his observations for the effects of ambient temperature on the length of his measuring tape, that observations made between reference frames differing in velocity and/or elevation require a correction for the effects of that difference on the size of the units of measurement he employs. Only after appropriate corrections have been made can valid conclusions be drawn as to what actually occurs between different velocity and/or elevation reference frames. The required technique for studying relativistic effects is Dimensional Analysis which was developed in the 19th century to facilitate experiments in hydraulic engineering. The Lorentz Transformations of both the Lorentz Transformation-Aether Theory and Special Relativity provide the information required to allow the use of Dimensional Analysis for a rigorous analysis of the effects of velocity. To examine the gravitational field, it was necessary for the author to devise a means of deriving gravitational equivalents of the Lorentz Transformations that did not include the loop of circular reasoning that caused General Relativity to be defective.

1.6- The real test of an expert's knowledge is his ability to make his subject matter intuitively understandable to an intelligent layman. If he cannot do so, there is only one possible reason. Regardless of his credentials and his acceptance by his peers, the reason is that he doesn't actually understand his subject matter. In the remainder of this text, the writer hopes to be able to pass that test of understandability and provide the reader with a useful insight into the nature of space, time, matter, gravitation, and cosmology in a manner which is consistent with the physical laws taught in undergraduate level physics. The author's tools are his simple minded belief that there is only one reality and everything we accept as true about that reality must be consistent with everything else we accept as true and his simple minded belief that Nature is constructed in the most straightforward manner possible. With that in mind, the author hopes that what follows will be both instructive and interesting to the reader and capable of raising the blood pressure of the academic community. Enjoy!

1.7- (Note: A rigorous derivation of the gravitational field and its effects on cosmology is provided in the author's text "Gravity" copyrighted in 1987 and which is included as an appendix to this book. This text was sent. at the time, to individuals identified as having a reputation in the field. Since then, the author has read some of the conclusions presented in "Gravity" in books subsequently written by a few of those individuals. Due to the controversial nature of this text, unless required by the subject matter, the names of individuals and publications have been omitted in order to prevent their possible embarrassment.)

Chapter 2 - Historical Background

2.1- By the last quarter of the 19th century, the Science of Physics was considered to be nearly complete. The electromagnetic equations of James Clark Maxwell had explained electromagnetic radiation and light was considered to be a vibrational wave propagating through a medium called the Aether in a manner similar to the propagation of sound through air. Using Maxwell's Electromagnetic Equations, J. J. Thomson derived the relationship between mass and energy, $E=M*C^2$, in 1888 when the alleged source of that relationship (Dr. Einstein) was still in knee pants. (The author has since received an E-mail which asserts that a Mr. Olinto D. Pretto of Italy published this relationship in 1903. This really doesn't matter too much, what is clear is that Dr. Einstein was not the original source of the relationship for which he was credited.) A difficulty which remained was that light was known to be a shear vibration acting in a plane perpendicular to the direction of propagation rather than a compressional vibration acting in the direction of propagation. Since shear vibrations cannot propagate through a fluid, it was recognized that the Aether must be solid. This conclusion raised the interesting question of how material particles could move through a solid without resistance. Even so, the concept of the Aether was so persuasive that the next logical step was an attempt to measure the effects of changes in the velocity of the Earth as it traveled through the Aether in its orbit around the Sun.

2.2- The most significant of the experiments was conducted by the the team of Michelson and Morley. They devised an experiment using optical interferometry which attempted to measure the difference in the velocity of propagation of light between two mutually perpendicular directions. To everyone's chagrin, the experiment produced a null result! No interference effects were observed as the Earth changed its orbital velocity through the hypothetical Aether by 36 miles per second over the course of a year even though the precision, accuracy, and stability of the experimental setup was more than adequate to reveal the anticipated effects.

2.3- Initial attempts at explaining the null result of the Michelson-Morley Experiment produced unsuccessful concepts such as the Aether Drift Theory in which the Aether was presumed to be carried along with the Earth, but by 1903 the Lorentz Contraction-Aether Relativity Theory was published. The key to this theory was the Fitzgerald Contraction which asserted that the length of material objects, in the direction of motion, was reduced as a function of the velocity of the object through the Aether in proportion to $(1-V^2/C^2)^{0.5}$ but were unaffected in directions perpendicular to that velocity. It was immediately recognized by Larmor that the Fitzgerald Contraction required an equivalent slowing in the rate of passage of time. Since, by that time Lorentz had used the known equivalence between mass and energy to provide the effects of velocity on mass, $1/(1-V^2/C^2)^{0.5}$, transformations involving the expression $(1-V^2/C^2)^{0.5}$ became known as Lorentz Transformations. Collectively, these transformations became known as the Lorentz Transformation-Aether Theory. A more meaningful name would seem to be the Aether Relativity Theory, and it will be referred to by this name where necessary to distinguish it from the Special Theory of Relativity. Under this theory, velocity through the Aether caused measuring instruments to change their calibrations in obedience to the Lorentz Transformations. Those changes in calibration were of exactly the amount required to insure that, in conjunction with the finite velocity of light, it was impossible to observe effects produced by our velocity through space (the Aether).

- **Redefinition:-** *Since the Lorentz Transformation, $(1-V^2/C^2)^{0.5}$, appears many times in the material which follows follows, the symbol 'B_V' will be substituted. Thus: $B_V=(1-V^2/C^2)^{0.5}$*

2.4- Since everything we experience, including the physiological sensations and behavior of our bodies, is the result of a measurement of some type, our absolute velocity with respect to space could never be observed. No matter what one's absolute velocity was, he could always assume himself to be at rest with respect to space and that everything that was not at rest with respect to him was moving through the Aether. With the effects on observations imposed by the Aether Relativity Theory, measurement of an observer's velocity with respect to the Aether was prevented by the fact that the finite velocity of light made it impossible to determine when two physically separated events were simultaneous. When the inability of an observer to communicate faster than the velocity of light is considered, it is simple but tedious to show, using elementary algebra, that the Aether Relativity Theory insures a null result of any attempt to determine an absolute velocity (velocity with respect to the Aether). An observer is therefore free to consider that any velocity reference frame between the limits of +/-C is valid as a base reference frame for making physical observations.

2.5- At the time, three difficulties seemed to remain with the Aether Relativity Theory. The first objection was that it did not account for the effects of velocity on electromagnetic phenomena. This objection was not a legitimate one. The three Lorentz Transformations allow the derivation of equivalent Lorentz Transformations for all physical parameters, including those of electromagnetics, by applying Dimensional Analysis to known physical equations. When these derived transformations are applied to electromagnetic phenomena, the Aether Relativity Theory is found to be valid for electromagnetic phenomena as well. The second objection was the question as to why, if the classical Aether is the absolute zero velocity reference for space itself, should Nature conspire to conceal our velocity with respect to it. That may have been a reasonable question at the time, but in the interim, quantum physicists have concluded that the forces between particles, such as between the atoms in a measuring stick, are electromagnetic in nature and are alleged to result from the exchange of virtual photons. A corollary to that conclusion is that since electromagnetic effects travel at the velocity of light, matter must adjust its parameters so that the velocity of light appears unchanged to a local observer. (It is fortunate that these adjustments occur. If they did not, travel at high velocity, such as the velocity of the Earth in its orbit or the velocity of the Sun in its galactic orbit, could be extremely hazardous to one's health.) The final objection to the Aether Relativity Theory is that if the Aether is a solid medium, as required for the propagation of light as a transverse wave, matter should not be able to travel through it without resistance. A means by which Nature may have resolved that objection is provided later.

2.6- In 1905 Dr. Einstein, apparently sensing an opportunity in the alleged failure of the Aether Relativity Theory to correctly predict the electromagnetic effects associated with velocity, published the Special Theory of Relativity. This theory was based upon Poincare's Principle of Relativity and asserted that any velocity between the limits of +/-C could be considered to be valid for use as a zero velocity reference for the purpose of physical observations. The Special Theory of Relativity provided the same transformations for mass, length, and time as did the Aether Relativity Theory published two years earlier. Under both approaches, any inconsistencies resulting from the effects of velocity on observations were concealed by the effects of the Lorentz Transformations and the fact that the finite velocity of light made the absolute synchronization of physically separated clocks impossible. As with the Aether Relativity Theory, regardless of one's velocity through space, he was free to consider himself at rest and apply the Lorentz Transformations to observations made in systems which were moving with respect to himself. It must be pointed out that Special Relativity did not have the difficulties with respect to electromagnetic phenomena alleged to be a weakness of the Aether Relativity Theory for the simple reason that, instead of resolving those difficulties, it arbitrarily defined them as non-existent.

2.7- When one compares the Aether Relativity Theory and Special Relativity objectively, one finds that they are identical theories and differ only in philosophical interpretation. Under the

Aether Relativity Theory, space is filled with a medium called the Aether which acts as the framework for the Universe and our velocity through that Aether is concealed by the effects described above. Under Special Relativity, the absolute velocity reference represented by the Aether is omitted since it does not appear in the mathematics. Both theories conclude that the effects of velocity on measuring instruments (including the physiological sensors of our bodies) and the finite velocity of light make it appear to any observer that he is at rest and that everything having a velocity relative to him is in motion. It should be obvious to all that the Aether Relativity Theory is a special case solution of the Special Theory of Relativity in which one of the infinite number of zero velocity references frames considered to be valid under Special Relativity is the correct one even though one cannot determine his velocity with respect to it. Special Relativity takes the position that, since our velocity with respect to an absolute spatial reference cannot be determined by observation, it is meaningless to consider the existence of an absolute velocity reference as part of physical theory. As we shall see, not only can our absolute spatial velocity be measured, asserting that it can't violates a basic rule. One should be extremely careful in declaring something to be impossible. Invariably as soon as such a declaration is made, some damned fool will come along and do it.

2.8- Recognition that our absolute velocity through space cannot be measured is a far cry from a proof that an absolute velocity does not exist. If it were proven that the absolute velocity reference represented by the Aether was not valid, then it would be proven that one of the velocities that Special Relativity allows to be considered as at rest can not be used as a basis for physical experiments. Such a proof would also be a proof that Special Relativity was invalid. Apparently, Dr. Einstein thought the interpretations associated with the Aether Relativity Theory to be correct since it has been reported that he maintained a belief in absolute simultaneity between physically separated events (a belief which requires the existence of the Aether) for 25 years after the publication of Special Relativity. He also is reported to have warned that "we have not proven that the Aether doesn't exist, we have only proven that we do not need it (for computations)".

2.9- Since the Aether Relativity Theory preceded the Special Theory of Relativity by two years and was in actuality the same theory in a different form, it was necessary to make a determination between them. That became a matter of belief rather than proof and, as the multitudinous deaths in religious wars over the centuries have amply demonstrated, the more unprovable a belief is, the more savagely men will fight to defend it. Such a savagery occurred in the discussions which followed. The Aether Relativity Theory was advocated by a cadre of physical scientists whose primary reliance was on their physical insights and who used their mathematical skills to quantify the results of those insights. Special Relativity was advocated by a different cadre of physicists who had mastered mathematics well but who had found that their use of physical insights, which, like art, requires an innate aptitude in addition to training, was unreliable. Since talent is scarce in any field, the advocates of Special Relativity won the battle. The proponents of the Aether Relativity Theory were ridiculed by having the Aether compared to the Emperor's Clothes in the fable of the same name. The general public was led to believe that the mystery resulting from the null results of the Michelson-Morley Experiment was resolved by Dr. Einstein even though Fitzgerald, Larmor and Lorentz had achieved that result two years earlier.

2.10- The author is seriously troubled by the historical accounts. The fact that the knowledge and insight to resolve the dilemma represented by the Michelson-Morley Experiment had already been provided by truly intelligent men (Thompson, Lorentz, Larmor, and especially Fitzgerald), degrades Dr. Einstein's contribution in this area from a work of brilliance to the rather trivial exercise of formulating the existing knowledge into mathematical terms for easier use in computational activities. [Dr. Einstein's famous equation, $\delta S^2 = \delta X^2 + \delta Y^2 + \delta Z^2 - C * \delta T^2$, which is accepted as the most succinct means of defining the effects of velocity, follows from the fact that the Lorentz Transformations for length and time are identical to the Pythagorean Theorem. The symbol "δ" indicates that an infinitesmally snall portion of the quantity defined by the symbol which follows is being considered. Thus δL refers to an infinitesmally small length.] However, the politics of the scientific community was not served by crediting Fitzgerald with the conceptual breakthrough since his approach did not suit its goals. As a result, Dr. Einstein was given that honor and was eventually proclaimed a deity of the new religion while the true contributors were relegated to footnotes in textbooks.

2.11- The strength of the feelings involved were brought home to the writer by personal experience. In the late 1950's, assuming that Special Relativity had been proven to validly represent our reality, the author began a good faith study of the subject for his own satisfaction. It was rather upsetting to learn the information provided in the previous paragraphs. Digging deeper, the author borrowed a technique from mathematics to show that the Aether Relativity interpretation must be correct because assuming the non-existence of the Aether led to an absurdity. In due course, this material was shown to a physicist whose specialty was Special Relativity. The man's reaction was astonishing. He did not take the intellectually reasonable although undiplomatic step of telling the author that he was an ignorant fool, instead he went into a rage and accused the author of being a "dangerous heretic who must be suppressed". (It is fortunate for the author that this isn't the 16th century.) His violent emotional reaction was akin to that of the Muslim Ayatollah who allegedly condemned the author of the "Satanic Verses" to death. The reaction could only have come from an individual whose quasi-religious beliefs were threatened. They were not the reactions of a man who accepted Dr. Einstein's dictum that the search for truth must take precedence over the teachings of established authority regardless of the prestige of that authority.

2.12- In 1915, Dr. Einstein published his General Theory of Relativity. In deriving this theory he combined a new and apparently original concept, the Principle of Equivalence, with the Principle of Relativity upon which Special Relativity was based. Simply stated, the Principle of Equivalence asserts that gravitational acceleration can be considered to be equivalent to inertial acceleration. Unfortunately, Dr. Einstein failed to recognize that Tensor Calculus cannot be used to derive a relativistic theory (as discussed later) and employed that mathematical technique in the theory's derivation. Its use for such a purpose introduced a mathematical error of a type which, if persistently made by a student of Elementary Calculus, would result in a failing grade for the course. As a result of this error, the derivation of General Relativity was impossible in terms of our observable three dimensional Euclidian Space.

2.13- Instead of recognizing and correcting the source of his difficulty, Dr. Einstein took the easy way out and arbitrarily added an extra degree of freedom by asserting that space was curved by the presence of mass and was properly described by the non-Euclidian geometry of Riemann. Objectively, his approach might be compared to that of a mechanic who installs the wrong part in a machine by hammering it into place instead of obtaining the correct part. Discussions of General Relativity at the time justified its validity by two rather questionable and irresponsible arguments. The first argument was that there was no reason not to accept the idea that space was curved "since no one could prove that it wasn't" (a proof that space is flat will be described later). The second argument was that, while General Relativity taught that the gravitational field created energy from nothingness, the Law of Conservation of Energy was not violated since the energy which was created could not escape from the field. It would seem, from this reasoning, that the Law of Conservation of Energy obeyed the Eleventh Commandment, "Thou Shalt Not Get Caught". (Newtonian Gravitational Theory also asserts that the gravitational field creates energy and allows that energy to escape from the field. That theory must be forgiven for this deficiency because, in the 16th century, it had not been recognized that energy must be conserved.)

2.14- As a result of the defect in its method of derivation, the relativistic corrections to the classical Newtonian Gravitational Theory provided by General Relativity were not rigorously correct but were only approximations. At the field strength existing at the surface of the Sun, these corrections revised the predictions of Newtonian Gravitational Theory by one part in a million. Due to the weakness of the Sun's field, General Relativity was able to predict, to within the limits of experimental accuracy, the anomalous precession of Mercury's orbit, the bending of the path of a ray of light as it passed close to the Sun and the slowing of time at the surface of the Sun as evidenced by the red shift of its spectral lines. The effects caused by residual errors in General Relativity resulting from its invalid method of derivation are about a million times too small to be observed within the Solar System.

2.15- It has been asserted that observations made of the red shift of the spectral lines in the light from extremely dense and/or extremely massive stars and the observed change in the orbital period of massive binary stars due to gravitational radiation provide the necessary verification for General Relativity in strong fields. However, in order for such observations to provide that verification, they must be combined with orbital observations made by an on-site observer. Until mankind has the equivalent of Star Trek's Warp Drive, observational validation of General Relativity in strong gravitational fields would seem to be impossible. At present, all that the spectral shift of light from massive stellar objects proves is that gravity is a relativistic phenomena. It does not prove that General Relativity is the correct description of that phenomena. To be fair, it must be pointed out that Dr. Einstein may have been aware of limitations in his derivation of General Relativity since it has been reported that he was uneasy about the extension of the theory to extremely strong fields.

Chapter 3 - The Nature of The Einstein Hoax

3.1- Since the Special Theory of Relativity and the Aether Relativity Theory which predated it are actuality the same theory and may be derived one from the other, the question arises as to the nature of Dr. Einstein's contribution to the solution of problems related to velocity. In a historical text published in the 1920's it was stated that his principle contribution was the demonstration that mathematics could be used to derive physical theory and, since mathematics could be taught to anyone, Science did not need to await the contributions of the "few great minds that arise in each century" (a category which most certainly includes Fitzgerald) to achieve progress.

3.2- The insidious end result of that philosophy was verified by a telephone call received by the author from a physicist at a highly respected Ivy League University whose status was sufficient to have had his work described in Time Magazine. The author was advised that the physicist's sole job was to search for mathematical relationships which provided predictions and to devise and perform experiments which determined whether those predictions agreed with observation. It was not considered to be the physicist's job to provide an understanding of the mechanisms by which Nature achieved its results. That task was the proper province of philosophers and meta-physicists and was beneath the dignity of physicists. This viewpoint is reinforced by one of the most respected theoretical relativistic physicists in the world in a statement in one of the most prestigious scientific publications in the world. He stated that he was unconcerned as to whether a theory correspond to reality because he didn't know what reality was, he only was concerned that a theory correctly predict experimental results. To place that high sounding philosophy in perspective, both a highly talented musician and a trained circus seal can play "Yankee Doodle" on a set of tuned bicycle horns and receive the applause of an audience. While the musician would probably want money and the seal would be content with a fish, the real difference is that the musician would understand the meaning of the music while the trained seal would have learned to play the tune through repetitive actions induced by a trainer. Unlike their counterparts of a century ago, it would seem that today many mainstream physicists do not feel the need to understand the phenomena upon which they are working but are content to do their work by manipulating mathematics and experiment using procedures they have learned only by rote. It is left to the reader to draw his own conclusions as to whether such an analogy is fair.

3.3- The publication of Special Relativity provided a golden opportunity for the majority of the academic community. Under the interpretations of Nature provided by Special Relativity the door was opened for the majority of its members who were without the talent required to understand its workings. At the same time, those with the necessary talent needed to understand reality instinctively recognized that contradictions were implicit in Special Relativity (described later) and could not accept the subject as it was presented. (Teachers of Special Relativity report that a significant percentage of intelligent and mathematically skilled students cannot master the subject.) The inability to accept Special Relativity, as presented, effectively eliminates individuals with a strong sense of reality (which by another name is called common sense) from the ranks of those who acted as advisors to PhD candidates and from the roles of those who perform the peer reviews which determine what is published in scientific journals. As a result, a selection process was gradually put in place which insured that only material which did not threaten the validity of Special and General Relativity was published. Material which appeared to be a threat, no matter how powerfully presented and how intellectually and observationally valid, was effectively squelched. On the other hand, material which supported Special and General Relativity, no matter how trivial or absurd, was readily published. Once this point was reached, it was possible to make the claim that the subject matter could not be understood in terms of common sense. It could only be understood in terms of mathematics and there were a limited number of minds in the world who could truly comprehend Dr. Einstein's work.

3.4- Early civilizations were based upon the invention of agriculture and the ability to determine the proper time for planting and harvesting crops was very important. In those societies, a small group of men studied the heavens and learned how to divine the seasons from the positions of the Sun, Moon, Planets, and Stars. Instead of passing along their knowledge, they kept it to themselves and became priests who provided life and death information for society as a whole. As a result of their monopoly of vital knowledge, more and more power and wealth flowed to them and in time they formed a religion. That religion eventually became the basis of all powerful states ruled by god-kings. In such a society, heresy was the most heinous crime imaginable, with revelation of the secrets of the religion to the masses a close second. From the vantage point of history, the motivation of these priests was obvious. They worked to achieve enormous power and luxury for themselves at the expense of the peasants. They did not work for the benefit of society as a whole.

3.5- The establishment of the relativistic effects as a mystery which could not be understood in terms of common sense placed the community of physicists into a position similar to that of those ancient priests. They possessed knowledge which could only be understood by those individuals who possessed the appropriate 'yup' in the form of a PhD in Physics. Naturally, no one whose innate sense of reality caused him to question the conclusions of Relativity ever received such a degree. (When the writer was interviewed for his first job, he was asked what kind of 'yup' he had. Puzzled, he asked for an explanation and was told that, when you spent the money it costs to go to college, you were not buying knowledge, you were buying a 'yup'. That 'yup' is required so that when you seek a job and the interviewer asks if you went to college, you can answer 'yup'. Of course, 'yup's from different schools are rated differently, but the prime purpose of schooling is to acquire that all important 'yup', it is not the acquisition of knowledge.)

3.6- The Einstein Hoax consists of maintaining the quasi-religious belief that the phenomena associated with velocity and gravitation cannot be understood by ordinary men using their common sense. It can only be understood in terms of mathematics performed by initiates who possessed the prerequisite 'yup's. Whether it is recognized or not, all of the essentials of a religion are present. There is a deity in the form of Dr. Einstein, who, like most of the men who have had that role thrust on them over the centuries, probably did not seek or even relish it. It has an established but unproven set of truths which were revealed by that deity. Finally, it is protected by selected defenders of the faith who, in this case, act through the peer review process to insure that heresy in any form is never published. The motive for the maintaining of the Einstein Hoax seems rather obvious, it's money. Society expends a large sums supporting this priesthood through tuitions paid by parents and grants by governments and industry. The donors believe they are paying for the teaching of the young, however, that teaching is mostly done by graduate students who are seeking their own 'yup's. The established possessors of the necessary 'yup's spend most of their time in research because, not only is that activity more interesting, it serves to advance their tenure protected careers. Should Special and/or General Relativity be shown to be fundamentally flawed, the careers of Relativists, most Cosmologists, and those working on Quantum Gravity and/or Unified Field Theory will have been wasted.

3.7- In the material which follows, the author will attempt to provide and justify the heresy required to remake Relativity Theory into a subject which both can be understood at the undergraduate or even high school level in terms of common sense and which will eliminate glaring defects in our current understanding of the fundamental principles of Nature and of Cosmology.

Chapter 4 - Does The Aether Exist?

4.1- Since the conclusion that the Aether does not exist is based upon unproven assertions on the part of key members of the community of physical scientists rather than upon evidence which would be admissible in a court of law, the subject must be examined with extreme care to guard against the possibility that evidence which would support its existence has not been knowingly or unknowingly suppressed. Remember, there are strong political reasons which act to bias the judgment of the academic community against the concept of the Aether. Determining whether the Aether exists requires asking questions which are readily answered by one interpretation, and which seem unanswerable by the other. Such questions encounter strong resistance when asked. In addition to those questions, experiments performed by quantum physicists have demonstrated the feasibility both of measuring the absolute velocity of an experimental setup through space and of communicating at velocities greater than the velocity of light. Achieving these results requires only minor modifications to their experimental arrangements.

4.2- How Does Light "Know" How Fast to Travel?:- This question arises from the fact that the velocity of light is independent of the velocity of its source. Under the interpretation of reality provided by Special Relativity, light is considered to consist of particles called photons which travel ballistically through empty space. Under the interpretation provided by the Aether Relativity Theory, light consists of packets (photons) of electromagnetic vibration transmitted through a medium called the Aether. Since one would expect the velocity of ballistic particles to be affected by the velocity of their source, Special Relativity would seem to be incapable of dealing with this question. The Aether Relativity Theory, on the other hand, has no difficulty. The velocity of propagation of a vibration in a medium is determined by the properties of the medium and is independent of the velocity of its source.

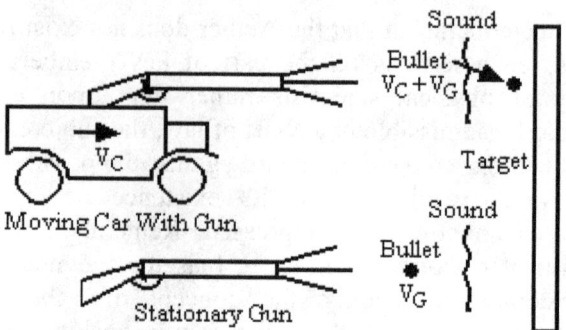

Stationary and Moving Guns Are Fired Simultaneously
Bullet from car arrives at target before bullet from stationary gun. Sounds of both gunshots arrive simultaneously.

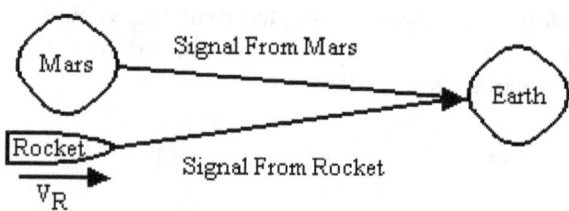

Rocket passes Mars at velocity V_R on way to Earth. Signals are sent from Mars to Earth and from rocket to Earth at instant that rocket passes Mars. Signals travel together and arrive at Earth at same time.

Figure 4.1 - The Propagation of Electromagnetic Radiation

4.3- Consider a car traveling down a road towards a target (Figure 4.1). On board the car is a man with a rifle. At the side of the road is another man with an identical rifle. At the instant that the man in the car passes the man at the side of the road, they both fire at the target. As expected, the velocity of the bullet fired from the car is increased by the car's velocity and it reaches the target before the bullet fired from the side of the road. The sounds of the shots travel together at the velocity of sound in air and reach the target at the same time. The bullets are material particles projected to the target and travel at different velocities. The sounds of the shots are vibrations traveling through a medium and travel at the same velocity. Consider next an analogous experiment (physically realizable) in which a rocket is passing Mars on its way to Earth. At the instant that the rocket passes Mars, a radio on the rocket and a radio on Mars send signals to the Earth. Along the whole path traveled by those signals, they remain side by side and they arrive at the Earth simultaneously. They travel together despite the fact that they were transmitted from sources having a velocity difference which could not be compensated at the transmitters since each transmitter was ignorant of the velocity of the other. Just as the simultaneous arrival of the bullets at the target in the preceding example would cause a reasonable man to suspect some form of chicanery, would not the same suspicions be aroused by the assertion of Special Relativity that photons travel ballistically through empty space at a velocity which is independent of the velocity of their source? It is difficult to envision a means by which the independence of their velocities from the velocities of their sources could occur unless photons were wavelike disturbances propagating through a medium. To date no explanations, other than the reliance on some form of magic, have been provided by the proponents of Special Relativity. Instead, they have shouted down the question whenever it was raised.

4.4- Why Does Light Travel at the Velocity of C?:- Current orthodoxy asserts that the Aether is not required to explain the propagation of light at its velocity of C. All that is required is the magnetic permeability, μ, and the dielectric constant, ε, of space. The velocity of light is then determined by the expression $C=(\mu*\varepsilon)^{0.5}$. As an analogy, if one strikes the end of a steel rod with a hammer, the sound of the blow propagates along the rod at a velocity, V, determined by the elasticity, e, and the density, d, of the rod in accordance with the expression $V=1/(e*d)^{0.5}$. If one accepts the above explanation for the velocity of light, consistency would require that he be willing to accept the conclusion that, since the propagation of sound at its velocity of V requires only the elasticity and density of the rod, the rod itself may be removed and only its elasticity and density retained to explain the propagation of the sound of the blow at the velocity V. While such a conclusion is obviously silly with respect to the rod, somehow it does not seem silly to Relativists when it is applied to Special Relativity's interpretation of the propagation of light.

4.5- How Does the Speed of a Clock After a Change in Velocity Compare with its Speed Before the Velocity Change:-

Consider, if your will, the following physically realizable experiment performed in compliance with the mathematical predictions of Special Relativity Theory and/or the Aether Relativity Theory. There are two locations, perhaps the Earth (reference frame A) and Mars (reference frame B) as shown in Figure 4.2, which are traveling at a significant velocity, V, with respect to each other. Observers at each location measure the velocity of the other location using Doppler radar as +V and -V respectively. There are identical clocks at each location which have had their calibrations compared by radio signals. Because of the finite velocity of light, the relative velocity between the observers causes each of them to observe that the clock at the other location is running more slowly than his clock. A rocket ship is at rest on the Earth and contains an observer and a third identical clock. Both the observer on the Earth and the observer in the rocket ship find that their clocks are running at the same speed and that the clock on Mars is running more slowly. The observer on Mars observes that both the clock on the Earth and the clock in the rocket are running more slowly than his. The rocket then takes off and lands on Mars. The observers on the Earth, on Mars, and on the rocket measure that the velocity of the rocket has changed by +V. The observer on the Earth measures that the clock on the rocket has slowed and it now runs at the same speed as the clock on Mars. The observer on Mars measures that the clock on the rocket has sped up and now runs at the same speed as his clock. The observer on the rocket observes no change of the speed of his clock, but, observing that his velocity has changed by V, concludes that its speed actually did change as a result of his change in velocity and concludes that the change was concealed from him by the effects of the Lorentz Transformation for Time.

4.6- With respect to the speed of the clock, the observer on the Earth asserts that the speed of the clock on the rocket slowed, the observer on Mars asserts that the speed of the clock on the rocket increased, and the observer on the rocket agrees that the speed of his clock has changed but recognizes that the change is concealed from him by relativistic effects. There is one test result upon which all three observers agree and which therefore must be accepted as observationally verified. The change in velocity of the rocket produced a change in the speed of its clock. Under the interpretations of the Aether Relativity Theory there is no conceptual difficulty. The change in the velocity of the rocket caused the speed of its clock to change uniquely, but the nature of that change is concealed from observation. The concepts of Special Relativity, however, produce an absurdity. One event, the change in the velocity of the rocket, has produced two different and mutually exclusive results. The change in velocity of the rocket has caused its clock to both slow down and to speed up, depending upon whether the Earth or Mars is considered to be stationary. Since there was a single event, the change in the velocity of the rocket, only a single result can have occurred. At this point, readers who have been trained in Special Relativity will object. They will state that the change in velocity of the rocket involves acceleration and the Special Theory of Relativity was not derived for accelerated systems. Such an

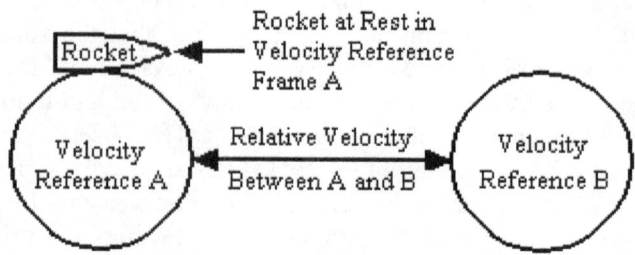

Conditions at Start of Experiment

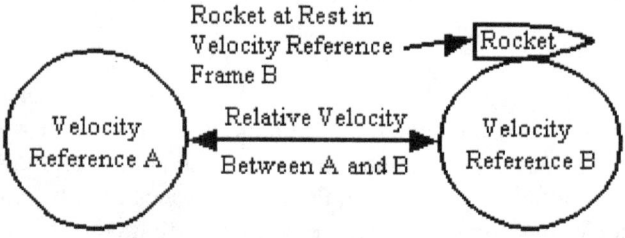

Conditions at End of Experiment

At the start of the experiment and again at the end of the experiment observers on the Earth, the rocket, and on Mars compare the speed of their clocks. All observations are taken under conditions of zero acceleration.

All observers agree that the rocket changed velocity. The observer at A asserts that the clock on the rocket slowed. The observer at B asserts that the clock on the rocket sped up. The observer on the rocket asserts that his clock changed speed due to the change in the rocket's velocity but that he cannot detect that change due to relativistic effects.

Figure 4.2 - Rocket-Clock Experiment

objection is irrelevant. All observations were made under conditions of zero acceleration and the Special Theory of Relativity is clearly applicable. A famous author has been quoted to the effect that the true measure of intelligence is the ability to hold two mutually exclusive ideas at the same time. *False!* The holding of two mutually exclusive ideas at the same time is evidence of a mind that is too lazy and/or incapable of resolving the inconsistency by correcting one or both of the ideas and who is too arrogant to admit the need for the correction and/or his inability to make it.

4.7- Enter the Tachyon:- In the 1960's it was recognized that the Lorentz Transformations did not prohibit velocities greater than the velocity of light. Instead they demonstrated that the velocity of light represented a velocity which material particles or objects could approach but not achieve because, at that velocity, the Lorentz Transformation became zero. At the velocity of light, kinetic energy (or mass if you prefer) became infinite and the rate of passage of time became zero. At velocities greater than the velocity of light, the infinities and zeros do not occur and it is theoretically possible for matter to travel at those velocities. Hypothetical particles which traveled at velocities greater than the velocity of light were postulated and given the name tachyons. For velocities greater than the velocity of light, the quantity within the square root sign in the Lorentz Transformation becomes negative and the Lorentz Transformation may be rewritten. It then becomes $i*(V^2/C^2-1)^{0.5}$, where i is equal to $(-1)^{0.5}$.

4.8- At this point, readers might question whether i has any physical meaning since $(-1)^{0.5}$ can exist only in one's imagination. Indeed, this was the viewpoint for several centuries until it was realized that the presence of i in a physical equation could be considered to represent a rotation of an effect into an axis which was perpendicular to the axis of the original coordinate system and was therefore unobservable. This concept has been found to be quite useful in the physical sciences and particularly in Electrical and Electronic Engineering since it allows phenomena which occur in two perpendicular axes to be represented in terms of the algebra normally used for a single axis problem. To observers confined to making their observations in the real axis, effects occurring in the imaginary axes are not directly observable and can only be inferred. Since i represents a rotation through 90 degrees from the real axis to the imaginary axis, as one might expect, equations containing i^2 represents a rotation of 180 degrees from the positive real axis to the negative real axis and produce observable effects which are reversed in sign.

4.9- While most of the properties of the hypothetical tachyon occur along the unobservable imaginary axis, it has at least one property whose Lorentz Transformation involves i^2 and therefore occurs in the real axis where observation is possible. That property is its velocity. (Since velocity is length divided by time, both of which are subject to Lorentz Transformations, the Lorentz Transformation for a tachyon's velocity contains i^2.) It is not surprising that the tachyon has never been observed as a particle, since in addition to the fact that many particles have been predicted long before they were observed, it would not be recognized as a particle because some of its key properties would be unobservable. It is possible, however, to draw a conclusion as to the "at rest" velocity of a tachyon. The "at rest" velocity of a particle traveling below the velocity of light is the velocity at which the magnitude of its Lorentz Transformation B_V is a maximum (V equals zero). By analogy, the "at rest" velocity of the tachyon would be the velocity at which the magnitude of its Lorentz Transformation is also at its maximum. Since this occurs when V is infinite, the "at rest" velocity of the tachyon should also be infinite. Experiments by quantum physicists have shown that mysterious effects called quantum numbers propagate at velocities which are significantly faster than the velocity of light, possibly at an infinite velocity. Quantum numbers would seem to have something in common with the hypothetical tachyon.

4.10- The very concept of the tachyon is devastating to the idea that Special Relativity is a valid representation of reality and it was vital to those who had built their careers around the Special Theory of Relativity that the idea of tachyons be discredited if the Aether Relativity Theory were to continue to be suppressed. To see why this should be so, consider the logic contained in the following statements:

- If I had a microscope, I would observe the existence of germs.

- The existence of germs does not depend upon the existence of the microscope.

The first statement asserts that, except for the case where microscopes cause germs, germs exist regardless of whether they have been observed. The second statement removes the escape clause from the first statement and it becomes equivalent to "germs exist". Now consider the following analogous statements:

- If I could communicate using tachyons, I would be able to establish absolute simultaneity between physically separated locations, measure my velocity with respect to space itself, and thereby verify the Aether Relativity Theory.

- The validity of the Aether Relativity Theory does not depend upon my ability to communicate using tachyons.

A little reflection should convince the reader that, if the last statement is true, the mere fact that I can conceive of communicating through the use of tachyons demonstrates that the limitation imposed on the Special Theory of Relativity by Aether Relativity Theory represents reality. ***The classical Aether must exist!***

4.11- Since it was vital that the idea of tachyons be suppressed and the fact that they had not been observed is not sufficient to accomplish that suppression, another approach was required. It was asserted that communication by tachyons would violate causality. (Causality is a very reasonable concept which asserts that a result cannot occur prior to its cause.) As an example, consider sending a signal by tachyon from the Earth to the Moon. If the time of transmission of the signal was 11:00:00 AM and the tachyon arrived at the Moon at 10:59:59 AM, it would seem to imply that it arrived at the Moon one second before it was transmitted. Literally interpreted, such a result would be a clear violation of causality. The argument falls apart when it is remembered that the clock on the Moon was synchronized with the clock on the Earth by an electromagnetic signal. If the Earth-Moon system were traveling through the Aether in a direction towards the Moon at a velocity of 0.81 times the velocity of light, the clock on the Moon would have a synchronization error causing it to be one second late with respect to the clock on the Earth. The apparent violation of causality would then be explained as being caused by the Earth-Moon system's velocity through the Aether. There are only two ways in which communication by tachyon can produce a violation of the Principle of Causality. The first possibility occurs if the tachyon arrives early by an amount of time greater than the observed time for light to make the trip. The second possibility is if tachyons were sent on a round trip from the Earth to the Moon and back and arrived on Earth before they were sent. While there is a school of thought which suggest that this can happen, it is based upon a misapplication of the Special Theory of Relativity.

4.12- The "Fictitious" Forces of Acceleration:- Newton's Second Law of Motion states that for every action there is an equal an opposite reaction. An exception to this rule seems to be the forces associated with inertial and gravitational accelerations. A force must be applied to an object to change its velocity, but there is no apparent opposing force to match the applied force. The same situation occurs when one considers the force of gravity. As you sit in your chair you are conscious of a force pressing you against it, but, as with inertial acceleration, there is no apparent opposing force matching it. As a result, the opposing forces required by Newtons's Second Law of Motion for both inertial and gravitational forces are referred to as fictitious. It is sometimes asserted that the General Theory of Relativity has shown that what appears to be the force of gravity does not occur but is a manifestation of the curvature of space associated with the source of the gravitational field. As we shall see later, General Relativity does not eliminate gravity as a force, it replaces it with the observable component of an enormous force acting along an unobservable fourth spatial axis. If one accepts the existence of the classical Aether, the fictitious forces present no conceptual difficulty since they are acting against the rigid medium of the Aether. Under Special Relativity, on the other hand, there is no medium for these forces to react against, and one is forced to accept the existence of exceptions to Newtons's Second Law of Motion.

4.13- Action at a Distance:- One of the reasons the concept of the Aether was accepted in the 19th Century was the need to explain the ability of forces to act between particles or objects which were separated in space. With the acceptance of Special Relativity, the Aether was banished from physical theories and another means of explaining the ability of these forces to act was required. The result was the introduction of the concept of virtual particles which bounced back and forth to produce the observed forces in a manner analogous to the production of force between two athletes throwing a medicine ball to each other. That concept has a difficulty. The mutual exchange of a medicine ball can only produce a **repulsive** force between the athletes. In order to produce an attractive force, the mass of the medicine ball would need to be negative. To date no theoretician seems to have raised the possibility that such is the case for the postulated virtual particles and one must conclude that, to supply an attractive force, the virtual particle must be under tension and consequently the distance over which it can act is limited by its size. If the exchange of virtual particles produces forces which act at a distance, they would themselves constitute a medium equivalent to the Aether and the need for the Aether to explain action at a distance would not have been resolved, it would merely been pushed down one level into the virtual particle. Since the advocates of Special Relativity have not addressed questions such as these, Dr. Einstein's statement that the need for the

Aether had been eliminated is not true, it only appears to be true because embarrassing questions have been swept under the rug.

4.14- The precept that electrostatic forces result from the exchange of virtual photons is readily tested. Such an experiment is diagramed in Figure 4.3. In this experiment, two metal plates are suspended parallel to each other in a large electromagnetically shielded and evacuated chamber. The plates are located symmetrically about the center of the chamber. Midway between the plates is a wire which is capacitively coupled to a suitable radio receiver. The plates are connected to high voltage D-C sources of equal amplitude and opposite polarity. The high voltages on the plates produces a strong electrostatic force of attraction between them which, since it acts over a distance, must either result from an electric stress in the Aether or, in accordance with the currently accepted concepts, from the exchange of virtual photons between the plates. (A

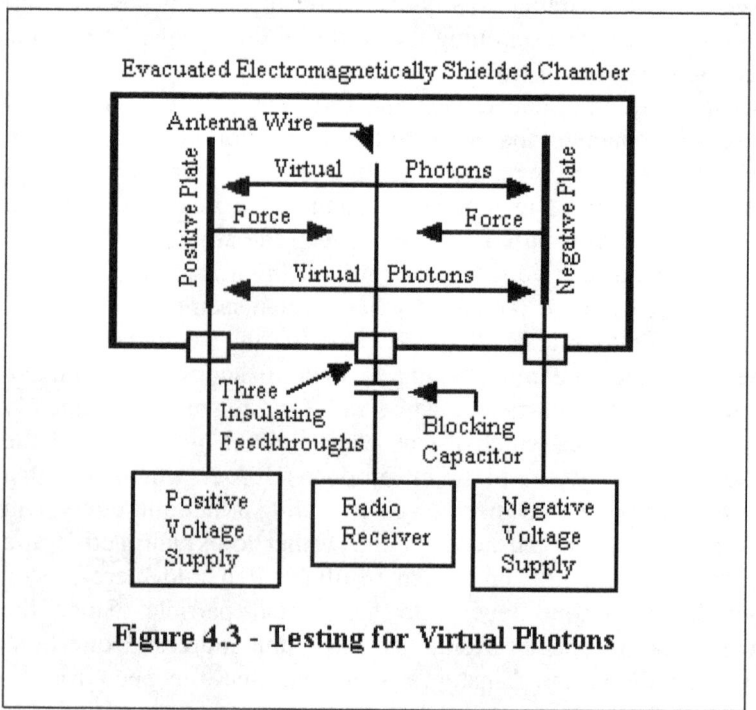

Figure 4.3 - Testing for Virtual Photons

magnetic equivalent of this experiment could be conducted with electromagnets substituted for the electrically charged plates.) The concept that the force results from an electric stress in the Aether is straightforward and needs no further explanation. The concept that the force results from the exchange of virtual photons is more complex and requires further examination.

4.15- If the force between the plates results from the exchange of virtual photons, then the relatively large spacing between the plates requires that the virtual photons have a fairly long wavelength and a frequency low enough to allow them to be sensed by a radio receiver operating at conventional television frequencies. As a result of that low frequency, a large flux of virtual photons is required to attain the force level that a high voltage can produce between the plates. A large virtual photon flux will, in turn, induce a high level of random noise in the wire located between the plates and will cause a high noise output in the radio receiver. (The evacuation of the chamber insures that the noise was not generated by ionized gas molecules.) While this experiment is readily executed, it is not necessary for it to be performed. If electrostatic forces resulted from the exchange of virtual photons, the electric fields which abound at various locations, such as between the surface of the Earth and clouds or the voltage on the picture tube of a TV set, would generate sufficient noise as to render electromagnetic communication impossible.

4.16- Since we regularly use electromagnetic waves as a means of communication and that communication seems unaffected by the presence of static electrical fields, it is safe to conclude that electrostatic forces do not result from the exchange of virtual photons. The only explanation which seems to remain is that they result from a stress in the Aether. If both our experience already denies, and a physically realizable experiment will also deny, that virtual photons act as carriers of the electrostatic force, they are most certainly ruled out as carriers of the magnetic force. If we are forced to abandon the idea that virtual photons carry the electromagnetic forces, consistency requires that we abandon the idea that virtual particles named gluons carry forces within the nuclei of atoms. It is time to go back to the subatomic drawing board. A possible nature of those forces will be discussed later.

4.17- The Radiometric Measurement of Our Velocity Through Space:- In November 1977, a paper read at an American Astronomical Society convention in Atlanta announced that measurements of the intensity of the microwave background radiation of space in different directions showed that the Earth was moving through space at a speed of about 700,000 miles per hour. The experiments were conducted by radiometers installed in a U-2 aircraft flown at an altitude of 70,000 ft. with a methodology which was apparently beyond challenge. The report of these experiments led to consternation on the part of cosmologists because it did violence to their existing concepts concerning the distribution of matter in space. An even more significant result of these experiments was not recognized. The observance of this velocity drives experimental nails into the coffin of the Special Theory of Relativity because the validity of that theory, in comparison to the more restrictive Aether Relativity Theory, depends on the fact that it is impossible for an observer to measure his absolute velocity though space. As often happens, as soon as one declares that something is impossible, some damned fool comes along and does it. The experimenters just didn't play fair. Inadvertently, by measuring the velocity of the Earth through space, they demolished the underpinnings of the Special Theory of Relativity and established that the Aether Relativity Theory was the correct interpretation of reality. *(Up with Fitzgerald, down with Einstein.)*

4.18- Dirac's "Sea" of Negative Energy:- In the 1930's Dr. P. Dirac considered the effect of the impact of a high energy photon (e.g.- 10^6 electron volts) against a more massive particle. He concluded theoretically that the impact would produce both an electron and a positron and had the satisfaction of having that prediction verified by observation. His theoretical treatment had one difficulty. It also concluded that all of the matter in the

Universe would vanish in a small fraction of a microsecond. Since the Universe continues to exist, it was necessary to revise the theory. The resulting revision was to consider that all of space was solidly filled with negative energy (whatever that is). The production of the electron-positron pair was considered to result when the impact of the photon knocked an electron from that sea of negative energy and left a hole where the electron had been. That hole represented a missing negative charge in the sea of negative energy and appeared to us as a positive electron. One might reasonably wonder how Dirac's concept of a sea of negative energy which pervades all of space differs significantly from the concept of the classical Aether.

4.19- The Characteristics of the Aether:- The discussions to this point, and more particularly those which follow in successive chapters, imply that, in order for "reality" to have the properties which we observe, the Aether must have, as a minimum, the following properties:-

- It must be a solid medium rather than a fluid. If it were not a solid medium, transverse electromagnetic disturbances (light) would not propagate since transverse disturbances cannot propagate through a fluid.

- It must have, as a minimum, a dielectric constant, a permeability, and occupy a volume since these properties are readily observed.

- It apparently is absolutely continuous rather than composed of minute particles. This continuity may well approach a zero size as a limit since it behaves as if it had a "Q" which approaches infinity.

  - A tuning fork made of steel will ring for a prolonged period after being struck since steel is a high "Q" material. One made of lead will merely "thunk" when struck since lead is a low "Q" material. The fact that disturbances in the Aether do not die out at a detectable rate while propagating through free space suggests that the Aether has a "Q" which is enormous and may well be infinite.

- Material particles must be constructed in such a way that it is possible for them to propagate through the solid Aether (Chapter 13).

4.20- Special Relativity and Occam's Razor:- It is sometimes asserted that Occam's Razor shows that the Special Theory of Relativity is to be preferred over the Lorentz Contraction-Aether Theory. Occam's Razor is a philosophical construct which asserts that, when there are two or more explanations of a phenomena, the simplest explanation should be chosen. In the case of these theories, Occam's Razor is ambiguous. Computations based upon Special Relativity are simpler than those strictly based upon the Lorentz Transformation-Aether Theory for the reason that it allows the observer's velocity reference frame to be used as the basis of computation rather than requiring the use of an infinite number of velocity reference frames that would seem to be required by the Lorentz Transformation-Aether Theory. The downside of the Special Theory of Relativity is that it requires the acceptance of the idea that the simultaneous existence of an infinite number of "real" velocity reference frames correctly represent reality.

4.21- The Lorentz Transformation-Aether Theory, on the other hand, asserts that there is only one absolute velocity reference frame, and it would superficially seem to require the use of separate calculations for each of the infinite number of possible velocities between the observers velocity reference frame and the absolute velocity reference frame. However, such a complication does not occur. As shown Figure 6.4 and the text which accompanies it, any reference to the absolute reference frame cancels from all calculations and observations and the mathematics of Special Relativity are completely applicable. (This happy effect results from the fact that the Lorentz

Transformations are multiplicatively commutative. - See Chapter 8.)

4.22- The Implications of Maxwell's Equations:- Based upon the discoveries of Faraday, Dr. Maxwell derived the famed equations which define the electromagnetic field. According to early texts, he did this by imagining "displacement currents occuring in space" even though he claimed not to have and understanding of what those currents could be and when Faraday requested an explanation of the theory in words, Maxwell is alleged to have been unable to provide it. If the author is to believe a recent communication, the current teaching is that Maxwell's Equations do not have a physical explanation, ***they just are!***

4.23- Dr. Maxwell derived his equations based on the idea of "displacement currents" in space. To understand the reason for considering these "currents", consider what happens when an A-C voltage is applied to a capacitor consisting of two parallel plates in a vacuum as shown in Figure 4.4. In this figure, the applied A-C voltage (which may be assumed to be identical at the supply and the plates) causes an A-C current to flow in the wires to the plates which is phased 90 degrees in advance of the voltage. Since this is a series circuit, the current in all parts of the circuit must be instantaneously the same. That means that the current must flow through the space between the plates, but, since the conventional carrier of electric current (e.g.- electrons, ions, etc.) cannot pass between the plates, the current must flow between the plates without the mediation of charged particles. In addition, a magnetic field which surrounds the electric field is generated in proportional to its rate of change. Similarly, an electric field is generated in proportion to the rate of change of the magnetic field. Since each of the fields is generated in proportion to the rate of change of the other, each leads the other in phase by 90 degrees. In combination, these two phase shifts add to 180 degrees and are thus capable of sustaining an oscillation by feeding energy cyclically from one field to the other *without the*

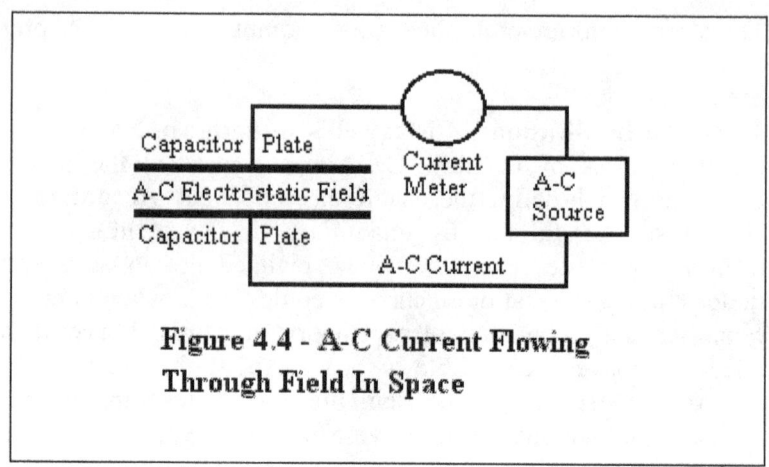

Figure 4.4 - A-C Current Flowing Through Field In Space

intervention of any other mechanism. Maxwell's Equations concisely describe the interaction and show that such an oscillation will propagate as a wave disturbance as part of the electromagnetic spectrum.

4.24- The difficulty in describing Maxwell's Equations in words does not derive from the equations themselves, the preceding paragraphs would seem to do that quite nicely. The difficulty arises when one tries to reconcile them with the "empty" space implied by Special Relativity. After all, how can electrical currents flow in empty space and how can magnetic forces exist in that space? On the other hand, if one considers that the space is filled with the classical Aether, the problem vanishes. One could consider that the Aether is stressed by the application of an electrical field. This stress reveals itself as a negative electrostatic potential at one plate and a positive electrostatic potential at the other plate and produces an attractive force between the surfaces applying the field (space has a dielectric constant). (One could also argue that the Aether contains two electrostatic omponents, positive and negative, in juxtaposition, and these components are pulled apart by the electric field.) The rate at which the electric stress is applied produces a hoop stress in the Aether which stores energy and which we observe as the magnetic field (space has a magnetic permeability). The rate of change of the magnetic field similarly produces an electric field which acts on the electric components of the Aether. These two effects are not mirror images of each other, we observe point electrostatic charges (electrons, positrons, etc.) but do not observe point magnetic charges (monopoles). (It should be noted that when the region of magnetic "hoop" stress propagates at the velocity of light, as it does in the photon, the "Fitzgerald Contraction" reduces the "hoop" from a circle to a line which is perpendicular to the velocity vector.) As we shall see later on, both the Velocity and Gravitational Transformations for permeability and the dielectric constant differ significantly. While viewing the Aether in this manner provides an explanation for an observed electromagnetic field, the possibility also suggests itself that if the stress is sufficiently intense, it can be relieved locally by a rupture which generates an electron-positron pair.

4.25- "Wormholes" and the Aether. There is a high probability that by now almost everyone who reads this text has been exposed to the idea, apparently current among relativists, of "wormholes in space". Passage through one of these "wormholes" would theoretically allow a traveller (or a message) to disappear at one point and "instantaneously" appear at a far distant point. (For example, a traveller might enter a "wormhole" in the Solar System and emerge at the Alpha Centauri System.) Such a transposition would allow the simultaniety uncertainty error between the Solar System and Alpha Centauri to be reduced from years to seconds or minutes, reducing, in the process, the applicability of Special Relativity from the limits of +/-C to a very small velocity limit. This result could only occur if the Aether actually existed and the adherents of Special Relativity Theory were wrong. It would seem that the academic community wishes to have its cake and eat it too. If they wish to think of "wormholes", to be consistent they must accept the idea of the classical Aether.

4.26- The Implications of the COBE Experiment:- Initial results from the COBE (Cosmic Background Explorer) experiments have revealed two items of information which are relevant to this discussion. The point of origin of our expanding universe has been observed, clearly establishing that the Universe has a fixed velocity reference frame (the velocity of that point of origin) thereby revealing the view of Space-Time resulting from Dr. Einstein's work to be false. The high resolution of the image of that point of origin reveals that our Universe exists in a three dimensional Euclidean geometry rather than in the curved Riemannian geometry asssociated with General Relativity. This conclusion is consistent with the observed precession of Mercury's orbit and the bending of starlight in the Sun's gravitational field, as shown in the Appendix "Gravity" copyrighted in 1987.

Chapter 5 - The Resurrection of Absolute Velocity By Quantum Experiments

5.1- While the preceding chapter is probably adequate to convince individuals whose thought processes include common sense that the Aether Relativity Theory and not Special Relativity represents reality, it will not convince most members of the academic community. In order to acquire their PhDs, they have, of necessity, allowed themselves to be brainwashed into submerging common sense reasoning in favor of reasoning by formal procedures. In so doing, not only they have they unknowingly suppressed the most powerful capability of the brain, its pattern recognition capability, and concentrated on the development of one of the brain's lesser capabilities, its ability to process logic, they have made it possible to ignore aspects of the problem which are not included in the postulate structure of the mathematics. To convince those individuals, experimental evidence is required. Fortunately, that evidence has been supplied by experiments in Quantum Physics.

5.2- An article in one of the world's most prestigious scientific magazines in the late 1980's described experiments which demonstrated that the polarization of paired photons (generated by a common source and travelling in opposite directions) was coupled in such a way that changing the polarization of one photon changed the polarization of the other. More significantly, they demonstrated that the velocity of the polarization coupling between the paired photons was at least 4 times the velocity of light. These results raise the question as to whether polarization, which quantum physicists designate as the photon's quantum number, could be considered to be coupled by an observable property of the as yet to be observed tachyon. If so, one would expect that the coupling of the polarization of paired photons would propagate at an infinite velocity. Since the energy content of a photon does not change as a result of its direction of polarization, the Special Theory of Relativity and/or the Aether Relativity Theory do not impose a velocity limit on the transfer of information by the coupling of the plane of polarization between paired photons. Although the authors of the article made no such claim, perhaps because they wished their work to be published, the experiments described in the article demonstrated both that our absolute velocity through space could be measured (validating the Aether Relativity Theory over the Special Theory of Relativity) and that communication at velocities greater than the velocity of light was feasible with minor modifications to the equipment.

5.3- In the article it was stated that attempts to use the apparatus to communicate at velocities greater than the velocity of light resulted in the transmission of noise instead of information. Their failure to achieve communication resulted from the fact that, while they were obviously good quantum physicists, they were not good communication engineers. In analogous electronic terms, they were attempting to communicate by phase modulating a randomly phased carrier. In such an arrangement, the randomly phased carrier injects white noise that prevents the information, which is actually present in the received signal, from being decoded. The experiment did demonstrate that, if the carrier had been coherent, the desired propagation of information at translight velocity would have been achieved.

5.4- The experimental setup is diagramed in Figure 5.1. [**Note:-** Limitations on page size forced this diagram, and the two which follow, to be rotated by 90 degrees. In actuality, the portion of the setup which is described as being at the bottom was at the left amd the portion described as being at the top was at the right.] A photon source is provided which contains excited atoms of a type which emit a pair of photons of the same polarization in opposite directions whenever one of them reverts to its unexcited state. Each of the photons of a pair (upper and lower) are sent to an optical switch which sends its photons in one of two directions in response to a command signal. Depending upon the setting of the switch, each photon passes through a horizontally or vertically oriented polarizer and is received by one of two photon detectors. (Two photon detectors are at each end of the

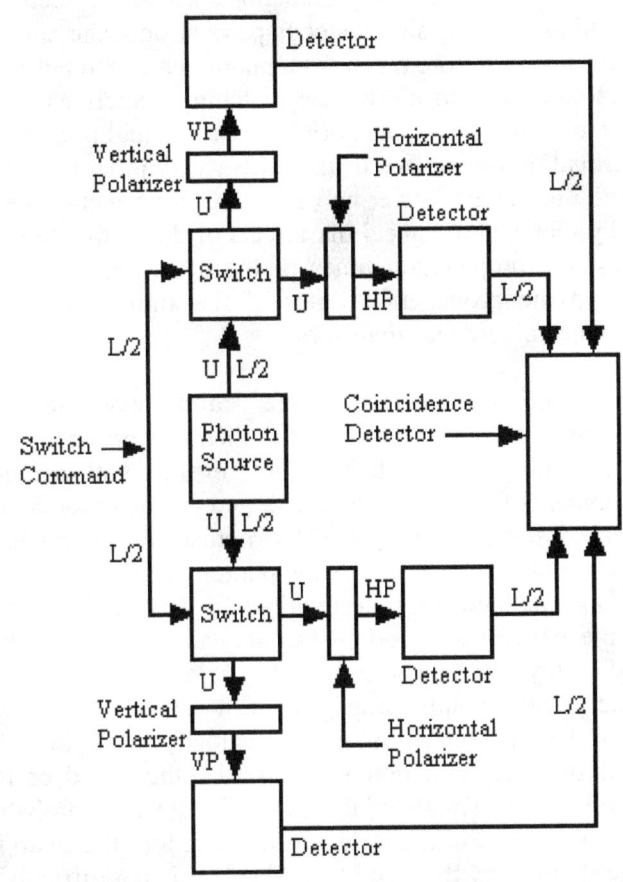

(See figure 5.3 for meaning of symbols)

Figure 5.1 - Block Diagram of Demonstration of Faster Than Light Polarization Coupling of Paired Photons

apparatus.) The outputs of the four detectors were compared in a coincidence detector. The purpose of the coincidence detector is twofold. It insures that the only detections which are recorded are those which occur as simultaneous pairs at opposite ends of the apparatus and therefore result from photon pairs and not from spurious photons. It also allows the matching detections to be sorted into four categories, horizontal/horizontal, vertical/vertical, horizontal/vertical, and vertical/horizontal. The lengths of the upper and lower halves of the experimental setup are carefully matched to cancel the effects of the transit time of the photons, the propagation times of the drive signals to the switches, and the propagation times of the signals from the detectors to the coincidence detectors.

5.5- The resulting detections were quite revealing. An overwhelming preponderance of them consisted of horizontal/horizontal and vertical/vertical events, with a much smaller number of horizontal/vertical and vertical/horizontal events. Since the response time of the optical switches which changed the polarization of one of the photons was on the order of 10 nanoseconds and the time required for light to travel the length of the experimental setup was 40 nanoseconds, such a result could only have occurred if the polarization coupling between the paired photons propagated at a velocity which was significantly larger than 4 times the velocity of light. The accuracy of this determination is limited by the speed of the switches and by the length of the setup. While it is probably quite difficult to improve the switches, the length of the setup is, in principle, limited by the size of the Earth. It is not difficult to envision an experimental arrangement which would demonstrate a propagation velocity for polarization coupling between paired photons greater than a million times the velocity of light.

5.6- Consider a modification to the experimental setup, as shown in Figure 5.2. In the experiment represented by Figure 5.1, care was taken to insure that the path lengths traveled by each of the paired photons and the length of the signal paths from the optical detectors to the coincidence detectors were equal. This was done

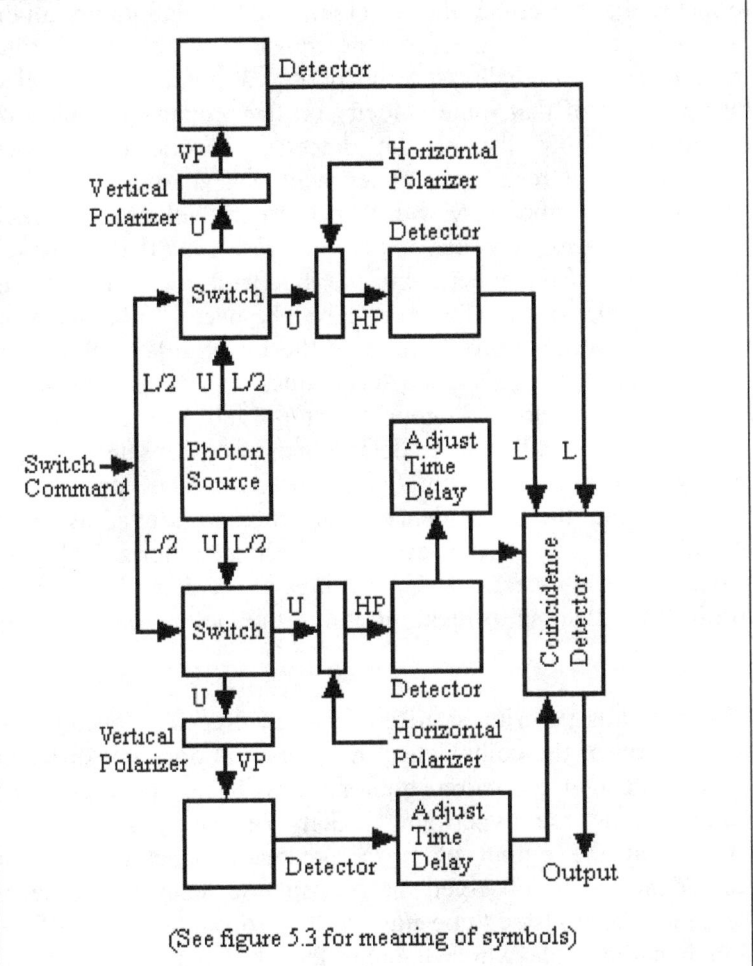

(See figure 5.3 for meaning of symbols)

Figure 5.2 - Block Diagram of Experiment to Measure Absolute Velocity Through Space

to insure that the effect of any velocity that the laboratory might have with respect to space on the transit times of the photons from the source to their respective photon detectors was canceled by the effect of that same velocity on the propagation times of the signals from the photon detectors to the coincidence detectors. As a result, the experiment was unaffected by the velocity of the laboratory with respect to any velocity reference frame arbitrarily chosen to be at rest. In the proposed modification of the experiment, the independence of the results on the velocity of the laboratory through space is eliminated by moving the coincidence detector to the lower side of the setup and substituting adjustable delay lines in the signal paths between the bottom photon detectors and the coincidence detectors. The adjustable delay lines compensate for the propagation delay of the signals between the top photon detectors and the coincidence detectors and are adjusted to provide the maximum level of horizontal/horizontal and vertical/vertical detections. The settings of the delay lines which result from that adjustment provides the output data for the experiment.

5.7- Since this experiment differs from the preceding one only in the location of the coincidence detectors and does not differ in the treatment of the paired photons, it will also be capable of demonstrating the hyperlight velocity of coupling of paired photons at the output of its coincidence detector. For that coincidence to be observed the propagation delay of the delay lines must be adjusted to be equal to the propagation delay of the signal in the cable which couples the photon detectors at the upper side to the coincidence detectors. The average of the settings of the delay lines provides the output data of the experiment. If one arbitrarily assigns a velocity of +V to represent the absolute velocity of the laboratory through space in a direction to the top, it should be possible to determine that velocity from the average delay, $T_{AV,}$ set into the delay lines. As observed in a velocity reference frame which is at rest, the velocity of propagation of the signal through the cable is increased by V and becomes C+V , and the setting of the delay line must be changed from its nominal value of $T_{AV}=C/L$ to compensate. The velocity of the laboratory through space is then given by $V=C-T_{AV}*C^2/L$.

5.8- Since the value of V is a number which may be broadcast, the measured velocity of the laboratory may be transmitted to a series of observers having velocities different from that of the laboratory and different from each other. If the Aether Relativity Theory correctly represents reality, the same number will be received by all of the moving observers and will represent the absolute velocity of the laboratory through space. If the Special Theory of Relativity correctly represents reality, the number which is broadcast from the laboratory will equal zero while the number received by each of the other observers will be equal to the velocity of the laboratory with respect to himself. All of those observers would receive a different number! (If any reader accepts such a result as possible, the author would like to meet him. There is a bridge over New York City's East River that he has been trying to sell for some time.) The success of the paired photon experiment of Figure 5.1 insures the impossibility of a result in which the delay line settings would remain unchanged at $T_{AV}=C/L$ as the Earth rotated on its axis and moved in its orbit.

5.9- The experiment of Figure 5.1 can be modified to eliminate its shortcomings as a communication system by providing a phase coherent carrier for the information to be transmitted by the paired photons. Figure 5.3 shows a means of providing phase coherency of the paired photons without degrading the signal to noise ratio by inserting additional polarizers at the photon source which are oriented parallel to each other and at an angle of 45 degrees to the horizontal and vertical polarizers already described. Since the purpose of this experiment is to demonstrate faster than light communication from top to bottom, the switch is eliminated from the bottom of the setup and the detectors at the top of the setup are replaced by photon absorbers. Data is inputted to the system through the switch at the top of the setup. In place of the switch at the bottom, a crystal is provided which divides the incident light into two polarized beams whose axes are perpendicular. The orientation of the polarizing crystal is chosen such that one of the beams is vertically polarized and the other beam is horizontally polarized. Each of these beams is sensed by a photon detector and the output of those detectors is decoded to provide the received signal.

5.10- Assuming that the addition of the coherency producing polarizers does not interfere with the polarization coupling of the paired photons observed in the experiment of Figure 5.1 (yet to be verified experimentally), the outputs of the decoder at the bottom of the setup will contain a signal identical to the signal inserted into the switch at the top of the setup. A computer simulation to determine the level of the signal received at each detector shows it to be more than adequate to discriminate against noise. As in the arrangement of Figure 5.1, 50% of the photons generated by the photon source will be received (assuming no spurious losses). Of these photons, 75% will be directed to the detector which was selected by the transmitting switch and 25% will be directed to the other detector. The resulting decoder output would then be expected to have a peak to peak amplitude (ideal case) equal to 50% of the photon output of one side of the paired photon source. It would be interesting to have such an experiment performed.

5.11- The Paired Photon Experiment described in Figure 5.1 clearly establishes that the Special Theory of Relativity only survived because the necessary experiments to validate its special case solution, the Aether Relativity Theory, were beyond the state of the art until several generations of physicists had been brainwashed into ignoring the fact that, unlike the Aether Relativity Theory, it contradicted common sense. The idea that quantum numbers, such as polarization, can propagate at an infinite velocity and exhibit many (if not all) of the properties of tachyons does not violate the concept that energy cannot be transmitted faster than velocity of light. The transmission of information does not necessarily require the transmission of energy. Since the energy of a photon does not change as a result

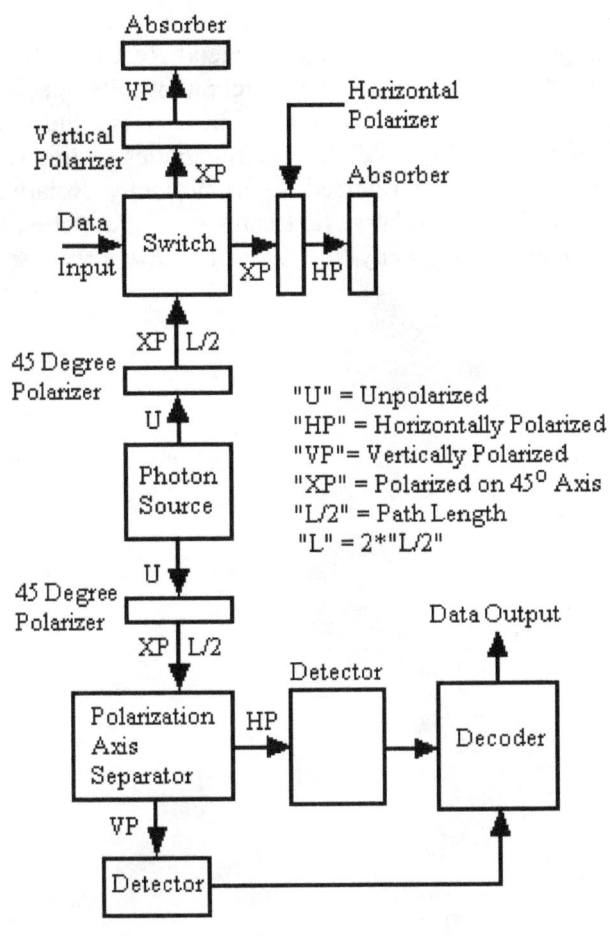

Figure 5.3 - Block Diagram of Faster Than Light Communicator

of a change in its direction of polarization, there is no prohibition, even in Special Relativity, against information represented by the direction of polarization of paired photons propagating at an infinite velocity. Special Relativity has been experimentally demonstrated to be incomplete. Its special case solution, the Aether Relativity Theory is the correct interpretation of reality, and, for the remainder of this text, its subject matter will be referred to as Velocity Relativity to distinguish it from the relativistic treatment of gravitation which will be designated as Gravity Relativity. ***The Emperor does have clothes after all!***

Chapter 6 - The Nature of Reality

6.1- Do The Effects Observed Between Velocity and/or Elevation Reference Frames Occur Because of a Change in Reality or Because of a Change in the Units of Measurement By Which That Reality is Observed?:- In the 19th century, it was universally assumed that there was an underlying reality which existed independently of the means by which it was measured. The Lorentz Contraction-Aether Theory of Relativity accepted this viewpoint, but, under the Special theory of Relativity, the measurement itself became the reality and the reality beneath the measurement became meaningless. To examine the difference between these philosophical approaches, let us consider the implications of a simple thought experiment as shown in Figure 6.1A. In this experiment, there are two rooms each containing identical clocks which may be interchanged without affecting the results. Also in these rooms are electrically operated buzzers activated by a common signal. An experimenter presses a button to sound the buzzers and observers in each room note the time at which the buzzer sounds. An hour later, the experimenter again sounds the buzzer, and the observers in each room again note the time. In room A, the observer reports that one hour has passed between the soundings of the buzzer. In room B, the observer reports that 50 minutes have elapsed. Since the clocks are identical, the philosophy associated with Special and General Relativity asserts that the observations show that time passes more slowly in room B than it does in room A. There is, of course, a problem with this assertion, the duration of time between the soundings of the buzzers was the same in both rooms since they were activated by the same signal and the propagation time for that signal between the rooms is much smaller than the difference in readings of the clocks. Repeating the experiment with the clocks interchanged between rooms yields the same result and shows that a difference between the clocks was not the cause of the difference in the observed durations.

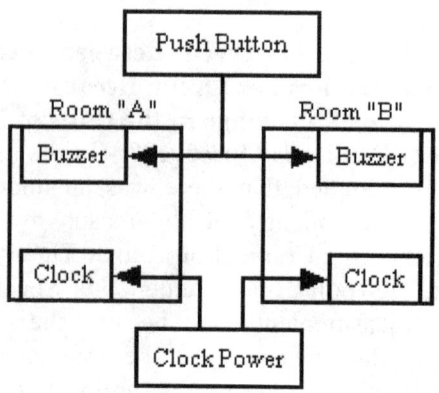

A:- Clocks in Separate Rooms

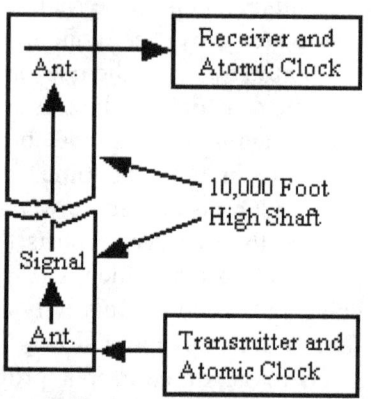

B:- Clocks at Top and Bottom of Shaft

Figure 6.1 - Clock Experiment to Show Change in Units of Time

6.2- The experimental results appear absurd until it is learned that the clocks are of the old fashioned type driven by synchronous electric motors with the clock in room A operated from a 60 Hz power source, as is conventional in the USA, and the clock in room B is operated from a 50 Hz. power source, as is conventional in Europe. The passage of time reported by each clock was determined by the room in which it was placed as well as by the actual passage of time. To measure the actual passage of time in each room, it is necessary to determine the effect of that room on the speed of its clock and to correct each observation for that effect with respect to an agreed upon universal standard. Without such a correction, the term GIGO applies to the experiment. (GIGO is an expression formulated in the early days of the computer. It means garbage in = garbage out.)

6.3- The preceding rather fatuous description was provided to show the need for recognizing that, when a quantity is observed using ideal instruments in reference frames which differ in velocity and/or elevation, it is necessary to distinguish between two effects. The effect of interest occurs as a result of a change of the quantity itself. Observation of that effect may be corrupted by a change in the calibration (size of the units of measurement) of the measuring instrument(s) which occurs between the reference frames. The predictions of General Relativity caused measurements to be made which demonstrated that the rate of passage of time slows as the elevation in a gravitational field is reduced. To be objective, one must determine whether the rate of passage of time actually decreases or whether clocks run more slowly (units of measurement for time are larger) at the lower elevation or whether both effects occur in combination. Fortunately, a physically realizable thought experiment can provide the answer.

6.4- Consider next an experimental setup in a vertical shaft drilled into a mountain, as shown in Figure 6.1B. A pair of identical atomic clocks are mounted at the top and bottom of the shaft. The height of the shaft is sufficient, perhaps 10,000 feet, so that the clocks can accurately measure the gravitationally induced difference in the rate of passage of time between the top and the bottom of the shaft. At the start of the experiment, a signal is sent from the bottom of the shaft to the top and the clocks are synchronized. At the end of the experiment, a second signal is sent from the bottom to the top of the shaft and the elapsed time is read on both clocks. Since the duration of the experiment will be affected by any velocity induced difference in the propagation time of the start and stop signals between the elevations, the effect of propagation time error is minimized by making the duration of the experiment equal to four years. This time span insures that the change in velocity of the shaft through space as a result of the orbital motion of the Earth and of its rotation is minimized. With this precaution, the difference in the duration of the experiment between elevations can be reduced to less than one picosecond. Since four years is $1.25*10^8$ seconds, the duration of the experiment is the same at both elevations to an accuracy of better than one part in 10^{20}. The slowing of time, as measured by the difference in readings of these clocks, over the elevation difference of 10,000 feet is about of one part in $3*10^{13}$. To an accuracy of better than 1 part in a million, the difference which will be observed in the duration of the experiment between the upper and lower elevation results from a difference in the speed of the clocks and not from a change in the rate of passage of time. We may conclude therefore that the rate of passage of time is an absolute which is independent of whether a measurement has been made and independent of the characteristics of any instruments might have been used to make such a measurement.

6.5- The preceding paragraph leads to the conclusion that a change of reference frame (elevation or velocity) causes a change in the size of the units of measurement for time (duration of time between the ticks of the clock) by which the clock gauges the passage of time while the rate of passage of time itself is unchanged. If such a conclusion applies to time in a gravitational field, consistency requires that it apply to all other measurements (force, mass, length, etc.) in which relativistic effects are involved. The relativistic theories must then actually be a means of keeping track of the consequences of changes in size of the various units of measurement which occur as a result of a change in velocity or elevation. They do not involve the size of those quantities in the absolute sense.

6.6- The Nature of Mass:- Throughout the science of physics, the concept of mass plays an important role. Subjectively, the meaning of mass was brought home to the writer quite vividly years ago as he stood on a dock while a large freighter was being moored. Apparently the captain of the ship and/or the tugboat crew were not sufficiently skilled, and the ship was pushed toward the dock at a speed equivalent to the crawl of a sleepy turtle. The ship contacted the group of pier supports, each composed about a dozen 12 inch diameter wooden pilings, and kept right on moving. Despite its extremely slow speed, before the ship came to rest it had pushed the massive pier supports about four feet sideways and made necessary a significant degree of dock repair. One only need to observe such an occurrence to appreciate the significance of the ship's inertial mass.

6.7- In the above experience, the author did not actually observe the mass of the ship, he observed the effects of the force which resulted when the pier supports attempted to lessen the ship's velocity. Observation of such a force is the only manner in which one can measure the inertial mass of any object or particle. Inertial mass cannot be observed directly. It can only be observed as the incremental impulse (force-time product) required to produce an incremental change in velocity (length/time quotient) and is more properly defined in terms of force, length and time. Similarly, gravitational mass is observable only in terms of force, length, and another known gravitational mass in accordance with Newton's Law of Gravitation. Finally, if one considers the equivalence between mass and energy, Thomson's $E=M*C^2$, one notes that only two of the three terms in that equation are independent. Since, unlike mass, both the velocity of light and energy are directly observable, mass must be a dependent variable without existence as an entity in its own right.

6.8- The misapplication of the concept of mass has led physicists to some weird conclusions. For example, the photon and the neutrino are considered to be massless particles despite the fact that they represent the presence of energy and have inertial and gravitational properties consistent with the level of that energy. (In the author's text "Gravity", it is shown that the gravitational mass represented by the energy of a photon or neutrino is twice that of the gravitational mass of the same quantity of energy in a the form of a material particle.) The reason that photons and neutrinos are called massless particles is that they do not possess mass when they are at rest. Since these particles only exist when they travel at the velocity of light, their designation as massless particles would seem to be rather frivolous. More damaging, the designation of these particles as massless obscures the fact that the gravitational mass of the background radiation in our universe exceeds the gravitational mass of its matter by a wide margin. This faulty definition has caused astronomers and cosmologists to spend a great deal of effort in searching for the dark matter needed to account for gravitationally induced behavior observed throughout the Universe. A rough calculation shows the so called massless particles easily contain enough gravitational mass to account for that behavior.

6.9- The classical concept of inertial mass is the incremental change of total energy, E, resulting from an incremental change in velocity, δV. Because $E=M*C^2$, it follows that $M=E/C^2$, and, in terms of that definition of mass, the effect of mass is more properly provided by the derivative of the conventional Lorentz Transformation for Mass with respect to velocity, the Lorentz Transformation for Incremental Mass. The curves of Figure 6.2 provide the relationship between the both the mass and the incremental mass of an object as a function of its velocity. It will be noted that the Lorentz Transformation for Incremental Mass is the slope (first derivative) of the curve for the Lorentz Transformation for Mass. When the Lorentz Transformation for

Incremental Mass is employed in the solution of relativistic problems, mass has the correct dimensional content and the discrepancy between its use in classical problems and its use in relativistic problems vanishes. Special Relativity's artificiality of substituting momentum for mass is no longer required.

6.10- The use of the Lorentz Transformation for Mass instead of the Lorentz Transformation for Incremental Mass led to the false conclusion that, since Special Relativity was derived for reference frames having relative velocity, it could not be applied to accelerated reference frames. That conclusion is incorrect. Both Special Relativity and the Aether Relativity Theory provide transformations for both length and time. Since velocity is the first derivative of length with respect to time and acceleration is the second derivative of length with respect to time, if Special Relativity and/or Aether Relativity can deal correctly with the effects of velocity, they must also be able to deal correctly with the effects of acceleration. Contrary to dogma, General Relativity is not required for that purpose.

6.11- The Mechanism Behind the Lorentz Transformations:-
The Lorentz Transformations provide a description of how matter behaves when its velocity is changed so that the Principle of Relativity is satisfied and observers moving with that matter can always consider themselves to be at rest. It is desirable to describe how these transformations come about in a manner which is consistent with common sense so that the effect can be readily understood and is not beclouded by mathematical obfuscation. In order to measure a length, one might use a yardstick (or meterstick if you prefer) as the unit of measurement. In the ideal case, the length of this yardstick is composed of a chain of atoms whose overall length is determined by the number of atoms in the chain and the distance between them. If the velocity of the yardstick is changed, the spacing between the atoms in the chain may change, but the number of atoms in the chain will not. The problem of determining length then boils down to determining how atoms control their spacing. For example, the two atoms in a hydrogen

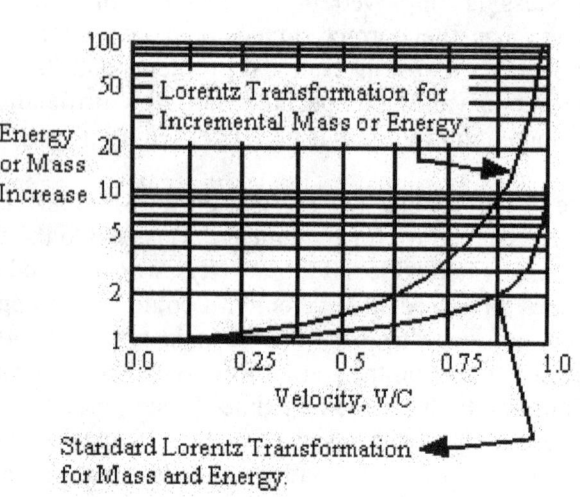

Standard Lorentz Transformation for Mass and Energy.

Lorentz Transformation for Incremental Mass or Energy Equals Slope of Lorentz Transformation for Mass or Energy.

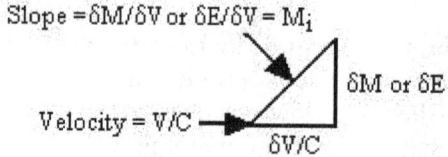

Figure 6.2 - Lorentz Transformations for Energy and Mass and for Incremental Energy and Mass

molecule maintain a separation of slightly over 10^{-10} meters and strongly resist any outside forces attempting to change that spacing. Since the spacing between the atoms represents a distance of 10^5 times the diameter of the proton in which 99.95% of the mass of the atom resides, it is obvious that the atoms have some means of measuring their separation and applying the forces needed to maintain that separation over what, to the atom, is an enormous distance.

6.12- In the macroscopic world, there are two means by which distances can be measured by electromagnetic means. They are triangulation and the radar principle. Triangulation, however, requires the pre-existence of a baseline of a known length, and, while it might seem to answer the question of how two hydrogen atoms might measure and maintain their separation, it is not a viable explanation because it requires that there be an independent means of defining the baseline. Triangulation does not resolve the problem, it merely moves it to another location. The other method, the radar principle, does not suffer from this limitation and is a reasonable candidate providing an entity exists which propagates at a fixed velocity through the empty space between the atoms. Electromagnetic radiation propagating through an Aether fills this role nicely.

6.13- To bring the process into the common sense world, consider an example in which two boats are station keeping with respect to each other and with respect to an indefinitely long straight bulkhead along the shore, as diagramed in Figure 6.3. The only instrumentation which these boats have to allow them to perform their station keeping function operates by sending sound waves through the water. On each boat, time is measured by a sonic clock whose unit of measurement is the round trip time of a sonic signal sent vertically from the bottom of the boat to a plate mounted a short distance below and reflected back to the boat. Each boat also sends a sonic signal to the bulkhead and measures the time, using its sonic clock, required to receive the reflection of that signal. It is then steered so as to maintain that time unchanged. Finally, a sonic signal is sent from the following boat to a retroreflector on the rear of the leading boat. The retroreflector returns that signal to the following boat. The rear boat adjusts its speed to maintain the time for the round trip signal, as measured by the sonic clock, unchanged.

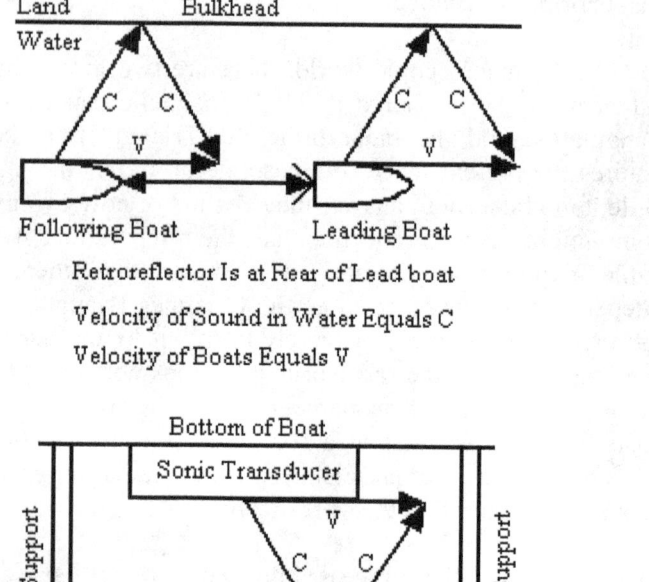

Figure 6.3 - Station Keeping Boats

6.14- The time required for sound to travel though water between two points is determined by three parameters, the distance between the points, the velocity of sound in water, and the velocity of the points of transmission and reception through the water. First consider the case of the round trip signal between the following boat to the leading boat. While the signal is traveling through the water, it propagates at the velocity of sound in the water, C. When the signal is sent from the following boat to the leading boat which is a distance L ahead, the receiving point is running away from the signal at the speed of the boat, V, and the time required for the outward trip is given by $T_O=L/(C-V)$. When the signal is returned, the following boat approaches the signal at the speed of V, and the time required for the return trip is given by $T_R*L/(C+V)$. The time for the round trip, T, is the sum of these times, $T=2*L*C/(C^2-V^2)$, or $T=2*L/B_V^2$.

6.15- When the signal is sent in a direction at right angles to the velocity of the boats, as occurs in the case of the signal of the sonic clock and in the signal reflected from the bulkhead, a different result occurs. While the signal is enroute, the receiving point moves laterally by an amount determined by the velocity of the boat and the time for the signal to make the round trip. As a result, the round trip signal has traveled a distance equal to the vector sum of twice the nominal distance to the target plus the distance that the boat has traveled during the round trip. For this situation, the distance the signal has traveled during the round trip has been increased by $C/(C^2-V^2)^{0.5}$, or $1/B_V$, in accordance with the Pythagorean Theorem for the sides of a right triangle. The effect causes the round trip time for the signal between the boat and the bulkhead to increase by the same ratio and also reduces the speed of the sonic clock by $(C^2-V^2)^{0.5}/C$, or B_V. With respect to the measurement of the distance between the boats and the bulkhead, the two effects cancel, and the distance the boats maintain from the bulkhead is independent of their velocity through the water. Unlike the situation occurring with the round trip of the signal to the bulkhead, the increase of the round trip time for the signal traveling between the following and leading boats is only partially canceled by the slowing of their sonic clocks. For this situation, the round trip time is increased in proportion to the square of the slowing of the sonic clock. In order to maintain correct separation between boats, as measured by signals sent through the water, it is necessary for the operator of the following boat to move closer to the leading boat. He must reduce the distance between the boats by a factor of $(1-V^2/C^2)^{0.5}$, or B_V.

6.16- To confine the analogy further, let us assume that the observers on the boats can only communicate between the two boats and between each boat and the bulkhead by means of sonic signals sent through the water. With this limitation, their only means of measuring the velocity of their boats through the water would be by timing a round trip signal between the boats using their sonic clocks. As a result, they would always measure their velocity through the water as zero regardless of their actual velocity. This is exactly analogous to what occurs in the processes described both by the Special Theory of Relativity and by the Aether Relativity Theory. If we allow the boats to communicate by radio as well as by sonic signals, they would quickly discover that clocks which were supposedly synchronized by the sonic signals sent through the water were not actually synchronized. The clock in the lead boat would be be set to an earlier time than the clock in the following boat. The amount of time that the leading clock is early would provide the information required to calculate the velocity of the boats through the water just as the ability to communicate at a velocity significantly greater than the velocity of light would allow us to determine our absolute velocity through space by establishing an absolute time reference. (Obviously, real clocks do not function in the manner of the sonic clock described above, but they obey the same Lorentz Transformation for Time as a function of their absolute velocity through space as does the sonic clock as a function of its velocity through the water. The Lorentz Transformation for Time will be discussed later.)

6.17- If we extend the station keeping analogy to a long convoy of boats, we obtain an analogy to a yardstick in which the separation of its atoms (and therefore its length), and the speed of its clock both are determined by the velocity though the medium (water or the Aether). The length of that yardstick obeys the Lorentz Transformations for Length in both axes. As a result, it is impossible for observers to determine their absolute velocity through space. Matter adjusts its size and clock speed to conceal that velocity. The concealment is made possible by the fact that the establishment of simultaneity between physically separated locations is limited by the finite velocity of propagation of information imposed by the speed of light. We cannot observe our absolute velocity through the Aether because Nature uses the velocity of light to determine the size of the matter which comprises our instruments.

6.18- The Impossibility of Measuring the Velocity of Light:- The velocity of light is a sacred and immutable quantity in the scientific community, no matter where or how or by whom it is measured, it always has the value of 186,236 miles per second. The quantity is so basic that its measurement is often part of the training process for PhD candidates. Now for the ultimate heresy. *The velocity of light has never been measured and it never will be measured!* When experiments which purport to measure that velocity are examined, it is found that they violate one of the basic rules of measurement. When making a measurement, it is necessary to compensate for any effect that the quantity being measured has on the scale factors of the instruments which are used. To the author's knowledge, this step has never been included in the measurement of the velocity of light.

6.19- Consider an attempt to measure the velocity of light in which a measurement is made of the time required for a pulse of light to be sent from the top of one mountain to a retroreflector on the top of an adjacent mountain and returned to its source. The experiment requires the use of a precise clock and a precise knowledge of the distance between the signal source and the retroreflector. Atomic clocks of extreme precision and accuracy are available and are readily transportable to the mountaintop. Determining the distance between the light source and the retroreflector is difficult since it involves precision surveying over a long distance of mountainous terrain. To overcome the difficulty of the survey, it is decided to measure the distance to the retroreflector by radar. The resulting experiment produces the correct value for the velocity of light. It concludes that C=C. Unfortunately, while such a result is correct, it is hardly useful.

6.20- While this example may appear frivolous, it is not. It is a valid analog of reality. Quantum physicists have concluded that the force between material particles is electromagnetic in nature and is asserted to result from the exchange of virtual photons. As a result, they are also asserting that the spacing between these particles is determined by the radar principle and therefore is in agreement with the predictions of Velocity Relativity Theory. Any attempt to measure the velocity of light must yield the meaningless conclusion that the velocity of light is equal to the velocity of light.

6.21- The making of a measurement requires, in effect, the writing of an equation in which the quantity to be measured appears only on the left side of the equal sign and all other quantities appear only on the right side. If the velocity of light, C, is to be measured, it is necessary that both the time, T, required for light to travel a distance, L, and the distance, L, be measured in a manner which is independent of C. The velocity of light may then be found by solving the equation $C=L/T$. The difficulty arises from the fact that both the measurement of time and the measurement of length involve the velocity of light. The length of the yardstick used to measure the length is asserted to be determined by the alleged exchange of virtual photons between atoms. The speed of all clocks (including atomic clocks) is determined by the resonant frequency of some type of oscillating spring-mass system. One of the factors which determines the frequency of such a system is the elasticity of the spring which is determined by the exchange of the alleged virtual photons. The other factor which determines its frequency is the mass of the oscillating system as determined by the energy represented by its mass divided by the square of the velocity of light. Obviously, the equation by which one would expect to use in measuring the velocity of light is not quite applicable.

6.22- While the author has not attempted the derivation, he is drawn to the conclusion that the correct equation for the measurement of the velocity of light is a rearrangement of the Fine Structure Constant, $\varepsilon * h * C / e^2 = 137$, where h is Planck's Constant, e is the charge of the electron, and ε is the dielectric constant of space. (ε is required if the equation is to be dimensionally correct. The current practice of omitting it from the equation of the Fine Structure Constant is erroneous.) With this rearrangement, the equation for the measurement of the velocity of light becomes $C=137*e^2/(\varepsilon*h)$. The reality of Relativity is that matter adjusts its size to satisfy this equation and any measurement which attempts to measure the velocity of light actually measures the Fine Structure Constant. Since this constant is dimensionless, it is the same at all velocities through space and at all elevations. Unless a physicist has been completely brainwashed during his education, he will recognize that the observed constancy of the velocity of light is not mysterious, it is inevitable and is completely unrelated to the actual velocity of light.

6.23- In Figure 6.4 three velocity reference frames are considered, "X", "A", and "B". It will be noted that, in accordance with the concepts of both the Special Theory of Relativity and the Aether Theory of Relativity, the relative velocity of "B" with respect to "X", V_{BX}, is provided in terms as the sum of the velocity between "B" and "A", V_{BA}, and the velocity between "A" and "X", as the sum of V_{AX} and V_{BA} divided by a factor, $1+V_{AX}*V_{BX}/C^2$. The denominator is required to compensate for the effect of the finite velocity of light on the measurement of the velocity differences. It is the effect represented by this term which prevents the direct addition of relativistic velocities and which prevents the observed difference of velocity between any two reference frames from exceeding the velocity of light. The denominator in the equation results from the limitation that the velocity of light imposes on the velocity of communication between reference frames. (The denominator becomes unity if the experimenter communicates at an infinite velocity, possibly through the use of paired photons.) If the product $V_{XA}*V_{AB}$ in the denominator is small compared to the square of the velocity of light, its effects can be ignored, non-relativistic mechanics are valid, and velocities may be added directly. This conclusion will become important when we examine the deficiency of Special Relativity with regard to accelerations.

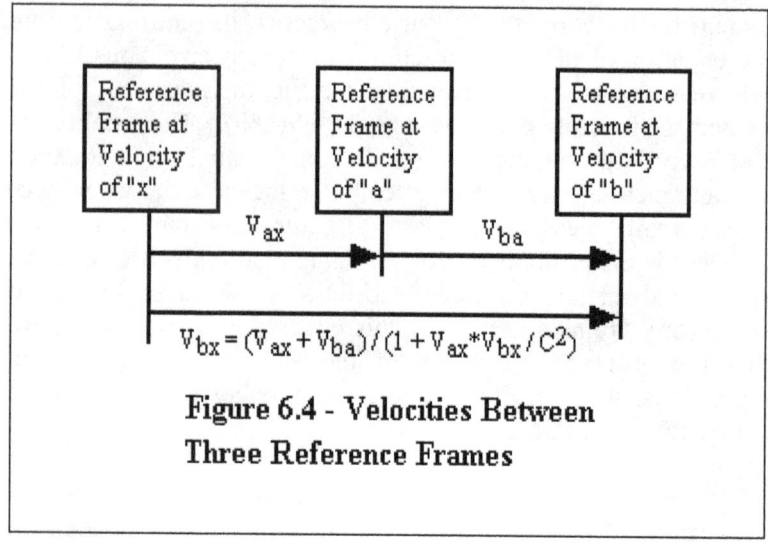

Figure 6.4 - Velocities Between Three Reference Frames

6.24- Consider that observers in reference frames "A" and "B" of Figure 6.4 wish to make observations between their reference frames. In order to insure that their observations are based upon the same reference frame, they agree to convert the results of their observations into observations as they would have been made in reference frame "X". After making observations within and between reference frames "A" and "B" they convert those results into the results which would have been observed in that reference frame and communicate those results to each other. After laborious manipulation of many pages of algebra they find that all of the terms in the equations which refer to reference frame "X" cancel and they are left with only those terms which refer to the effects occurring in and between velocity between reference frames "A" and "B". Reference frame "X" does not exist in the solution, and, allowing that reference frame "X" represents the velocity reference frame of the Aether, it becomes obvious why our velocity with respect to that Aether cannot be observed.

6.25- The addition of velocities as defined by the Special Theory of Relativity and by Aether Relativity produces a conflict with common sense. Under Special Relativity, when one adds the velocity of one reference frame to the velocity of another reference frame, the denominator in the velocity addition equation described above insures that the velocity difference between those reference frames is not equal to the algebraic difference. When the velocities approach the velocity of light, the effect is so pronounced that, when the algebraic velocity difference approaches twice the velocity of light, the observed velocity difference remains less than the velocity of light. (See Figure 6.5A.) The strange nature of this curve results from the fact that, under Special Relativity, observers in reference frames "A" and "B" are both free to consider themselves at rest and that the other observer is moving. Both conclusions cannot be true, and as above, the only conceptually valid means of dealing with the observations is for the observers in both reference frames to agree upon a reference frame which they accept as stationary and to use the mathematics of either Special or Aether Relativity to convert their observations to the results which would be obtained if the observations had been made in the stationary reference frame. When that step is taken, the addition of velocities between reference frames "A" and "B" becomes consistent with common sense. (See Figure 6.5B.) Under the Aether Relativity Theory, the confusion about the adding of velocities does not occur, the absolute velocity reference represented by the Aether forces the observers to make the necessary corrections to their observations.

6.26- At this point, a conventional relativist will ask why it should be necessary to assume the existence of an artificially selected velocity reference frame when that reference frame does not appear in the data. Perhaps the simplest response to that question is to direct the reader's attention to the problem of navigation on the surface of the Earth. A position on the Earth's surface is defined in terms of its latitude and longitude. The observation of latitude presents no problem, the equator provides an observable absolute reference for zero latitude at a location 90 degrees away from the spin axis of the Earth. The observation of longitude does present a problem. There is no absolute reference

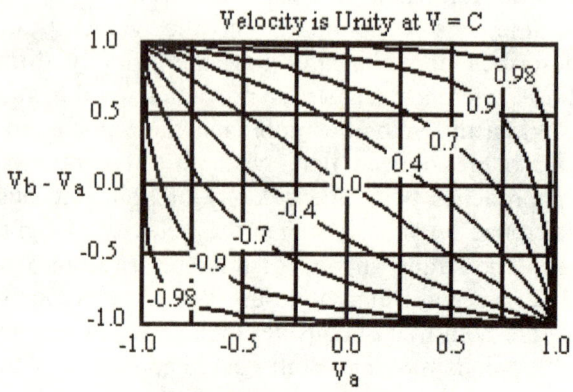

**A:- The Relativistic Addition of Velocities.
This is what we observe.**

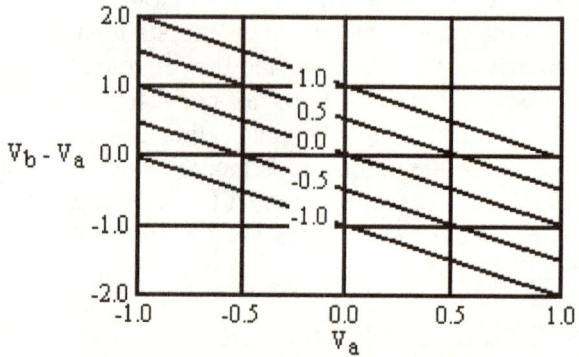

**B:- The Actual Addition of Velocities.
This is what actually occurs.**

Figure 6.5 - The Relativistic and Actual Addition of Velocities

for zero longitude. Our ancestors, however, were practical men. They arbitrarily defined the zero meridian of longitude as passing through Greenwich England and based all observations of longitude upon that artificially chosen absolute longitude reference. ***In order for an observation to produce rigorously correct results between reference frames, compensation of the effects of the difference in reference frames on the units of measurement is required. This, in turn, requires that a reference frame be arbitrarily chosen as a standard. If this step is not taken, GIGO prevails..***

6.27- The Location of Kinetic Energy:- When a bullet is fired from a gun, kinetic energy is added to the projectile by the expanding gases in the gun barrel. That kinetic energy is eventually imparted to the target, but during the time of the bullet's flight, it travels with the bullet. The Lorentz Transformations may be used to determine the location of that kinetic energy. That determination is readily made with the use of a thought experiment. Consider that energy has been stored in an ideal massless spring by compressing and tying it, as shown in Figure 6.6 (Due to space limitations, Figure 6.6 is divided into two parts which should be read together as one diagram.) Since the action of the spring is one dimensional, energy is stored along a single axis. Consider next that the bullet consists of three springs with identical energies of compression mounted orthogonal to each other with one of the springs being parallel to the bullet's projected path. Since, for our thought experiment, the springs themselves are considered to be massless, the only mass the bullet possesses is the energy stored by the compression of the three springs.

6.28- When the springs acquire a velocity, the acquisition of that velocity causes kinetic energy to be added to the energy of compression which had been stored in them. Applying the Lorentz Transformation for Mass and multiplying by the square of the velocity of light shows that the total energy of the springs has been increased in proportion to $1/B_V$. The product of the Lorentz Transformation for Transverse Length times the Lorentz

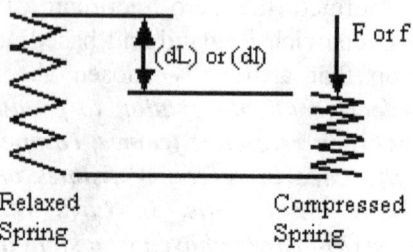

Relaxed Spring   Compressed Spring

Spring is considered to be ideal and massless with only the energy stored in the spring having inertial mass.

$e_T = f_T * (dl_T)/2$ $e_P = f_P * (dl_P)/2$

$E_T = F_T * (dL_T)/2$ $E_P = F_P * (dL_T)/2$

$E_T = e_T * (1 - V^2/C^2)^{0.5}$ $E_P = e_P * (1 - V^2/C^2)^{0.5}$

Object consisting of energy stored in three mutually perpendicular compressed springs as observed in local reference frame. When local reference frame is stationary, $e_T = E_T$ and $e_P = E_P$

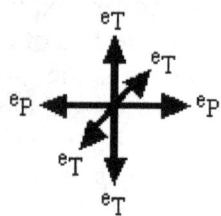

"Stationary" Reference Frame

Figure 6.6 - The Location of Kinetic Energy
(Continued on Next Page)

Contiuation of Figure 6.6 showing the same object "moving" with respect to observer at a velocity of 0.86*C.

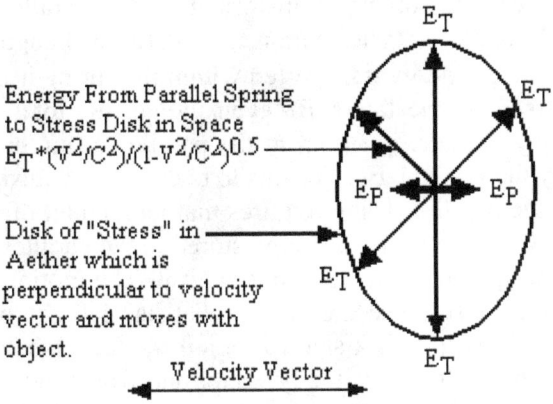

Energy From Parallel Spring to Stress Disk in Space
$E_T*(V^2/C^2)/(1-V^2/C^2)^{0.5}$

Disk of "Stress" in Aether which is perpendicular to velocity vector and moves with object.

Velocity Vector

"Moving" Reference Frame

In "moving" object, energy stored in the transverse springs is increased in proportion to the sum of the stored energy plus the kinetic energy. The energy stored in the parallel spring is reduced in proportion to the Lorentz Transformation and equals zero at $V=C$. The remainder of the parallel spring's stored energy and kinetic energy, equal to V^2/C^2 times its total energy, is transported as a perpendicular shock wave in a disk like region of the Aether which moves with the object.

Figure 6.6 - The Location of Kinetic Energy
(Continued From Previous Page)

Transformation for Transverse Force shows that the energy stored in the transverse springs has been increased by the same factor. For these transverse springs, therefore, it is apparent that their kinetic energy is stored as an increase in their energy of compression and that stored energy is returned when the bullet is brought to rest. In the parallel axis, the situation is more subtle. The product of the Lorentz Transformation for Parallel Force times the Lorentz Transformation for Parallel Length, and therefore the total energy transported within the spring itself, has been **decreased** by the factor B_V even though the total energy transported by the energy stored in the parallel spring has been increased by the factor $1/B_V$. For this to occur, the acquisition of velocity by the parallel spring requires that an amount of energy equal to $V^2/(C^2*B_V)$ times the energy stored in the parallel spring leave that spring and yet travel along with it. The motion of the parallel spring must cause the energy that has left the parallel spring to be stored in a disk shaped region of space located in a plane perpendicular to the velocity vector and which moves with the parallel spring. For that energy to be stored, the Aether must be distorted and stressed in the region involved. For Newton's Laws of Motion to be valid, the interchange of energy between the source/sink of kinetic energy and both the springs and the disk shaped distortion and stress in the Aether must occur at 100% efficiency. It also follows that the inertial forces associated with a change in velocity are not fictitious as modern physicists would have you believe, they are, like the force of gravity, a real force exerted against the Aether.

6.29- A Model for the Neutrino? In the previous paragraphs it was pointed out that the Lorentz Transformations for Force and Length require that part of the rest mass energy and **all** of the kinetic energy associated with energy stored in a direction parallel to the velocity vector must be transported in a disk shaped region of stressed space moving with that energy. One would conclude, however, that, at $V=C$ if the original rest mass energy were greater than zero, the energy stored in the disk would become infinite, exactly as would the energy stored in a direction perpendicular to the velocity vector.

6.30- Let us consider a situation in which the rest mass energy is stored only in a direction parallel to the velocity vector. As shown in Figure 6.6, the energy stored in the spring is given by $E_S = e_P*(1-V^2/C^2)^{0.5}$ and the energy stored in the disk of stressed space is given by $E_D = e_P*(V/C)^2/(1-V^2/C^2)^{0.5}$. Now let us consider that the amount of the original rest mass energy is reduced as a function of the velocity, V_F, to be attained (a series of experiments is required, one for each of the final velocities, V_F) so that the energy in the disk will approach the nominal rest mass energy, e_P, as the V_F of the individual experiments approaches C. This can be achieved by reducing the initial rest mass energy, e_P in proportion to $(1-V_F^2/C^2)^{0.5}$. The expression for the energy transported by the spring for each of the values of V_F becomes $E_S = e_P*(1-V_F^2/C^2)$ and the energy transported by the disk of stressed space for each of the values of V_F becomes $e_P*(V_F/C)^2$.

6.31- Let is now consider values of V_F which approach the value of C as a limit. As V_F approaches C, the portion of the total energy stored in the spring approaches zero as a limit while the portion of the energy which is stored in the disk approaches the original rest mass energy as a limit. A direct determination of the energy in the spring when V_F equals C is obvious, it is zero. A direct determination of the energy in the disk would appear to be meaningless since it would involve the multiplication of zero by infinity and, as a result, can have any value between the limits of +/-infinity. The actual value can be determined however by using the same procedure as is used in integral calculus, allowing V_F to approach infinitesimally close to C and determining the amount energy in the disk when the velocity of C is approached as a limit. This method allows us to conclude that, when V_F is equal to C, all of the energy is stored in the stressed disk in space and none of the energy is stored in the spring. In the case under consideration, the amount of the energy stored in the disk is equal to the energy e_F. We have now described a particle (Figure 6.7) which does not have a charge, possess no rest mass, travels at the velocity of light and transports energy and momentum. Has a neutrino been described? The author believes it has.

6.32- How Could Such a Neutrino be Launched and/or Absorbed? The mechanism of launching or absorbing a neutrino which was constructed as described in the proceeding two paragraphs might seem to pose a conceptual problem. This would be the case if the formation of the neutrino involved an acceleration from rest to the velocity of light and/or if the absorption of the neutrino involved a deceleration from the velocity of light to the rest velocity. There is no conceptual difficulty however if the neutrino were formed or absorbed already traveling at the velocity of light. Such an emission/absorption characteristic is already known. When photons are emitted or absorbed, the process involved behaves in exactly this manner, at least when observed for a time which is longer than the period of the photon. One would expect such neutrinos to have a discrete frequency just as the photon has a discrete frequency and one would expect it to be emitted and absorbed in discrete quasi-spectral

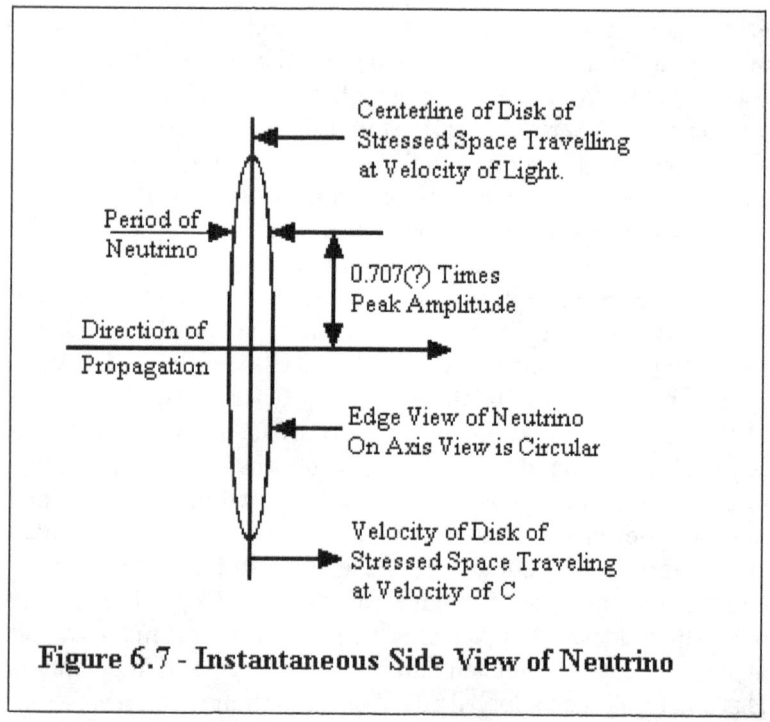

Figure 6.7 - Instantaneous Side View of Neutrino

lines analogous to the absorption and emission of the spectral lines of photons from atoms. If neutrinos are emitted in that manner by nuclear processes in a star, frequency shifts due to thermal Doppler and of gravitational time dilation should reduce the observable emission of neutrinos by a star. At present, experimenters are trying to account for the fact that the neutrino emission from the Sun is about a third of what they expect. Perhaps the emission spectrum of the neutrinos is sufficiently broadened so as to prevent their detection by present methods.

6.33- How is the "Aiming Problem" Implicit in Figure 6.3 Resolved?- In Figure 6.3 sound waves are shown being projected from the ships to the bulkhead in the direction required to allow them to arrive at each ship's new position after they have been reflected. The question has been posed to the writer as to how the projecting mechanism can provide the aiming necessary to achieve this result when it is obviously directed perpendicular to the bulkhead. The answer is in the simultaneity effect. In this system, the time delay between the forward edge of the laterally directed projector and its trailing edge causes the leading edge of the sonic wave fronts being projected to lag behind their trailing edges. Since waves travel in a direction perpendicular to their wave fronts, the sonic signal is projected forward as required. It is easy to show that the amount of this "lead angle" is precisely the amount required to satisfy the requirements of Figure 6.3. The same effect applies to light traveling though an "Aether Wind".

Chapter 7 - Applying the Lorentz Transformations Properly

7.1- In the gravitational field, it has been predicted and experimentally verified that differences in elevation result in differences in the observed rate of passage of time. Earlier in this discussion, it was shown that the difference occurs because a change in elevation changes the calibration of clocks and does not, in contrast to present dogma, affect the absolute rate of the passage of time. Since a change in reference frame (elevation) changes the speed of a clock, it is not only reasonable to expect that the scale factor of other types of instruments also be affected, the rules of Dimensional Analysis and the Principle of Relativity require that such be the case. If one is to examine the effects of velocity and/or elevation therefore, it is necessary for him make observations within each of the reference frames using the units of measurement of that reference frame (local units of measurement). He must then choose a standard velocity reference frame and a standard elevation reference frame and use the Velocity and/or Gravity Transformations to convert the results of his observations into those which would have been obtained with the units of measurement of the standard reference frame. The procedure is analogous to the practice employed by surveyors who adjust distances measured with steel tapes for the error caused by the difference between the ambient temperature and the temperature at which the tape was calibrated. Somehow, it seems reasonable that physicists should be expected to adhere to the level of rigor required of the more mundane field of surveying. (It is the change in the speed of clocks resulting from a change in reference frames that produced the Twin Paradox of Special Relativity. There is no paradox. A twin who returned from a high speed trip would be younger than his sibling who had remained stationary because his biological clock would, on average, have run slower.)

7.2- In order to insure that the basic system of units represent directly observable entities, it is necessary to convert the familiar

Mass-Length-Time system of units to a Force-Length-Time system. (For a definition of the basic units of this system see Table 7.2.1.) Unlike force, which is directly observable and is a fundamental entity in its own right, mass has no independent existence and cannot be directly observed. The mass of an object (or particle) can be observed in one or more of three ways. It can be observed in terms of its inertia as the incremental impulse (force-time product) required to produce an incremental change in velocity (length/time ratio). It can be observed in terms of the gravitational force between the energy (force-time product) equivalents of two masses divided by the square of their separation (length-length product). Or it may be determined by the energy (force-length product) released when the object is converted to radiation. Accepted texts on Special Relativity correctly provide the Lorentz Transformations for Time, Parallel Length, Transverse Length, and Parallel Force. Unfortunately, an unrecognized error was been made in the derivation of the Lorentz Transformation for Transverse Force. The transformation provided in texts is $1/B_V$ whereas the correct transformation is B_V. The existence of this error is revealed by the Right Angle Lever Paradox discussed below. (A rigorous derivation of the Lorentz Transformations for Parallel and Transverse Force and the Right Angle Lever Paradox is included in Appendix 2.)

Table 7.2.1:- Revised Lorentz Transformations

Quantity	Symbol	Parallel	Transverse
Force	F	1	B_V
Length	L	$1/B_V$	1
Time	T	B_V	B_V

Note on Table 7.2.1:- Current texts erroneously provide $1/B_V$ as the Lorentz Transformation for Transverse Force.)

$B_V = (1 - V^2/C^2)^{0.5}$

7.3- Two errors in present Velocity Relativity Theory have been eliminated. The first error was eliminated by the recognition that inertial mass is properly represented by its incremental mass, M_I, because inertial mass refers to effects which occur under conditions where the change in velocity is small. This correction allows the mathematics of Velocity Relativity Theory, as Special and/or Aether Relativity Theory will designated from this point on, to be valid for accelerated reference frames. Indeed they must be valid for accelerated reference frames since acceleration is the second derivative of length with respect to time and both terms are subject to their respective Lorentz Transformations. The second error is eliminated by providing the correct Lorentz Transformation for Transverse Force. With these corrections, it is possible to provide the Lorentz Transformations based upon a Force-Length-Time system of units rather than the conventional Mass-Length-Time system.

7.4 A paradox, known as the Right Angle Lever Paradox, was discovered early on in discussions of the Special Theory of Relativity. As is the case with all paradoxes, its existence revealed that an error has been made. The error occurred in the initial derivation of the Lorentz Transformation for Transverse Force and correcting that error eliminates the paradox. (The correct transformation is readily derived and is provided in "Special Relativity Corrections" in Appendix B.) However, instead of recognizing its existence and working to find the source of the error, the academic community elected to accept the erroneous Lorentz Transformation for Transverse Force as correct and seek an esoteric method of resolution. After all, a religion had been established and it would not do for priests of that religion to admit to having been in error.

7.5- The Right Angle Lever Paradox is illustrated in Figure 7.1. Consider the condition where a right angle lever having equal length arms in its own reference frame is moving with respect to a reference frame which is considered to be stationary. A force is applied to the end of one of the arms of the lever which is prevented from rotating by an equal force applied to the other arm. Since the lever is observed not to rotate in either the stationary or the moving reference frame, it follows that the net torque applied to it in each of the reference frames is zero. From the diagram we may write, for the moving reference frame, $F_{TM}*L_{PM} = F_{PM}*L_{TM}$ and, since the arms are of equal length in that reference frame, it follows that $F_{TM} = F_{PM}$. We should also be able to write $F_{TS}*L_{PS} = F_{PS}*L_{TS}$, but, in the stationary reference frame, the length of the parallel arm, L_{PS}, is reduced by the effects of the velocity in accordance with the Lorentz Transformation for Length, B_V. For the net torque to be zero in the stationary reference rame, F_{TS} must equal to F_{PS}/B_V. Minkowski

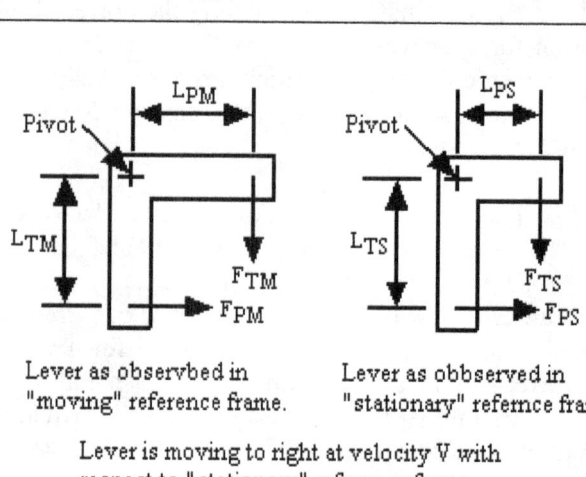

Lever as observbed in "moving" reference frame.

Lever as obbserved in "stationary" refernce frame.

Lever is moving to right at velocity V with respect to "stationary" reference frame.

$L_{PS} = L_{PM}*(1 - V^2/C^2)^{0.5}$; $L_{TS} = L_{TM}$; $F_{PS} = F_{PM}$

Under Special Relativity:-

$F_{TS} = F_{TM}*(1 - V^2/C^2)^{0.5}$
Torques are unbalanced in both reference frames.
T_P does not equal T_T.

After Error Correction:

$F_{TS} = F_{TM}/(1 - V^2/C^2)^{0.5}$
Torques are balanced in both reference frames. $T_P = T_T$

Figure 7.1 - The Right Angle Lever Paradox

correctly provided the Lorentz Transformation for Parallel Force as equal to unity but the currently accepted Lorentz Transformation for Transverse Force is the reciprocal of its correct value. This error forces one to conclude that the observed angular acceleration of the Right Angle Lever of Figure 7.1 cannot be zero in both the stationary and moving reference frames. The lever, not knowing this to be impossible, does not undergo angular acceleration in either reference frame and the Right Angle Lever Paradox results.

7.6- To by-pass the need to admit that an error had been made and correcting the Lorentz Transformation for Transverse Force, relativistic theorists devised a rather imaginative explanation. This explanation has appeared in more than one postgraduate text and conflicts so severely with common sense that many teachers of Relativity by-pass the topic despite the brainwashing they underwent in the process of attaining their positions. In this explanation, the rate at which the torque unbalance of the lever, as observed in the stationary reference frame, increases its angular momentum is countered by the rate that energy is added to the lever by the force, F_{PS}. In undergraduate Physics 101 (Mechanics), which is a prerequisite to receiving a PhD in Physics, one is taught that the existence of a moment requires the existence of two equal and opposite forces separated by a distance. In this case, the reaction force components existing at the hinge pin provide the second forces. As a result, any energy added at the end of the lever is immediately removed at the hinge pin and the rate of change of energy in the lever is zero. One is also taught in Physics 101 that the angular momentum of an object is the product of its moment of inertia and its angular velocity. Since the angular velocity of the lever remains zero in both reference frames, the rate of change of its angular momentum is also zero. The supposed resolution of the Right Angle Lever Paradox degrades to the statement that zero equals zero. This conclusion is most certainly true, but it is hardly very useful. The only means of resolving the Right Angle Lever Paradox is to correct the error in the Lorentz Transformation for Transverse Force.

7.7- Dimensional Analysis as Applied to Relativistic Phenomena:- While it is not commonly recognized, Dimensional Analysis is the most effective tool available for the investigation of the effects of a change in velocity and/or elevation. Observations are translatable into equations, such as $V=\delta L/\delta T$. [This equation states that the velocity at which an object is moving is equal to the incremental distance it travels, δL, divided by the incremental time, δT, required for it to travel that distance.] Usage of Dimensional Analysis is simplified by the fact that only three independent dimensional entities are required. (More than three dimensional entities are found to be redundant.) The dimensional content of every parameter encountered in an observation may be derived from those three entities because each term in an equation describing a physical process or phenomena must have the same content of dimensional entities. Apples must not be equated to oranges.

7.8- While Dimensional Analysis obviously applies within a given reference frame, the Principle of Relativity adds the requirement that, with the appropriate transformations, it must also apply between reference frames which differ in velocity and/or elevation. (For reference frames differing in velocity, these transformations are the Lorentz Transformations which will be termed Velocity Transformations from this point on. For reference frames which differ in elevation, an equivalent set of transformations termed Gravity Transformations is required.) Combining the rules of Dimensional Analysis with the Principle of Relativity allows the phenomena associated with relativistic effects to be unpeeled so that they may be understood at the common sense level. To facilitate such a use of Dimensional Analysis, Table 7.8.1 provides the dimensional content of various physical quantities. Based upon the earlier discussion of the meaning of mass, the table is based upon the more rational Force-Length-Time (FLT) system of units rather than upon the conventional Mass-Length-Time (MLT) system.

Table 7.8.1 - The Dimensional Entities Contained in Various Physical Quantities

Quantity	Symbol	Dimensional Content
Force, F	F	F
Length, L	L	L
Time, T	T	T
Energy	E	$F*L$
Planck's Constant	H	$F*L*T$
Velocity	V	L/T
Acceleration	A	L/T^2
Incremental Mass	M_I	$F*T^2/L$
Momentum	U	$F*T$
Angular Momentum	J	$F*L*T$
Gravitational Constant	G	$L^4/(F*T^4)$
Ergo-gravitational Constant	D	$1/F$
Temperature	ϕ	$F*L$
Charge	Q	L
Dielectric Constant of Space	ε	$1/F$
Permeability of Space	μ	$F*T^2/L^2$

Notes on Table 7.8.1:-

- Current texts erroneously provide $1/B_V$ as the Lorentz Transformation for Transverse Force.)

- The dimensional content for velocity is unaffected by the relativistic correction term for the addition of velocities since that correction term is dimensionless.

- The dimensional content of the gravitational constant is determined from the expression for Newtonian gravitational force, $F = G*M_{I1}*M_{I2}/L^2$.

- The ergo-gravitational constant is the conventional gravitational constant defined in terms of the energy equivalents of the gravitating masses. It is identical to Dr. Einstein's Cosmological Constant and is equal to G/C^4.

- Temperature is kinetic energy per available degree of freedom and has the dimensional content of energy.

- Evaluation of the dimensional content of the expression for the electrostatic force between charges provides $Q^2/\varepsilon = F*L^2$.

- Evaluation of the expression for the electromagnetic force between moving charges provides $Q^2*\mu = F*T^2$. The velocity of light is given by $C = (\varepsilon*\mu)^{0.5}$.

If one accepts the precept that the exponent of a dimensional entity must be an integer, there are two possibilities.

- The first possibility is that the dimensional content of Q is equal to L, the dimensional content of ε is equal to $1/F$, and the dimensional content of μ is equal to $F*T^2/L^2$.

- The second possibility is that the dimensional content of Q is equal to unity, the dimensional content of ε is equal to $1/(F*L^2)$, and the dimensional content of μ is equal to $F*T^2$.

The dimensional content for charge, Q, must equal to the dimensional content for length, L, in order for the Principle of Relativity to apply, and this value, and its related values for ε and μ is chosen for Table 7.8.1.

7.9- At this point, it is time to consider the observation of the velocity of light in a reference frame different from the one in which the observer finds himself. Everyone who has been exposed to an undergraduate level course in physics has been exposed to the idea that the velocity of light is constant, yet, as has already been discussed, the velocity of light has never been measured and can never be measured because the calibration of instruments is affected by changes in the velocity of light. The velocity of light is constant only when it is measured by a local observer using local units of measurement.

7.10- Consider the case of an observer in the stationary reference frame in communication with an observer in a moving reference frame. The observer in the moving reference frame measures the velocity of light in a direction along the relative velocity vector using local units of measurement and reports its value as its standard value of C. The observer in the stationary reference frame, knowing that the instruments used to make the measurement in the moving reference frame have been distorted by the effects of its velocity, compensates the reported value of the velocity of light in the moving reference frame using the appropriate Velocity Transformations and obtains $c=B_V^2*C$, which may be rewritten as $B_V=(c/C)^{0.5}$. It follows that the Velocity Transformation term, B_V, is the square root of the ratio of the velocity of light in the other reference frame divided its velocity in the base reference frame. As will be seen, when the behavior of the gravitational field is correctly described, this definition holds for gravitational transformations between elevations as defined in terms of B_G.

7.11- It has been suggested that mass could also be determined by counting particles, and, in response, the following clarification is provided:

- In attempting to use mass as a fundamental observable entity, care must be taken to insure that the observation of mass does not require knowledge of another mass(es), unless that mass(es) has been determined by an independent means. Otherwise, such an observation defines mass in terms of itself, an obvious absurdity.

- When one determines mass by "weighing", the necessary independent means is available. The laws of orbital mechanics allow the determination of the mass of a primary object to be determined in terms of the orbital period of a satellite and the gravitational constant. The gravitational constant, in turn, can be measured in terms of inertial mass by means of the Eotvos Experiment of Figure 8.1.

- In order to determine the mass of an object by counting its particles, a similar independent means of measuring the mass of the particles is required. This process is complicated by the fact that the mass of a compound particle (e.g..- an atom having an atomic weight greater than 1, or a molecule) does not equal the sum of the masses of its component parts (e.g.- the mass of a helium atom is less of the mass of four hydrogen atoms due to the energy released during its fusion). A mass spectrometer could be employed to determine the mass of every particle, atom, or molecule of interest, but, as is the case of "weighing", such usage would define mass in terms of its inertial effects.

It would appear, then, that a refinement of our understanding is required. There are only two primary means of observing mass. It may be observed by the energy released during annihilation and it may be observed by "shaking". "Weighing" or "counting" are a secondary means of observation which rely on the primary means for their validity.

Chapter 8 - Generating the Gravity Transformations

8.1- Enter the Principle of Equivalence:- Dr. Einstein made an enormous contribution to our understanding of physical reality by incorporating the Principle of Equivalence into gravitational theory. Proper use of this principle results in the assertion that the effects of velocity and gravity are equivalent and that gravitation is a relativistic phenomena. If that principle correctly applies to the gravitational field, it must be possible to generate Force-Length-Time Transformations between gravitational reference frames which are equivalent to the Force-Length-Time Transformations between velocity reference frames. Once derived, these gravitational transformations allow the actual effects of a change in elevation to be observed and, as shall be seen, will reveal far more about reality than one might reasonably hope.

8.2- It must be pointed out that, in most texts (including the writings of Dr. Einstein), the meaning of the Principle of Equivalence is overstated. It is commonly asserted that there is no observable difference between a gravitational acceleration and an inertial acceleration. This conclusion is not quite true. The force observed as a result of inertial acceleration is always accompanied by an observable change in velocity since inertial acceleration is the rate of change of velocity with respect to time. The force due to gravitational acceleration, on the other hand, does not produce a change in velocity, but it is accompanied by a gradient in acceleration due to the curvature associated with all gravitational fields. (The force of gravity is less at the ceiling than it is at the floor.) The force which is observed is the sum of these forces, greatly complicating the design of the inertial navigation systems employed in aircraft and submarines by requiring that they be able to distinguish between the two effects. The only conclusion that can rigorously be drawn from the Principle of Equivalence is that the gravity transformations must be exactly analogous to the Lorentz Transformations. These transformations are readily derived and are provided in Table 8.9.1.

8.3- Disillusion sets in, however, when one realizes that the use of the Principle of Equivalence in the generation of General Relativity was based on the observation that inertial and gravitational masses were identical when evaluated in terms of the force produced by acceleration, as expressed by the equation $F=M*A$, and the force produced by gravitational attraction, as expressed by the equation $F=G*M_1*M_2/L^2$. The most precise determination of the gravitational constant, G, was made by the Eotvos Experiment illustrated in Figure 8.1. In this experiment, three identical metal spheres were used. Two of these spheres were mounted in a dumbbell configuration on a horizontal rod suspended from a fine torsion wire and the resonant frequency of their suspended masses was determined. The third test mass was placed a known distance from one of the suspended masses and the torsional deflection of the suspended masses resulting from the gravitational attraction between the masses was also determined. From these observations it is possible to unequivocally evaluate the gravitational constant, G. So far so good, but then circular reasoning was employed to determine the relationship between gravitational mass and inertial mass using the same equations and, surprise surprise, both types of mass were found to be identical in magnitude. It was then concluded that both types of mass were identical. Of course they were observed to be identical, the difference between them was compensated by the gravitational constant, G. The gravitational mass of an object is actually $G^{0.5}$ times as large as its inertial mass. Disillusionment set in even deeper when the author read in several texts that the solution of the equations which led to General Relativity also provided the value of the gravitational constant. Apparently the writers of those texts never realized that the gravitational constant, G, was an input to the theory resulting from the false assumption that inertial and gravitational masses of objects were numerically equal rather than merely proportional to each other.

8.4- The first step in generating the Gravity Transformations is to define the gravitational equivalent of the velocity ratio, V/C, which is the basis of the Velocity Transformations. That equivalent is the gravitational potential, Θ, which exists between elevations. Using the upper elevation units of measurement as a reference, Θ is equal to the energy, δE, released as the energy of fall by an object of rest mass equivalent energy, E, as it is lowered from the upper to the lower elevation. [Its value is provided by the relationship $\Theta = \delta E/E$.] It is demonstrated in

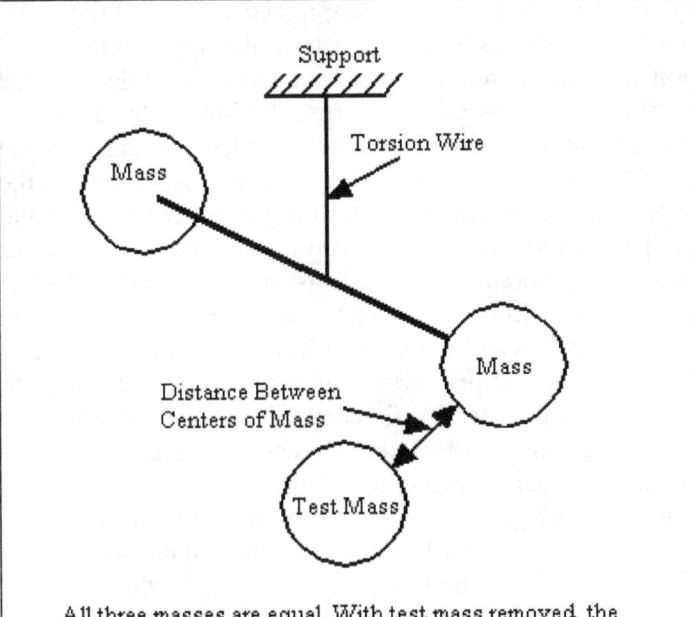

All three masses are equal. With test mass removed, the torsional resonant frequency of the pair of masses about the vertical axis is measured. With test mass in position, the torsional deflection of pair of masses about the vertical axis due to gravitational attraction of the test mass is measured. Also measured is the distance between the centers of the test mass and the adjacent mass.

Figure 8.1 - The Eotvos Experiment

"Gravity" that the Gravity Transformations are independent of direction (horizontal or vertical) greatly simplifying their determination since only those resulting from elevation changes need be considered.

8.5- As derived in "Gravity", the gravitational transformation for time, T, is readily determined in terms of the gravitational potential, Θ, and the Gravitational Transformations for force, F, and for length, L, using the ideal thought experiment illustrated in Figure 8.2. In this thought experiment, mechanical energy is stored in a spring which is then compressed and tied at the upper elevation. The spring is then moved from the upper to the lower elevation and the stored energy is recovered by releasing the spring. Along with this stored energy, the lower elevation also receives the energy of fall of that stored energy along with the energy of fall of the relaxed spring. The energy of fall of the relaxed spring is then used to return it to the upper elevation and plays no part in the thought experiment. Remaining at the lower elevation is the stored energy and its energy of fall. The net energy transported from the upper to the lower elevation is then converted to photons and transmitted from the lower elevation to a receiver at the upper elevation and converted back to mechanical energy at 100% efficiency. Since there are no losses in this hypothetical closed cycle, the energy recovered from the photons at the upper elevation must equal the energy originally stored in the spring. If this were not the case, it would be possible, in principle, to build a perpetual motion machine which created energy from nothing. In "Gravity", this thought experiment is used to derive the Gravity Transformation for Time as $T=1/(1+F*L*\Theta)$. It should be noted that if the transformation for the $F*L$ product is arbitrarily made equal to unity, the time dilation provided by General Relativity is obtained.

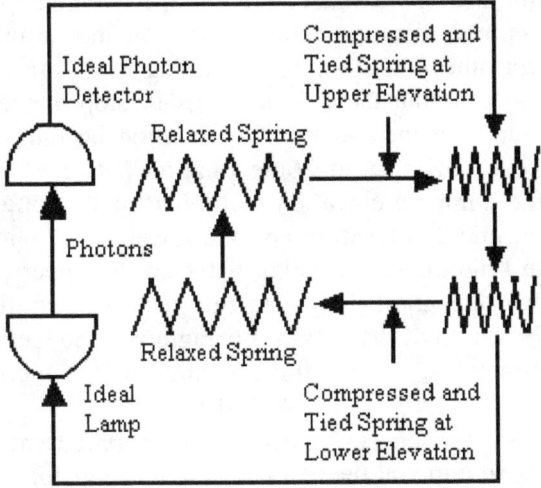

Energy released by lowering mass of relaxed spring equals the energy required to raise the relaxed spring. The spring itself acts as a passive working fluid and its mass effects may be ignored.

The energy stored in the spring plus the energy of fall is converted to photons by an ideal lamp and is projected without loss to an ideal photon collector at the upper elevation where it is used to comprerss the spring.

Since this is a closed loop cycle, it must satisfy the First Law of Thermodynamics. Energy cannot be created or destroyed within a closed system.

Figure 8.2 - Spring-Lamp Thought Experiment

8.6- The next requirement which must be met is that the Gravity Transformation for Time must have a property which the author designates as multiplicative commutivity. As an example, when one goes from the first floor of a building to its third floor, it does not matter if the elevator happens to stop at the second floor. The requirement that the time dilation be multiplicative commutative means that the time dilation between the upper elevation and a middle elevation multiplied by the time dilation between that middle elevation and the lower elevation must be equal to the time dilation existing between the upper elevation and the lower elevation. Imposing this requirement allows the time transformation of the previous paragraph to be factored into two transformations, one for time, T, and one for energy, $F*L$, providing $T=(1-\Theta)$ and $F*L=1/(1-\Theta)$. The resulting transformation for energy meets another requirement of a satisfactory gravitational theory. The Gravity Transformation for Energy, $F*L$, brings the gravitational field into compliance with the Law of Conservation of Energy and, in so doing, eliminates a serious flaw of both the Newtonian Theory of Gravitation and of General Relativity. The Gravity Transformation for Energy shows that the rest mass equivalent energy of an object, as measured with upper elevation units of measurement, is reduced by an amount equal to the energy released by lowering it to the lower elevation. The total energy in the system remains unchanged as required by the Law of Conservation of Energy. The requirement for multiplicative commutivity is not an invention of the author, it is characteristic of all continuous fields. It is taught in undergraduate courses in Field Theory that the difference in characteristics observed between two points in a continuous field is independent of the path which is traveled between those points and, in order for this requirement to be met, the field must be "multiplicatively commutative". Perhaps more significant is the fact that the property of "multiplicative commutivity" is necessary if the Principle of Relativity is to be valid. It is the fact that the Lorentz Transformations are "multiplicatively commutative" which allows the Special Theory of Relativity to work.

8.7- At this point, the determination of the Gravitational Transformations could be completed by a literal application of the Principle of Equivalence and asserting that the relationship between the Velocity Transformations for Parallel Length and for Time must bear the same relationship to each other as the Gravity Transformations for Length and Time. One would then conclude that the Gravity Transformation for Force was equal to unity and the Gravity Transformation for Length, L, was $1/(1-\Theta)$. As shown in "Gravity", the application of this triad of transformations to the gravitational field around the Sun leads directly to the observational results (i.e.- time dilation, anomalous precession of Mercury's orbit, bending of starlight) which supposedly have verified General Relativity. Unfortunately, General Relativity introduced the idea that the presence of mass(energy) causes the geometry of space to be distorted and it was this distortion which produced the gravitational effects. If the presence of mass(energy) does distort space, some or all of the observed bending of light and of the observed orbital precession would result from the curvature resulting from that distortion rather than from the effects of the Gravity Transformations for Length and Time. The introduction of non-Euclidian geometry into the concept of the gravitational field thus allows the possibility of an infinite number of geometries in the vicinity of a gravitating object. The only limitation on these geometries is the requirement that the product of the Gravity Transformations for Force and for Length compensate each other so that the Gravity Transformation for Energy remains correct. While General Relativity introduced the idea of curved space, it does not meet this requirement and cannot be a valid description of the gravitational phenomena.

8.8- Since General Relativity has generated a dogma in which space is non-Euclidian, it is necessary to disprove the concept. To do so, consider the ideal thought experiment diagramed in Figure 8.3. In this experiment, a pair of ideal retroreflectors are mounted on opposite sides of a massive object by ideally rigid and massless booms. Photons are reflected back and forth between the retroreflectors in a non-divergent beam which passes close to the object. The object's gravitational field causes the trajectory of the photons to bend as they pass by the object and strike the retroreflectors at an angle to the geometric line between them. Since photons posses inertial mass, the deviation of their path causes a reaction force to be exerted on the retroreflectors as they are reflected. In the horizontal direction (of the diagram), the component of that force on one of the retroreflectors is canceled by an equal and opposite component of that force exerted on the other. In the vertical direction (of the diagram), the force components on the retroreflectors do not cancel, they add and produce a net downward force on the object. This downward force must be balanced by an upward force acting on the object which can only result from the gravitational attraction of the beam of photons. The requirement for a force balance in the vertical direction is basic. If a net force were to exist in the vertical direction, it would be possible, in principle, to construct a perpetual motion machine. This machine would allow the system to continuously export energy without any change in its internally and externally observed states. This is a NO-NO and cannot occur.

8.9- Conceptually, the bending of the path of the beam of photons can result from one or both of two effects acting separately or in combination. One of these effects is conventional refraction. The other effect is the apparent bending of their path because they are traveling in a curved three dimensional non-Euclidian space and they are actually traveling along a straight line. To the degree that the observed bending of their path is caused by refraction, the bending of the path of the photons requires the force of gravity to act between the object and the photons. To the degree that the observed bending of their path is caused by the curvature of non-Euclidian space, no change in direction actually occurs and the force of gravity is not acting. In "Gravity", the writer derived the gravitational force acting upon such a beam of photons and the net forces they apply to the retroreflectors. By showing that these forces are equal, the derivation demonstrates that observed bending of the path of light

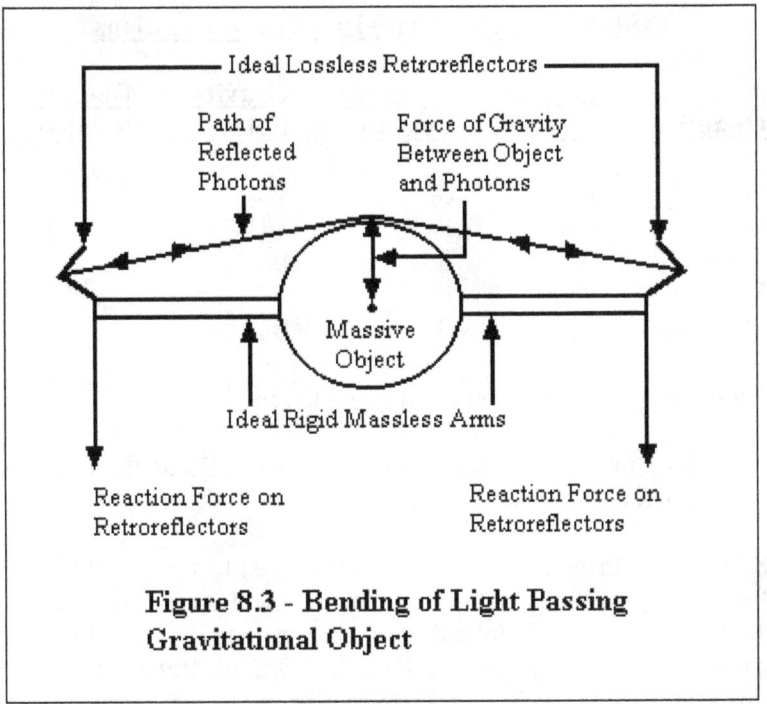

Figure 8.3 - Bending of Light Passing Gravitational Object

in a gravitational field results entirely from gravitational attraction. A postulated curvature of space makes no contribution to the photon's path and space in the gravitational field must be three dimensional Euclidian, exactly as common sense would require. It is now possible to provide Gravity Transformations to match the Velocity Transformations already provided. Defining B_V as the velocity transformation, B_G as the Gravity Transformation, and B_{GE} as the gravity transformation of General Relativity, the Relativistic Transformations for Velocity, Gravity, and for General Relativity may be provided in the same form, as shown in Table 8.9.1.

Table 8.9.1:- The Relativistic Transformations

Quantity	Parallel Velocity	Transverse Velocity	Gravity Relativity	General Relativity
Force, F	1	B_V	1	1
Length, L	$1/B_V$	1	$1/B_G$	1
Time, T	B_V	B_V	B_G	B_{GE}
Space, S	1	1	1	B_{GE}
Stiffness, k'	B_V	B_V	B_V	1

Note 1:- B_V, B_G, and B_{GE} are all equal to $(c/C)^{0.5}$.

Note 2:- Current texts erroneously provide $1/B_V$ as the Lorentz Transformation for Transverse Force.

8.10- If the Principle of Equivalence apples to the gravitational field, a proper gravitational theory will provide Gravity Transformations identical in form to the Lorentz Transformations for Parallel Velocity. Gravity Relativity meets this requirement. Its basic transformations are:

Force $F=1$
Length $L=1/(1-\Theta)$
Time $T=(1-\Theta)$

The equivalent transformations for General Relativity are provided for reference. It will be noted that, under General Relativity, the transformations for both Force, F, and Length, L, are equal to unity. Applying these values to the results of the Spring-Lamp Experiment of Figure 8.2, one obtains the time dilation, B_{GE}, of General Relativity as its accepted value of $T=1/(1+\Theta)$. The basic transformations of General Relativity become:

Force $F=1$
Length $L=1$
Time $T=1/(1+\Theta)$

Space $S = 1/(1+\Theta)$

The above listing includes an additional line for the Gravitational Transformation for Space which Dr. Einstein required to allow him to complete his derivation of General Relativity. It represents the alleged curvature of space made necessary by the mathematical error in its derivation and which arbitrarily forced the transformation for length to equal unity. It will be noted that these transformations are inconsistent with the Velocity Transformations for Parallel Force and for Parallel Length. As a result, General Relativity contradicts one of its postulates, the Principle of Equivalence. Furthermore, because its transformations for force and for length are both equal to unity, the effects of the transformation for time are not compensated between reference frames and General Relativity also contradicts its other basic postulate, the Principle of Relativity. Somehow, it seems reasonable to suspect the validity of a theory which contradicts the postulates upon which it is based.

8.11- Deriving the General Theory of Relativity presented Dr. Einstein with considerable difficulty. He could not make it work in terms of three dimensional Euclidian space. After about a year and a half of failure, he overcame his difficulties by resorting to the non-Euclidian geometry described by Riemann in the middle of the 19th century and added the extra degree of freedom that curved space provided. The effect of that curvature is to add additional space as elevation is reduced, as defined by "S" in Table 8.9.1. This modification allowed him to solve his mathematical equations in a self consistent manner. The resultant theory provided predictions for the behavior of the gravitational field which are more accurate than the predictions of Newtonian Gravitational Theory because they allow the theory to be relativistic, but the theory contains an error on the order of Θ^2. That error is about 5 orders of magnitude too small to be detected in the extremely weak gravitational fields existing within the Solar System or by observations of the spectral lines of distant stars. Unfortunately the small size of this error allowed Dr. Einstein to predict the bending of starlight, the time dilation, and the precession of orbits caused by the gravitational field which are now accepted as proof of the validity of General Relativity without fear of contradiction by the results of observation. There is an indication that he suspected that his theory was defective since he is reported to have had concerns about its extension to extremely strong fields such as those associated with neutron stars. Unfortunately, direct observation cannot be used to distinguish between General Relativity and Gravity Relativity. Both approaches yield the same predictions to the foreseeable limits of observational accuracy. The difference between the two approaches can only be observationally determined by close-up observation of objects having extremely strong fields, such as a neutron stars. Such observations probably require the invention of Star Trek's Warp Drive.

8.12- There is an observational test which can be applied to General Relativity and to Gravity Relativity using current technology. We live in a Universe which approximates a gravitationally collapsed object. At present, its observed radius is several times larger than the radius of the Event Horizon resulting from its estimated mass. However, astronomical observations indicate that the Universe started expanding from its origin as a point source about 15 billion years ago. At its start, the Universe was significantly smaller than its Event Horizon. Apparently, between 3 and 5 billion years ago, the Universe expanded through and is now several times larger than its Event Horizon. Since General Relativity predicts that nothing can escape from within the Event Horizon, the astronomical observations directly deny the validity of General Relativity. Gravity Relativity, on the other hand, is more than merely compatible with astronomical observations, the cosmology which follows from the Gravity Transformations seems to agree quite well with observation. Most recently, however, the observations of the COBE (Cosmic Background Explorer) provided extremely strong evidence. It observed radiation from the original "big bang" from which the Universe sprang. This observation revealed that space was three dimensional Euclidean and that the Universe had an absolute velocity reference. These observations throw the Space-Time concepts of Special Relativity and the curved space concepts of General Relativity into a cocked hat for what should be very obvious reasons.

8.13- As derived in "Gravity", the gravitational potential, Θ, is equal to R_H/R, where R_H is designated as the Horizon Radius and is equal to the radius of the Event Horizon and R is the actual radius of the object, both as observed externally. Table 8.13.1 provides the Velocity Relativity and Gravity Relativity Transformations of the dimensional entities listed in Table 7.8.1 evaluated in terms of the velocity ratio, V/C, in terms of gravitational potential, Θ, and in terms of the externally observed ratio of the Horizon Radius to the actual radius, R_H/R.

Table 8.13.1:- Dimensional Content of Various Physical Quantities

Quantity	Symbol	Parallel Velocity	Transverse Velocity	Gravity
Force	F	1	B_V	1
Length	L	$1/B_V$	1	$1/B_G$
Time	T	B_V	B_V	B_G
Energy	E	$1/B_V$	B_V	$1/B_G$
Planck's Constant	H	1	B_V^2	1
Velocity	V	$1/B_V^2$	$1/B_V$	B_G^2
Acceleration	A	$1/B_V^3$	$1/B_V^2$	$1/B_G^3$
Incremental Mass	M_I	B_V^3	B_V^3	B_G^3
Momentum	U	B_V	B_V^2	B_G
Angular Momentum	J	1	B_V^2	1
Gravitational Constant	G	$1/B_V^8$	$1/B_V^5$	$1/B_G^8$
Ergo-gravitational Constant	D	1	$1/B_V$	1
Temperature	ϕ	$1/B_V$	B_V	$1/B_G$
Charge	Q	$1/B_V$	B_V	$1/B_G$
Dielectric Constant of Space	ε	1	$1/B_V$	1
Permeability of Space	μ	B_V^4	$1/B_V^3$	B_G^4
Stiffness	k'	B_V	B_V	B_G

Note 1:- $B_V = (1 - V^2/C^2)^{0.5}$

Note 2:- $B_G = (1-\Theta)$ or $(1 - R_H/R)$

Note 3:- Velocity Transformations are modified by $(1 + V_1 * V_2/C^2)$ in the denominator as described in Chapter 6.

8.14- A school of thought exists which is associated with the Inflationary Theory of the origin of the Universe. This school teaches that, although the components (stars, galaxies, etc.) of the Universe are observed to be separating at a high velocity, they are actually stationary in space. It is the space that is expanding instead and outside of the Universe there is no space! Such an argument might be acceptable but for two reasons. Firstly, for such to occur, energy would be required to be continuously added to that Universe from some unnamed source. This requirement is evidenced by the fact that, if the matter present in the Universe were to elect to fall back in on itself, it would have further to fall and would release more energy as time passed. This concept suffers from the same difficulty, but in reverse, as does the Universe of the preceding paragraphs. It cannot reconcile the observed Universe with the Law of Conservation of Energy and must be dismissed unless a strong justification can be provided. Secondly, such a Universe would not provide a reciprocal relationship between the gravitational energy and time transformations, and, without that reciprocal relationship, the Principle of Relativity would not apply and physics would change as elevation changed. It would seem that such a concept must be dismissed. The required effects do not seem to occur.

8.15- The Source of Gravitational Energy:- When the author was in high school, he asked his physics instructor where the energy was stored when a weight was lifted from the floor to the ceiling. The answer given was that "the energy was stored in the gravitational field". The instructor believed that his answer explained everything when, in reality, it was an admission that the scientific community did not know how or where gravitational energy was stored and it did not have the intellectual integrity to admit that it did not know. ("We don't know" would have been a reasonable answer, but hiding the fact that the answer was unknown is inexcusable.) The author had, and still has, what might be considered to be the unreasonable conviction that the primary purpose of a gravitational theory should be to provide a description of the relationship between elevation and energy which is consistent with the Law of Conservation of Energy. It is the energy of fall, and the force through which that energy is manifested, which is the only first order effect of the gravitational field. The precession of planetary orbits, the time dilations, and the bending of the path of light rays are second order effects which pale into insignificance in comparison. As we have seen, and as rigorously shown in "Gravity", when they are properly applied, the combination of the Principles of Relativity and Equivalence yield a description of the gravitational field which is both consistent with Law of Conservation of Energy and the observed second order effects which are erroneously supposed to have validated General Relativity.

8.16- The gravitational transformations provided in Table 8.13.1 show that the release(absorbtion) of energy as an object changes its elevation results from the release(absorbtion) of a portion of its mass equivalent energy. (Because these transformations are multiplicatively commutative, any elevation may be considered to be the upper elevation, a second elevation closer to the center of the field may be considered to be the lower elevation, and the gravitational potential, Θ, considered to be defined in terms of the upper elevation units of measurement.) Since the locally measured energy represented by the mass of the falling object is the same at both elevations, as required by the Principle of Relativity, its mass equivalent energy as measured by upper elevation units of measurement has been reduced by Θ at the lower elevation. This reduction in absolute mass equivalent energy equals the energy released by gravitation and the Law of Conservation of Energy is satisfied. To release this energy of fall, the gravitational field causes the force of gravity to act over the distance of fall.

8.17- Returning to Table 8.13.1, it will be found that the transformation for velocity is $1/(1-\Theta)^2$. This transformation shows that the velocity of light, which remains unchanged when measured locally, is reduced in the absolute sense at a rate equal to the square of the reduction which occurs in the mass equivalent energy of the falling object. Effectively, the proximity of energy modifies the Aether and produces a local reduction in the velocity of light. (The Aether has already been shown to be the underlying structure of space.) In this sense, the gravitational field is actually a velocity well for light. This velocity well, in turn, allows an object within it to shed some of its internal energy by moving to a lower elevation. It does this by pushing itself downward against the Aether. This push, which we cannot directly observe, is the fictitious force of gravity currently described in texts. We can only observe the equal and opposite reaction to this fictitious force required by Newton's Second Law of Motion (for every action there is an equal and opposite reaction) as the force of gravity. The Gravity Transformation for Energy requires that the falling process cease when, as observed with upper elevation units of measurement, the energy released by falling equals the original energy content of the object. This cessation of falling does in fact occur as the end state of gravitational contraction and will be discussed in detail later.

8.18- The reduction of the velocity of light in a gravitational field not only produces the force of gravity by allowing energy contained within an object to be released, it refracts the path of a ray of light (and all radiation) passing through the field in the same manner as the slowing of light in a lens refracts its path and allows your eyes to focus on this sentence. When a photon is in a gravitational field, it experiences the same impulse to release its energy as does matter. However, unlike a material object, a photon cannot transform its energy into kinetic energy as its elevation is lowered because it must always travel at the local velocity of light. As observed with upper elevation units of measurement, the energy (frequency) of that photon remains unchanged. As observed at the lower elevation, where the units of measurement for time are larger and the units of measurement for energy are smaller, its frequency and energy have increased. Gravitational refraction of the path of light is accompanied by a gravitational force of attraction. As derived in "Gravity", the gravitational force acting on photon is twice the gravitational force acting on a material particles of the same energy. What does not seem to have been recognized is that the gravitational attraction of the photons (and neutrinos) associated with the background radiation level of space can easily represent the dark matter currently sought by astronomers and cosmologists.

8.19- The Effect of the Gravitational Field on the Velocity of Light:- From the time that Special Relativity was published until the early 1980's, it was accepted as an absolute truth that the velocity of light in a vacuum was a constant that was unchanged by any change in velocity or elevation reference frame. Along with this viewpoint was the idea that a straight line was defined by the path of ray of light between two points. The author was rather startled to read in a book entitled "Was Einstein Right?" by Dr. Will that the velocity of light was no longer considered to be unchanged in a gravitational field, but was reduced in proportion to that theory's time dilation. Even more startling was the fact that the change in viewpoint was made by the academic community without the slightest embarrassment even though its change undermined the philosophical foundations of both Special and General Relativity.

8.20- The correct effect of changes in elevation on the velocity of light is provided by the Gravity Transformation for Velocity in Table 8.13.1 as a function of B_G. B_G is defined as $(1-\Theta)$ or $(1-R_H/R)$ where Θ is the gravatational potential and R_H/R is the ratio between the distance, R, to the center of a gravitationally attracting object and its Horizon Radius, R_H, both as measured with the units of measurement existing at an infinite radius. The departure of the velocity of light from its nominal value of C when R approaches R_H is provided in Figure 8.4. It will be noted that, because of the minus sign in the transformation, as the radius approaches the Horizon Radius, the velocity of light abruptly approaches zero. The effect of this drop in the velocity of light causes the path of a ray of light to be refracted by the gravitational field for the same reason that the lenses in your eye refract the light to bring the image of this text to a focus on the retina of your eye. The velocity of light is slowed down by the nature of the space through which the light is passing. This change in the velocity of light is concealed from a local observer who, of necessity, measures the velocity of light with units of measurement which have been altered by the gravitational field.

8.21- If one could observe, from a distance, an object which has collapsed to less than three times its Horizon Radius, he would find that the effects of refraction causes it to appear to have a radius of 6.75 times the Horizon Radius. This occurs because the line of sight from the observer to the surface of the object is refracted towards the object and causes it to appear larger than it actually is. The effect is shown in Figure 8.5. A hypothetical observer on the surface of the object would find that the angle between the zenith and the horizon became smaller than 90 degrees as the object contracted to a radius smaller than three times the Horizon Radius and became equal to zero when the object had contracted to the Horizon Radius. The possible paths of rays of light leaving the surface of a collapsed object are illustrated in Figure 8.6. Surprisingly, gravitational refraction has no effect on the ability of the object to radiate energy to space. As the object contracts, the approach of the horizon towards the zenith reduces the solid angle from which radiation can escape from its surface to space, but the effect is exactly counterbalanced by the fact that the surface from which the radiation is effectively emitted to space has the optical size of 6.75 times the Horizon Radius. In terms of the ability of the object to radiate energy to space, refraction may be ignored.

8.22- As shown in "Gravity", the slowing of light in a gravitational field not only refracts the path of light, it refracts the paths of moving objects. This refraction of path is in addition to the orbital effects of gravitational attraction and is responsible for the anomalous precession of planetary orbits which supposedly prove the validity of General Relativity. The basic Law of Motion of Newtonian Physics which states that "an object having a velocity will retain that velocity unless it is acted upon by an outside force" is still true, however, the velocity which obeys this law is the velocity of the object as measured in terms of the locally observed velocity of light.

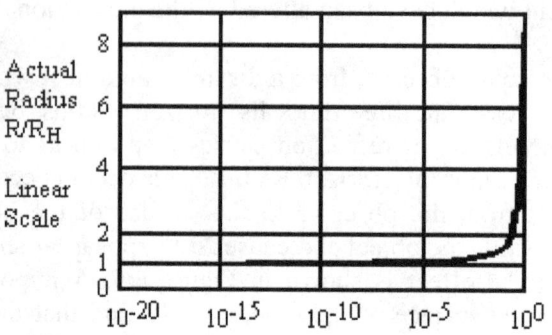

Actual Velocity of Light, Compared to Nominal Value

Exponential Scale

Note:- Actual radius and velocity are measured with units of measurement which exist at an infinite radius. As measured with local units of measurement (existing at radius R), light travels at its nominal velocity of C.

Figure 8.4 - Velocity of Light Reduction in Region of Horizon Radius

8.23- Experiments have been proposed, and may already have been performed, to provide additional verification of General Relativity. These experiments consist of placing extremely accurate gyroscopes in orbit and observing the precession of their axes as a result of their orbital velocity. Such experiments will indeed verify that gravity is a relativistic phenomena since the relativistic gyroscopic effects which will be observed are a direct consequence of the gravitational refraction of the velocity vector. They will not show that General Relativity is the correct relativistic gravitational theory.

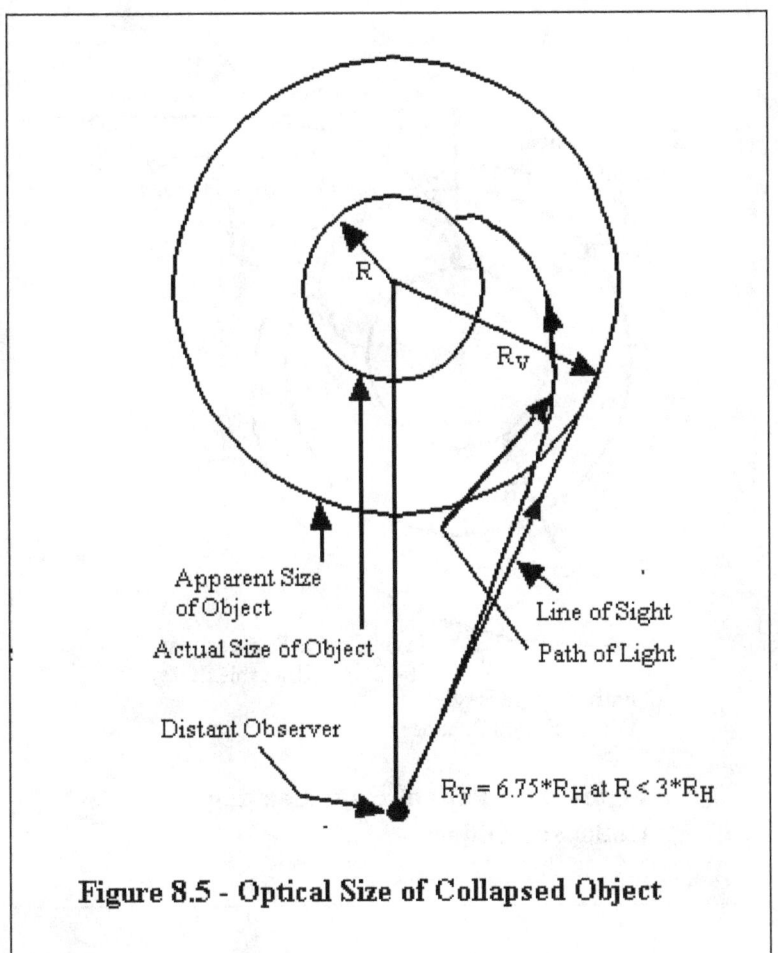

Figure 8.5 - Optical Size of Collapsed Object

8.24- *Gravity results from the fact that the velocity of light is slowed by the proximity of energy. This slowing of the velocity of light results in the changes in the size of the units of measurement illustrated in Table 7.2.1, Table 7.8.1, Table 8.9.1, and Table 8.13.*

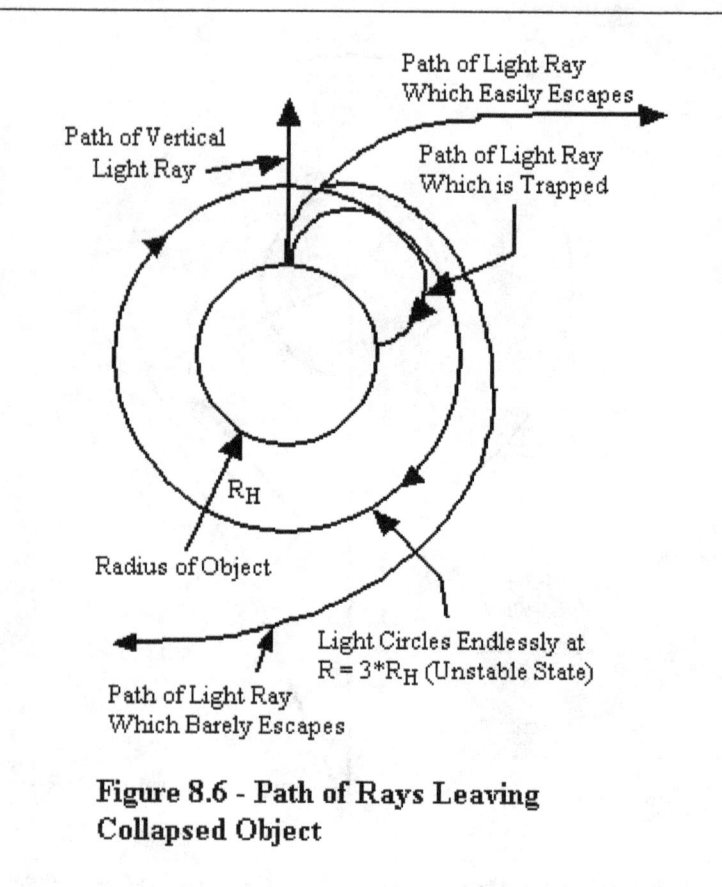

Figure 8.6 - Path of Rays Leaving Collapsed Object

Chapter 9 - Dr. Einstein's Error and the Introduction of Curved Space

9.1- Where did Dr. Einstein go wrong? - Since the General Theory of Relativity yields results which are in conflict with the postulates upon which it is based, it is apparent that at least one significant mathematical error was made in its derivation. In order for the existence and nature of that error not to have been recognized by Dr. Einstein and his contemporaries, it must be of a type which would not be obvious to individuals of sufficient stature and scientific sophistication to be in a position to question his work. It is likely that the error was recognized from time to time by bright undergraduates, but their objections would not be accepted by an academic priesthood dedicated to preserving the true faith. Questions from such sources would be brushed aside because obviously they could only have resulted from a lack of understanding on the part of questioners who were not sufficiently trained (brainwashed?) to understand the subject matter and who did not as yet possess the proper 'yup's.

9.2- In deriving the General Theory of Relativity, Dr. Einstein employed a mathematical tool called Tensor Calculus. Properly applied, this tool is extremely useful and normally insures that the effects of all possible variables are considered. It does have a limitation however, it cannot be used for deriving a relativistic theory. Tensor Calculus, in its simplest form, is a process in which partial derivatives of the variables of interest are arranged in a set of simultaneous equations similar to those encountered in conventional algebra. A typical group of Tensor Calculus equations is illustrated below:

$$A_{11}*\delta X + A_{12}*\delta Y + A_{13}*\delta Z = W_1$$
$$A_{21}*\delta X + A_{22}*\delta Y + A_{23}*\delta Z = W_2$$
$$A_{31}*\delta X + A_{32}*\delta Y + A_{33}*\delta Z = W_3$$

9.3- In this group of equations, the symbols A_{11} through A_{33} are constants determined by the problem, the symbols X, Y, and Z are the variables whose value is to be determined, W_1, W_2, and W_3 are the sums of each of the equations, and the symbol δ indicates that the term is a partial derivative. Solution of equations of this type requires a mathematical procedure known as integration. Therein lies the rub. To perform that integration, it is necessary to know that the coefficients (eg:- A_{11} through A_{33}) of the partial derivatives are independent of the variables. To understand why, consider the use of the rules of Elementary Calculus to integrate the expression K*δX. If K is independent of the value of X, the result of the integration is K*X+C", where C" is the constant of integration. However, if K is equal to X, the result is $X^2/2$+C", an entirely different result. A basic rule of all types of Calculus, including Tensor Calculus, is that the variable to be integrated must be completely defined in the expression itself and not be hidden within other terms, in this case, K.

9.4- It is because the solution of a problem in Tensor Calculus requires the performance of mathematical integration that Tensor Calculus is unsuitable for the derivation of a relativistic theory. The partial derivatives in the Tensor Calculus matrix used to derive the General Theory of Relativity involve length. It has been demonstrated that the units of measurement of length change between reference frames which differ in velocity. Since the derivation of General Relativity is based upon the Principle of Equivalence, it follows that the effect of a change in elevation on the units of measurement for length must be known in order for a meaningful, mathematically valid, integration to be performed. (The size of the units of measurement are analogous to the "size" of K in the previous paragraph.) Unfortunately, the effect of a change in elevation on the units of measurement for length cannot be known until the integration has been performed correctly and a valid integration cannot be performed until the effects of a change in elevation on the units of measurement for length are known. Consequently a valid derivation of gravitational theory is not possible by this method. When the attempt is made, it arbitrarily forces the Gravity Transformation for Length to equal unity regardless of its correct value. Until the Tensor Calculus equations have been solved, the required information needed to solve them is not available. It is difficult to understand, however, why Dr. Einstein did not recognize that, in employing Tensor Calculus to derive General Relativity, he was incorporating an erroneous loop of circular reasoning (unless he did not actually understand the nature of the relativistic effects). *If an undergraduate student of elementary calculus persistently made an equivalent error he would receive a failing grade for the course.*

9.5- Dr. Einstein's mathematical error made it impossible to achieve a solution of the General Relativity Tensor in a manner which is consistent with Euclidian geometry. He is reported to have struggled with this difficulty for about 18 months and finally resolved his impasse' by adding another, otherwise superfluous, degree of freedom. He did this by incorporating the curved space described by Riemann Geometry. This addition permitted the mathematical equations to be solved, but the results were clearly not rigorously correct since, as Table 8.9.1 shows, the resultant General Theory of Relativity clearly violates the Principles of Relativity and Equivalence upon which it is based. Unfortunately, the fact that gravity is a relativistic phenomena insured that General Relativity contained enough truth to enable it to gain acceptance by refining the predictions of Newtonian Gravitational Theory to agree with the observed anomalous precession of Mercury's orbit, the observed red shift of Solar spectral lines, and the bending of the path of starlight by the Sun's gravitational field to within the limits of observational error.

9.6- During the Solar eclipse of 1919 when the bending of the path of starlight by the Sun's gravitational field predicted by General Relativity was verified, Dr. Einstein's confidence in the outcome surprised many of his contemporaries. In expressing this confidence, he was on safe ground. He knew that gravitation was a relativistic phenomena and that it didn't matter whether General Relativity was rigorous. The fact that General Relativity made its predictions based upon relativistic effects insured that any residual error would be about a million times too small to be revealed in the Sun's puny gravitational field. There is a suggestion that Dr. Einstein recognized that General Relativity was not rigorous since he is reported to have expressed misgivings as to its application to the strong fields such as exist around neutron stars. The question that can never be answered is whether Dr. Einstein believed that the curved space solution of General Relativity was valid, or whether he engaged in scientific fakery with the confidence that it could not be unmasked. One must remember the comment made in a meeting in which General Relativity was evaluated, "why shouldn't we consider space to be curved, nobody can prove that it isn't". The author finds it difficult to differentiate Dr. Einstein's use of curved space to allow him to complete his derivation of General Relativity from the actions of a mechanic who installs the wrong part into a machine by hammering it into place.

9.7- The Meaning of Curved Space:- The most significant property of the gravitational field is not revealed in the effects represented by the precession of orbits, the time dilation, or the bending of the path of a ray of light as it passes the Sun. These are second order effects. The only first order effect of the gravitational field is typified by the force which holds you to your chair as you read this sentence and the energy of fall which you will experience if you fall from the chair. This force, and the energy of fall it implies, is the primary reality of the gravitational field. There is no way in (expletive deleted) that a curvature of space can account for that force and that energy without the presence of an attractive gravitational force. It is irresponsible to assert, as proponents of General Relativity must, that "the apparent ability of the gravitational field to create energy from nothingness does not constitute a violation of the Law of Conservation of Energy because the energy which is created cannot climb out of the field". Any proper gravitational theory must explain the reason for the force which holds you to your chair and the source of the energy which is released when you fall in a manner which is consistent with the requirement that energy be conserved. The author was once reminded by a physicist that some texts assert that General Relativity has eliminated gravity as a force. For some reason that physicist became strangely silent when it was suggested that he step through an adjacent second story window so that we could discuss the subject across the window sill.

9.8- Under Newtonian Theory, gravity is an attractive force acting between two concentrations of energy (masses), E_1 and E_2, in inverse proportion to the square of their separation, R, in accordance with $F=D*E_1*E_2/R^2$. The theory makes no attempt to explain the source of that force and the energy it represents. Since it is a descriptive law derived several centuries before the Law of Conservation of Energy was recognized, there is no need for it to apologize for that omission unless, of course, one intends to retain it as the theory which represents reality. Its conclusion that the force of attraction between objects varies inversely with the square of their separation reasonably follows from the fact that the area of a spherical surface varies as the square of its radius. General Relativity, on the other hand, does not have the luxury of ignoring the source of gravitational force and gravitational energy since it claims to represent reality and was derived at a time when the Law of Conservation of Energy was well established.

9.9- At first glance, the non-Euclidian geometry of Riemann which was used by Dr. Einstein in formulating General Relativity appears sophisticated. When one examines Dr. Riemann's geometry in detail, he finds that it is a subset of conventional Euclidian geometry in which a non-Euclidian geometry of three spatial dimensions is used to describe the properties of a "surface" which can be considered to be contained in a Euclidian geometry of four spatial dimensions. As the author pointed out in "Gravity", any non-Euclidian geometry of N spatial dimensions can be contained in a Euclidian Geometry of N+1 spatial dimensions. (An example of non-Euclidian geometry as a subset of Euclidian Geometry is the two dimensional non-Euclidian geometry which describes the surface of our three dimensional Euclidian Earth.).

9.10- For a reasonable and conscientious man to accept the concept that gravity results from the distortion of our familiar three dimensional Euclidian space into a four dimensional non-Euclidian space, he must be provided with a reasonable description of the nature of that distortion as it appears in the four dimensional Euclidian space. The closest approach to such a description seems to be speculation as to whether our three dimensional space is positively curved, as is a sphere, or whether it is negatively curved, as is a saddle. It is commonly asserted that if our observable three dimensional space is positively curved, it encloses a four dimensional space of finite volume, while if its curvature is negative, the volume of the four dimensional space enclosed is infinite. Actually, a common sense understanding of the concept easily reveals that there is no connection between whether the four dimensional space is enclosed by the three dimensional surface and therefore has a finite volume and whether the curvature of the three dimensional space is positive or negative. As an example, the two dimensional non-Euclidian surface analogous to the reflector of an automobile headlight is positively curved but the volume of the three dimensional space it encloses is infinite. On the other hand, the two dimensional non-Euclidian surface represented by the inner portion of an automobile inner tube is negatively curved, but it encloses a finite volume. The factor which determines whether the geometry of the surface contains an infinite or a finite volume is determined not by the polarity of its curvature but by whether the curvature along each of its axes is greater or less than the curvature of a parabola. For some unexplained reason these same mathematical physicists assume that whether or not our Universe reverses its expansion and collapses in upon itself to produce a cosmic crunch is determined by whether space is positively or negatively curved. The word 'assume' provides valuable advice. Divided into syllables, it is a reminder that when you ass*u*me, you run the risk of making an 'ass' of 'you' and 'me.'

9.11- In terms of Newtonian Theory, gravity is a rather straightforward phenomena. As Figure 9.1 illustrates, an object suspended above a central gravitational mass and having no orbital velocity experiences a force impelling it towards that mass. This is the force you feel applied to the seat of your pants as you sit in your chair. Under General Relativity, as illustrated in Figure 9.2, an object suspended above a central gravitational mass and having no orbital velocity experiences the same attractive force. However, that force is of enormous magnitude, is inversely proportional to the size of the central mass, and acts in a direction at right angles to our observable three spatial dimensions. The force holding you to your chair is the component of that enormous attractive force which is observable in our three dimensional space as a result of its distortion into the fourth spatial axis. Under General Relativity, the actual gravitational force which produces the observable force component holding you to your chair is on the order of 2^{18} tons. Under both Newtonian Theory and General Relativity, gravity is an attractive force acting at a distance. The difference between the two concepts is that, under Newtonian Theory, gravity is an attractive force acting towards the central mass responsible for the field, while General Relativity asserts that the presence of a central mass creates an enormous attractive force aligned with an unobservable fourth spatial axis and simultaneously distorts our familiar three dimensional flat space towards that axis to allow a component of that enormous force to appear as the force of gravity. General Relativity does not eliminate gravity as a force, it converts it into an incredibly large force acting along an unobservable fourth spatial axis towards a source which does not seem to have a physical existence and which produces infinite energy from nothingness. Come on fellows, give us a break.

9.12- The only reason for considering that space is curved by the presence of mass into a fourth spatial axis and the existence of that enormous attractive force acting along that axis is that mathematical error made by Dr. Einstein in deriving General Relativity. The concept of curved space has survived because of

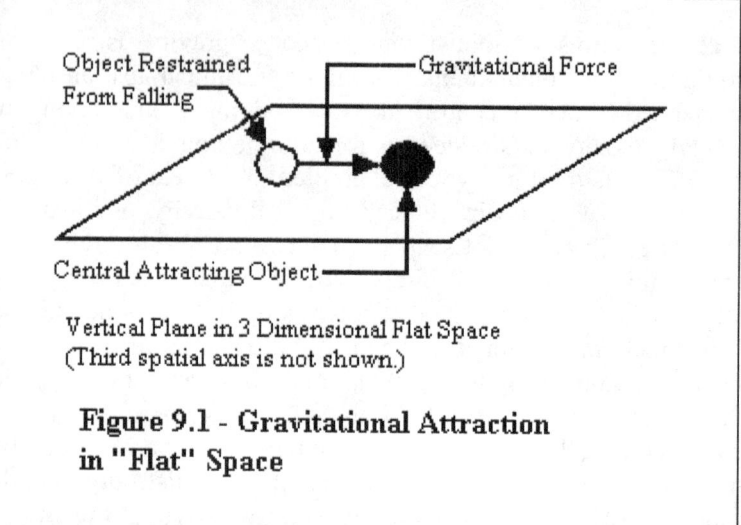

Figure 9.1 - Gravitational Attraction in "Flat" Space

the effectiveness of the defenders of the faith in suppressing the questioning by heretics who would dare to challenge the revealed truth. Lately, however, there seems to some hedging on the question of whether space is curved among senior members of the academic community. For example, in a recently published book, a respected authority in the field (who had received a copy of "Gravity" in 1988) states that there is no difference in the results obtained when one considers space to be curved by the gravitational field and when one considers lengths to shrink as elevation is lowered.. (Needless to say, the author takes issue with that assertion.) Remember, as mentioned earlier, it has been demonstrated in "Gravity" that, if space is curved, it is possible, in principle, to build a perpetual motion machine of the first kind. That machine would be capable of exporting energy forever without any change in its internally or externally observed states. Again, if any reader believes that such a machine is possible, there is a bridge that the author has been trying to sell.

9.13- The author's contention in "Gravity" (1987) that any non-Euclidian geometry of N dimensions can be contained in a Euclidian geometry of N+1 dimensions or higher seems to have

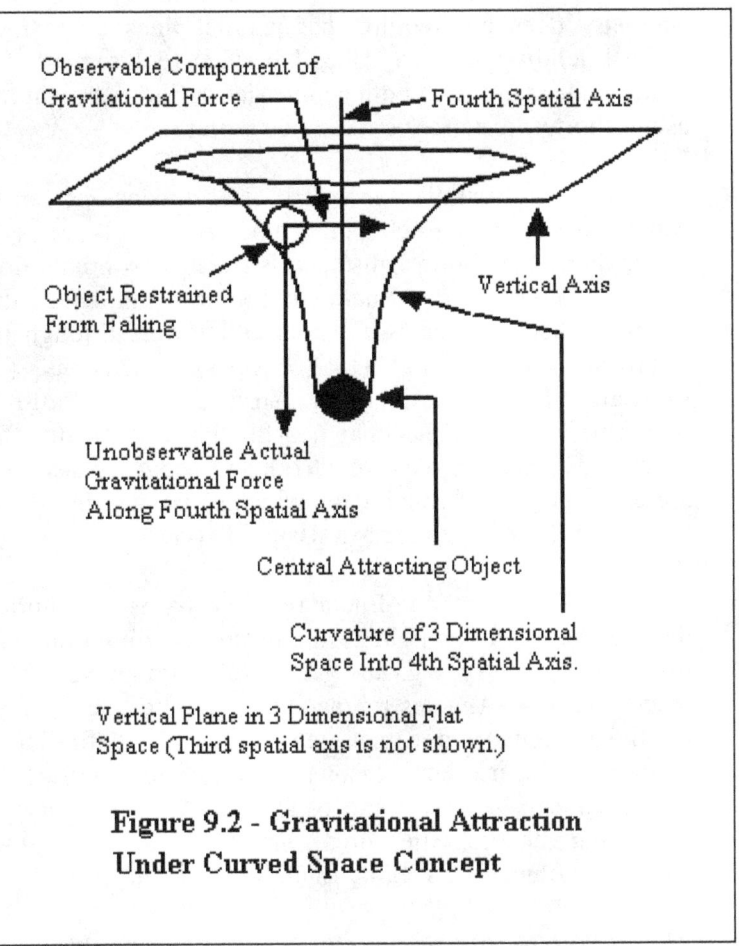

Figure 9.2 - Gravitational Attraction Under Curved Space Concept

been acceptable to the academic community since it has subsequently appeared in books written by at least two individuals who have received copies of that text, although one of these individuals asserted that a Euclidian geometry of N+3 dimensions was required. It is desireable, therefore, to re-examine the geometry of Riemann.

- The primary difference between the axioms of Riemann geometry and Euclidian geometry is that Riemann's geometry does not require that parallel lines never meet while Euclidian geometry does have that requirement. The writer understands that both geometries define a straight line as the shortest distance between two points.

- Since the non-Euclidian geometry of N dimensions can be contained within a Euclidian geometry of N+1 (or N+3) dimensions, the shortest distance between two points does not lie within the non-Euclidian geometry. It leaves that geometry between the two points and travels through the Euclidian geometry of N+1 (or N+3) dimensions. (To illustrate, if one considers the surface of the Earth to represent a two dimensional non-Euclidean geometry, the shortest distance between New York and Los Angleses is not a great circle, it is through a tunnel which passes almost two hundred miles beneath the Mississippi River.)

- The straight line of non-Euclidan geometry is the shortest distance between two points which **remains** within the non-Euclidian geometry (e.g.- the great circle path between New York and Los Angeles.) Since the straight line of non-Euclidian geometry does not meet the rigorous definition of a straight line, it seems reasonable to question the rigor of non-Euclidean geometry except as a convenient means of describing the properties of a curved "surface" contained within a "volume" defined by Euclidian geometry.

- *The author asserts that a rigorous theory should not be based upon Riemannian geometry without an adequate and relavent treatment of the higher order Euclidian geometry it implies.*

9.15- The idea that a Euclidian geometry of at least six spatial dimensions (N+3) was required to contain a three dimensional non-Euclidian geometry is supported by the assertion that each plane in a three dimensional non-Euclidean space, (X-Y), (X-Z), and (Y-Z), requires a separate degree of freedom in the higher order space. The requirement for the additional dimensions

(N+3) would be reasonable if one considered the three planes to be independent. However, these planes are not independent, they are interlocked into the three dimensional non-Euclidean space as a unit. Consider three dimensional Euclidean space to be composed of a series of, $(X-Y)_K$ planes, $(X-Z)_L$ planes and $(Y-Z)_M$ planes (where K, L, and M are integers between one and infinity). These planes meet at vertices "K,L,M". If the three dimensional Euclidean space is curved into a non-Euclidean space about any or all of its principle axes, the vertices of the planes must remain coincident, point "K,L,M" must remain point "K,L,M". As a result, it seems reasonable to conclude that only one extra degree of freedom is required and the three dimensional non-Euclidean geometry may be validly considered as a hyperplane contained in a four dimensional Euclidean geometry.

9.16- Whether the writer's assertion that a Euclidian space of (N+1) dimensions is adequate to contain a three dimesional non-Euclidian space correct is unimportant. It is rigorously shown in "Gravity" that our universe is a three dimensional Euclidian space. The curvature of that space that is currently accepted as representing reality results from Dr. Einstein's attempt to overcome the error introduced by his misuse of Tensor Calculus in the derivation of General Relativity.

Chapter 10 - Gravitational Contraction and Collapse

10.1- The Formation of a Gravitational Object:- Common experience reveals that an accumulation of matter creates a gravitational field which attempts to compact that matter into a mathematical point. We experience that field as a force which impels us towards the center of the Earth and would fall to that center if it were not for the material of which the Earth is composed. Each layer of that material is attracted towards the center and is supported by increased pressure in the underlying layers until the pressure maximum is reached at the Earth's center.

10.2- For an object the size of the Earth, normal matter is able to withstand the gravitationally induced pressure and nothing dramatic occurs. For larger objects, gravitational compression is more complicated. Typically, such objects have the same composition as the interstellar gases (99% hydrogen and helium) from which they were formed. As the object contracts, its gases are compressed and the temperature at its center increases due to the temperature rise of adiabatic compression. (Adiabatic temperature rise is the phenomena which causes the temperature of the Earth's atmosphere to decrease with increasing altitude and causes the surface of Venus to be hot enough to melt lead.) If the mass of such an object is more than 10 times that of Jupiter, adiabatic temperature rise and gravitationally induced pressure cause the temperature and pressure at its center to reach a level where hydrogen fuses to form helium. This fusion releases large amounts of energy and raises the temperature, and therefore the gaseous pressure, at the core to the point where it is capable of resisting gravitational pressure and the contraction stops. The object has then become a star in which fusion energy released at the core produces the temperature needed to allow gaseous pressure to balance gravitational pressure. Simultaneously, energy flows from the hot core to its surface, is radiated to space, and must be replaced by the fusion of more hydrogen.

10.3- Eventually, enough hydrogen has fused to helium to make the hydrogen burning process ineffective and the core contracts to release gravitational energy to replace the energy flow no longer provided by hydrogen burning. Core contraction continues until a temperature and pressure is reached where helium can fuse into still heavier elements and again supply the energy flow needed to resist gravitational pressure. This stage in a star's evolution is dramatic. The switch from hydrogen burning to helium burning requires a large increase in the temperature of the core and produces a correspondingly large increase in the rate of heat flow to the surface. To radiate that larger rate of heat flow, the star requires a much larger surface area and it expands to become a red giant. (In about five billion years, our Sun will reach this stage and its surface will encompass the orbit of Mars.) When the helium fuel is exhausted, the successively heavier elements which are the ashes of the lighter elements already burned are themselves burned into even heavier elements. Their burning, however, does not induce a significant increase of core temperature and burning rate and the star contracts from its red giant state to approximately its original size. The generation of energy in the core by the fusion of progressively heavier elements ceases when the core has been burned to iron because the formation of elements heavier than iron absorbs energy instead of releasing it. The future of the star from this point on depends upon its mass.

10.4- If the star is smaller than about 1.4 solar masses, the matter in its core is capable of resisting the pressure exerted by the overlying material and it eventually cools to the temperature of space. If the star is larger, normal matter in the core is not capable of resisting gravitational pressure and atomic electrons are squeezed into the atomic nuclei to form neutrons. Since neutrons do not repel each other, the core collapses almost instantaneously from the density of compressed atoms to the density of neutrons. This collapse produces a shock wave which raises a significant portion of the hydrogen overburden to fusion conditions and the star explodes. If the star is smaller than about 3 solar masses, the gaseous overburden is not able to contain the explosion and a supernova results. The remnants of such an explosion are a neutron star at the center of an expanding shell of gas. If the star is between 3 and 4.5 solar masses, the overburden is sufficient to limit the explosion and the object sheds a portion of its gaseous envelope. That explosion, known as a nova, is much less severe, and may occur several times as the star reduces its mass to the point where its neutron core can support the overburden of stellar gasses.

10.5- Gravitational collapse occurs when the size of the collapsing star exceeds 4.5 solar masses. Such stars are not capable of reducing their mass by exploding. The gaseous overburden is too great and a collapse of the star which results from the inability of the neutrons in its core to resist gravitational pressure goes to completion. Under both Newtonian Gravitational Theory and General Relativity, such an object has no choice but to collapse almost instantaneously to a mathematical point called a Singularity. (Recently a suggestion was made that electrons will collapse to form kaons under sufficient pressure. Even though kaons, like photons and neutrinos, compress rather than collapse under pressure, the mathematics associated with both theories show that kaons are not capable of preventing the final gravitational collapse.)

10.6- The fact that both Newtonian Gravitational Theory and General Relativity predict that large objects eventually contract to a mathematical point at a velocity greater than the velocity of light to form Black Holes, Singularities and Wormholes should have led to the conclusion that the theories were defective and/or incomplete. Unfortunately, in this area at least, science has been converted into a religion. As a result, physicists are prevented from raising embarrassing questions either because of the lemming effect characteristic of all religions or because challenging the true faith would end their careers. Since the author is neither a lemming nor is he dependent upon the goodwill of the defenders of the true faith, he is free to challenge that faith and assert that a proper gravitational theory must yield the following results:

- It will predict the cessation of contraction at the radius where the velocity of fall from an infinite distance is equal to the velocity of light.

- It will be consistent, in the absolute sense, with the Law of Conservation of Energy.

- It will be consistent with the Principle of Relativity.

- It will be consistent with the Principle of Equivalence.

- It will yield predictions which are consistent with observation.

Of the above, General Relativity only satisfies only the last requirement and that satisfaction is superficial. Gravity Relativity, on the other hand, meets all of the requirements.

10.7- Modeling the Gravitational Object:- In "Gravity", the author has examined gravitational collapse by assuming a highly artificial model of the gravitational field to allow its inherent characteristics to be examined while retaining sufficient simplicity to allow easy solution. This model is diagramed in Figure 10.1. The assumptions which were made are tabulated below:

- The mass of the object is contained in an infinitesimally thin shell which is at a constant radius from the center. All of its gravitational mass and its entire internal volume is at the same gravitational potential.

  - This model is structurally unstable and will collapse in response to the slightest deformation, just as a plastic soft drink bottle will collapse under a slight external pressure even though it can withstand considerable internal pressure. This instability does not limit its usefulness in analyzing the gravitational field. (This instablity also applies to the "Dyson Sphere" concept.)

- Where it is necessary in the discussion to eliminate the effects of energy loss by radiation, the surface of the object is considered to have zero emissivity.

10.8- These simplifications allow the object to be treated as if it had a single radius, R, instead of requiring it to be an infinite series of nested shells of decreasing radii which are at an increasing gravitational potential, density, temperature and pressure as the center of the object is approached. The simplified model does not accurately represent the interior conditions of a gravitationally contracted object, but it is rigorous outside of the object. While conclusions based upon such a model will quantitatively differ from what actually occurs, they will be a reasonable approximation. In "Gravity", this model is solved for the two end limit cases. One case results when none of the energy released by gravitation is radiated to space. The other results when all of the energy released by gravitation is radiated to space. (Actual gravitational contraction follows a path between these end limit cases and is determined by the portion of the initial total energy which has been radiated to space.) To provide clarity to the discussion, observations made with the units of measurement existing at a quasi-infinite distance from the object are described as "actual" or "actually observed". Similarly, observations made with the units of measurements existing near or within the object are described as "local" or "locally observed". It must be reiterated that the "actual" units of measurement remain unchanged as a result of a change of elevation while the "local" units of measurement change in a manner which satisfies the Principle of Relativity both within and between elevations, as provided by the Gravity Transformations of Table 8.13.1.

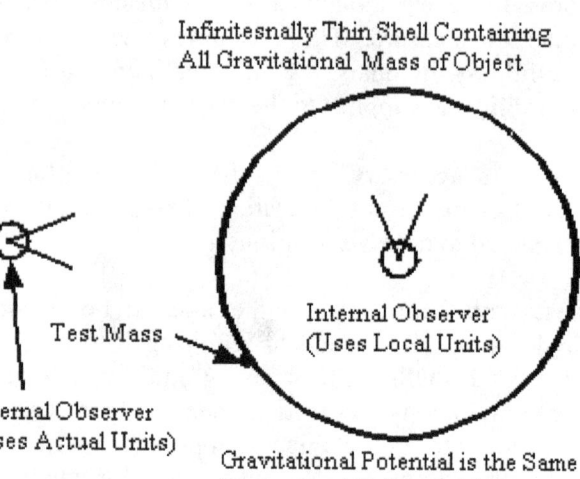

Surface of Shell Has Zero Emissivity to
Prevent Radiation of Energy to Space

External Observer Uses Units of Measurement
Existing at an Infinite Radius. His Results are
Termed "ACTUAL"

Internal Observer Uses Units of Measurement
Existing at Gravitational Potential Within Object
His Results Are Termed "LOCAL

Note:- Computation of Energy Density Violates
This Model as it Assumes Energy is Uniformly
Distributed Throughout Volume of Sphere

Figure 10.1 - Idealized Model of Gravitationally Contracted Object

10.9- In "Gravity" it is shown that the gravitational potential, Θ, may be determined from the actual radius, R and the Horizon Radius (designated as the Schwarzchild Radius in "Gravity"), R_H, using the expression $\Theta = R_H/R$. (To provide a reference for the reader, the Horizon Radius for an object the mass of the Sun is about 1.38 kilometers.) Substituting this expression into the gravitational transformation $(1-\Theta)$ provides the basic gravitational transformation as also being equal to $(1-R_H/R)$. The Horizon Radius is determined, in turn, by the total energy content, E_T, of the gravitating object, the Ergo-gravitational Constant, D, and the portion, ς, (the range of ς is 0 to 1) of the total energy content which is in the form of radiation. Its value is defined by the equation:

$$R_H = (1+\varsigma)*D*E_T$$

The term, ς, is added to the expression for the Horizon Radius derived in "Gravity" because, as that text shows, energy in the form of radiation gravitates at twice the rate as energy in the form of matter. The portion, ς, of the total energy of a contracting object which is in the form of radiation changes as the object contracts due to the release of gravitational energy as radiation or the conversion of radiation into matter. The material which follows has been simplified by normalizing the radius of the object in terms of R/R_H.

10.10- Since the gravitational field results from a reduction in the velocity of light in the Aether caused by the proximity of energy, the first effect to be considered is the effect of gravitational potential on the velocity of light. That effect is shown in Figure 8.4. Unless the actual radius, R/R_H, is less than about 10, the actual velocity of light, is essentially unchanged from its standard value of C. As the radius, R/R_H, approaches unity, the actual velocity of light rapidly approaches zero. As observed locally, of course, the velocity of light remains unchanged at its nominal value of C.

10.11- The Actual and Locally Observed Size of a Gravitationally Collapsing Object:- Since matter controls its size and its separation from other matter by a method akin to measuring the local velocity of light, the reduction of that velocity by the presence of the energy represented by a gravitating object reduces the size of all of the units of measurement by which distances are observed. At first, as an object contracts from a large radius, the contraction, as both actual and locally observed, proceeds in the manner one would anticipate from classical physics. As the actual radius, R/R_H, approaches unity, the minus sign in the basic Gravity Transformation causes the local units of measurement for length to shrink more rapidly than the actual observed radius. To an internal observer, the effect causes the velocity of contraction to slow to zero when the locally observed radius, R/R_H, has fallen to 4. (The actual value of R/R_H is 2 at this point.) As the actual value of R/R_H of that object contracts from 2 towards its limiting value of 1, the resultant reduction in size of the unit of measurement for length causes the locally observed value of R/R_H to approach infinity. The effects are shown in Figure 10.2.

10.12- The simplified model of the gravitational field described above and illustrated in Figure 10.1 is useful in providing insight into the nature of gravitational collapse. For this purpose, the author has written a program for use on a PC to allow him to approximate the gravitational contraction of an object having a selected mass, as expressed in units equal to the Solar Mass, between the radii of 10^{10} and $1+10^{-10}$ times its Horizon Radius.

- As the object contracts, the velocity (as a fraction of the local velocity of light) and the kinetic energy of a particle falling to the object from an infinite radius increases.

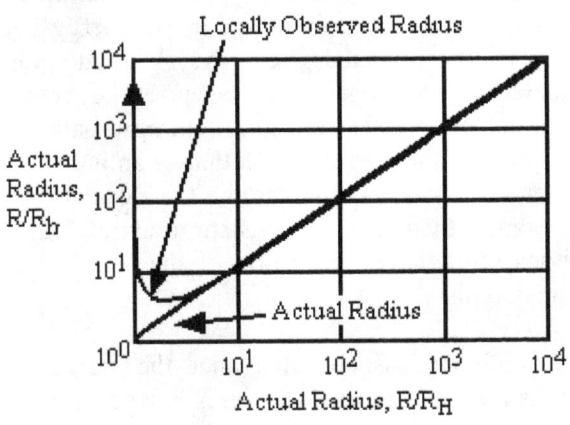

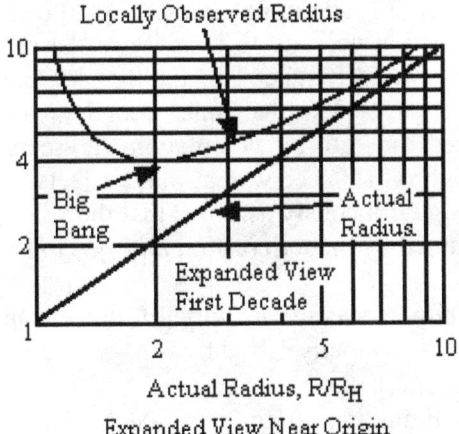

Figure 10.2 - Locally Observed vs. Actual Radius of Object

- The increase of kinetic energy of the falling particle raises its temperature until its kinetic energy becomes equal to its rest mass energy. At that point, its temperature no longer rises because the creation of additional matter is a higher entropy path for the absorption of the energy of fall than is an increase in the temperature of existing matter. The locally observed temperature then remains constant at about 4.25 trillion Kelvins until the final portion of the locally observed expansion phase.

- The simulation assumes that once the matter creation stage is reached, internal pressures balance gravitational pressures and the velocity of fall no longer increases. The simulation was also run without allowing the creation of additional matter by allowing the actual contraction velocity to increase to the velocity of light. The only difference in the results obtained was a reduction of less than one second in the time for the process to go to completion. It seems reasonable, therefore, to accept that any effects resulting from having erroneously assumed that the velocity of collapse is limited by matter creation may be ignored.

- The locally observed velocity of the expansion stage is limited to the velocity of light.

  - This limitation is not imposed by Velocity Relativity because the locally observed expansion is not caused by velocity. The particles which are observed to be separating from each other are essentially stationary. It is the reduction in the actual velocity of light which makes them appear to be separating at a high velocity.

  - This limitation would result if the propagation velocity of gravitational effects is limited to the velocity of light. Since the gravitational force producing collapse propagates through the interior of the object, the force causing the collapse would then cease to act once the locally observed expansion velocity became equal to C.

- Since gravitational energy is released from the energy contained in the matter and radiation within the field rather than from the field itself, there is no reason to believe that the gravitational field itself contains energy in any form. Velocity Relativity then does not impose its velocity limits of +/-C on the propagation velocity of a gravitational field. To the author's knowledge, there has been no experimental determination of the velocity of propagation of gravitational effects.

- The computer simulation was run both with and without a limitation on the velocity of propagation of gravitational effects. The only significant difference was, that without this restriction, the expansion phase, as locally observed, is so rapid that nucleons begin receding from each other at faster than light velocities within a second after the expansion phase started. Since our Universe is most certainly an ancient gravitational object which is expanding and which contains a large number of observable nucleons, the author accepts that gravitational effects propagate at the local velocity of light.

10.13- The Actual and Locally Observed Rate of Contraction of a Freely Contracting Gravitational Object:- The first result of interest provided by this simulation is the actual velocity of contraction of the object as a function of the ratio between its actual radius and its Horizon Radius, R/R_H, as shown in Figure 10.3A. As the object contracts from a large radius ($R/R_H=10^{10}$), the velocity of contraction increases to slightly less than the velocity of light and remains at that level as the energy of fall begins to create additional matter instead of increasing the temperature of the matter that already exists. Beyond this point the effects of the gravitationally induced

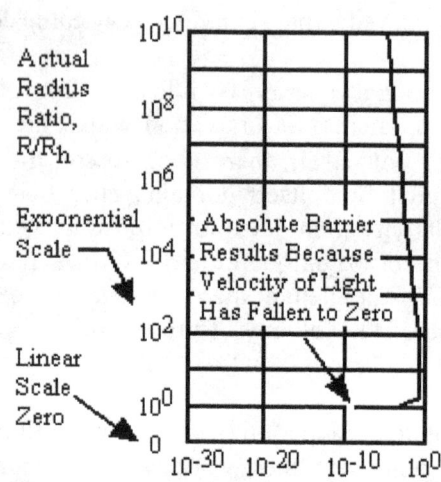

A:- Externally Observed Velocity of Contraction

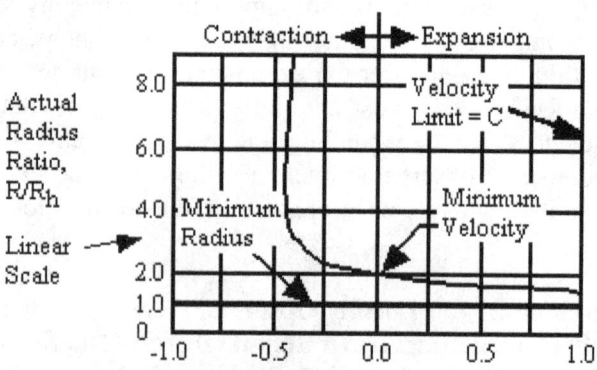

B:- Locally Observed Velocity of Ccontraction, V/C

Figure 10.3 - Externally and Internally Observed Velocity of Collapse

reduction in the actual velocity of light begin to become significant. The actual velocity of contraction reduces abruptly when the actual radius falls below $4*R/R_H$ and runs into a figurative "brick wall" when the actual radius, R, has fallen extremely close to R_H because, at that radius, the actual velocity of light has become very close to zero. The locally observed velocity of contraction behaves differently as the actual radius, R, approaches R_H, as shown in Figure 10.3B. At large radii, the locally observed and actual velocities are nearly identical, but at an actual ratio, R/R_H, of about 5, the locally observed velocity of contraction begins to decrease from its value near -C (contraction), comes to rest at an actual radius ratio, R/R_H, of 2, and increases to its maximum value of +C (expansion) at an actual radius ratio of about 1.5. It maintains the locally observed expansion velocity of +C indefinitely as the actual radius, R, approaches the Horizon Radius, R_H, and the locally observed radius approaches infinity.

10.14- The Effect of Gravitational Collapse on the Observed Rate of Passage of Time:- The next result provided by the computer simulation is the relationship between the locally observed and actual time for the object to collapse, as shown in Figure 10.4. Since the mass of the object affects the time required by the collapsing process, the results are provided for the smallest object which astrophysicists predict as being subject to collapse, 4.5 solar masses. (The plot starts one microsecond after the start of the expansion phase in order to permit the use of an exponential scale.) It will be noted that the passage of time, as observed with local clocks, increases in proportion to the square root of the passage of time as observed with external clocks. Thus, the passage of one year inside the object requires a passage of time outside of the object which is greater than the age of our Universe. It must be emphasized that the difference in the observed passage of time does not result from a change in the actual rate of passage of time, it results from the slowing of "clocks" within the object.

10.15- The Actual and Locally Observed Energy of a Nucleon During Gravitational Collapse:- The actual slowing of the velocity of light to zero at the Horizon Radius insures that gravitational collapse will not allow the radius of an object to decrease to less than R_H in a finite time. One would hope, however, that Nature had a less namby-pamby means of terminating the process, and indeed it does. The termination of gravitational collapse occurs because, as the actual radius approaches the Horizon Radius, the radiation pressure of photons and neutrinos increases sufficiently to balance gravitational pressure. As the object contracts, its locally observed radius, as diagramed in Figure 10.2B, decreases from the radius of the star from which it was formed to a minimum of four times the its Horizon Radius. From that point on, the locally observed unit of measurement for length decreases more rapidly than the actual radius of the object. This causes the object, as locally observed, to expand rather than contract, with the locally observed radius approaching infinity as the actual radius approaches R_H. During the locally observed contraction, the energy density (mostly in the form of matter) of the object increases from that of normal

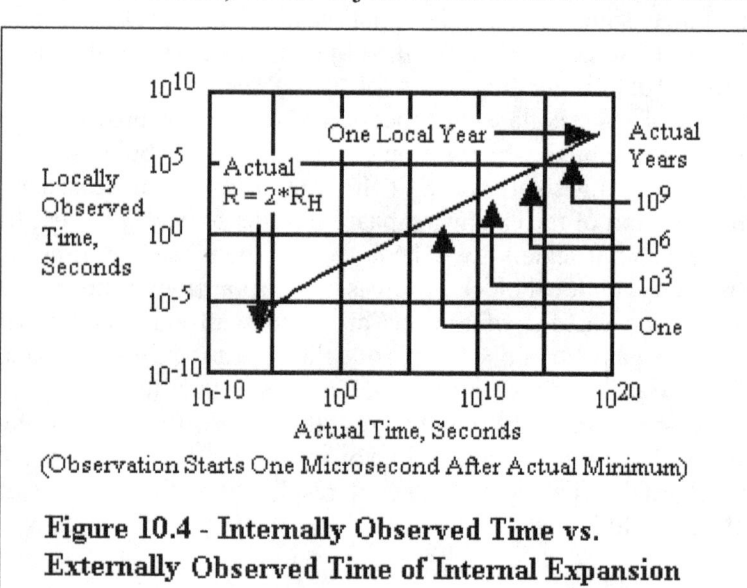

(Observation Starts One Microsecond After Actual Minimum)

Figure 10.4 - Internally Observed Time vs. Externally Observed Time of Internal Expansion

matter to that of neutrons and finally to an energy density which, if the object is not too large, far exceeds the energy density of nucleons. When the inflection point shown in Figure 10.2B has been passed, the locally observed energy density of the object begins to decrease because of the locally observed increase in volume and eventually falls to the energy density of a nucleon. During this portion of the expansion process, the energy which had been stored as a result of the compression of the original nucleons is released by the expansion to form additional nucleons at an internally observed temperature of 4.25 trillion Kelvins. The process generates a large increase in the locally observed energy content of the object, both in terms of the number of nucleons present and its radiational energy, as shown in Figure 10.5A. (The reduction of the actual energy present in nucleons is shown in Figure 10.5B.) As the locally observed expansion continues, the object becomes normal matter which then expands to become a gas. The expansion associated with this phase acts to reduce the temperature of the object.

10.16- As the internally observed expansion phase proceeds and the temperature drops below its limit of 4.25 trillion Kelvins, matter can no longer reduce the absolute energy represented by individual particles of matter, as required by the Principle of Relativity, by creating additional matter. After this point has been reached, matter has no choice but to release energy in the form of radiation. For the small portion of its energy which is electromagnetic in nature (electrons/positrons, charge of protons/antiprotons, and the electromagnetic energy associated with the orbiting of electrons/positrons) there is no problem. The resulting radiation consists of photons and adds to the observed background electromagnetic radiation of space or, if the matter is contained within an object such as a planet or star, serves to warm that object. (Planets, for example, are observed to radiate more energy than they receive from the Sun.) The 99.95% of the energy content of matter that is contained within atomic nuclei is a different matter. Based upon the arguments provided in Chapter 13, it would seem that the required shedding of energy must be accomplished by broad spectrum radiation of neutrinos

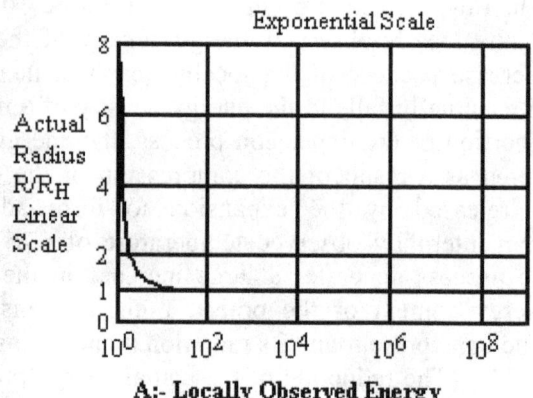

A:- Locally Observed Energy Increase Per Original Nucleon

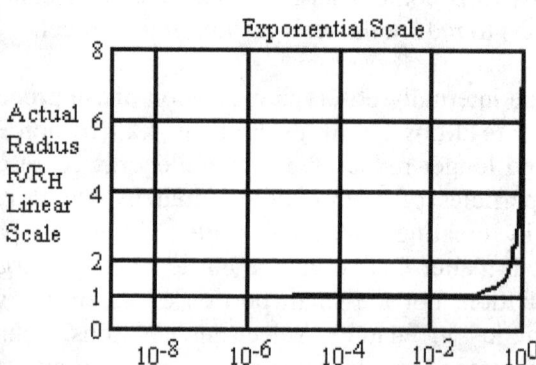

B:- Actual Energy Decrease Per Nucleon

Figure 10.5 - Locally Observed and Actual Total Energy Change of Nucleon Due to Actual Contraction of Object

similar to the black body radiation of electromagnetic energy. Such a broad spectrum radiation of neutrinos, however, is not directly observable within the current state of the art and, indeed, may never be observable. Neutrinos are currently detectable only by their absorption by atomic nuclei. Since that interaction most likely occurs at sharply defined neutrino frequencies, analogous to the electromagnetic spectral absorption lines of atoms, it is probable that only an extremely small percentage of the emitted neutrino radiation traversing space can be detected. (Gravitationally induced time dilation between the center of the Sun and the Earth can easily provide enough frequency shift to account for the failure of experiments designed to detect the expected emission level of neutrinos produced in Sun.) Unlike photons, neutrinos are not trapped by matter and they escape directly to space without heating the object through which they are passing. It seems reasonable to expect that the distribution of energy between photons and neutrinos in the background radiation of space would closely match the ratio of electron mass to nucleon mass of atoms. If this is the case, 99.95% of the background radiation level of space would consist of undetectable neutrinos. While this conclusion may seem extreme, in terms of background radiation temperature it is not hard to accept. Since the energy of radiation varies as the 4th power of absolute temperature, the required energy ratio is achieved if a very reasonable background neutrino radiation temperature of 22.3K coexisted with our Universe's background electromagnetic background temperature of 3.5K. (Since neutrinos and photons do not interact, they can co-exist at different temperatures.)

10.17- The Endpoint of Gravitational Collapse:- "Gravity", using the above model, provides the gravitationally induced pressure and the energy density of an object having a total energy of E_T, as observed with actual units of measurement and in terms of the gravitational potential, Θ (alternatively expressed as R_H/R). The derivation of these expressions did not consider that energy in the form of radiation gravitates at twice the rate of energy in the form of matter. When this effect is considered by the inclusion of the factor ς, these expressions, with no energy radiated to space, become:

Gravitational Pressure = $R_H^4/[4*\pi*D^3*(1+\varsigma)^2*R^4*E_T^2]$

Radiation Energy Density = $3*R_H^3/[4*\pi*D^3*\varsigma^2*R^3*E_T^2]$

And, since radiation pressure is proportional to the energy density of the radiation, the radiation pressure, P_R, is given by:

Radiation Pressure = $3*R_H^3/[4*\pi*D^3*\varsigma^2*R^3*E_T^2]$

Equating the expressions for the gravitational pressure and the radiational pressure enables the relationship between the gravitational potential, R_H/R, and the portion, ς, of the object's total energy which is in the form of radiation enables the value of ς at which a pressure balances occurs to be determined. The result is plotted in Figure 10.6 as a function of the gravitational potential. The curve has been adjusted to include the effects resulting from the fact that energy in the form of radiation gravitates at twice the rate of energy in the form of matter.

10.18- As a result of radiation pressure, actual gravitational contraction comes to a halt at a radius, R, which is close to the Event Horizon radius, R_H, because further actual contraction increases the locally observed radiation pressure. Actual contraction from this point on can only continue as a result of the radiation of energy to space. For this to occur, the actual temperature of the object must be warmer than the space in which it exists. As provided in "Gravity", the gravitational equilibrium temperature between space and the collapsed object imposed by the Gravitational Transformation, $(1-R_H/R)$, in terms of the temperature of the external space, Φ, in Kelvins, is given by $(1-R_H/R)=4.85*\Phi^{0.5}*10^{-7}$. The characteristics of a collapsed object in which all of its energy has been radiated to space has also been derived in "Gravity". Actually, the object has shrunk to the size of a mathematical point which contains zero energy. As

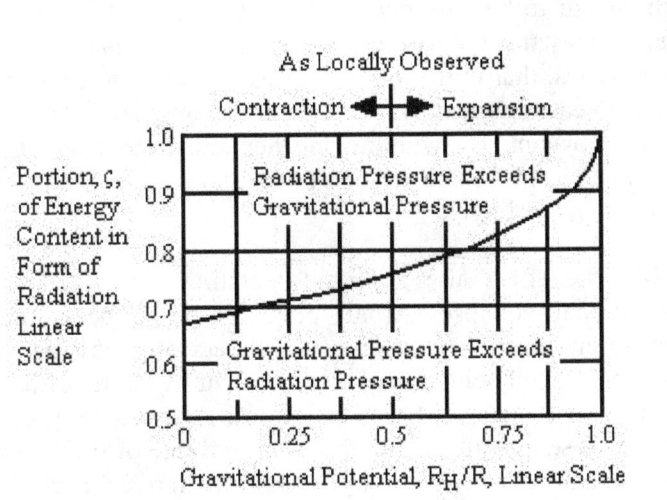

Figure 10.6 - Pressure Balance Between Radiation and Gravity vs. Gravitational Potential

observed locally, the object has shrunk to the radius of its original Event Horizon and contains an amount of energy which is equal to the energy originally contained in the matter from which it was formed. This energy, however, will be entirely in the form of radiation. *Gravitational collapse of an object will not go to completion as long as the temperature of the space surrounding the object exceeds absolute zero.*

10.19- The Conservation of Angular Momentum During Gravitational Contraction:- When a figure skater is spinning on the toe of a skate, her rate of spin is increased dramatically by the simple act of bringing her extended arms to her side. The effect is one of the most familiar examples of the Law of Conservation of Angular Momentum and results from two effects. In retracting her arms, the skater increases the kinetic energy of her arms by doing work against the centrifugal force of the spin. This energy, along with the original kinetic energy stored in her arms must be conserved and, in order for this requirement to be met, her angular velocity must increase. The effect obeys the Law of Conservation of Angular Momentum which states, that in the absence of external torques, the product of an object's moment of inertia and its angular velocity must remain constant. By withdrawing her arms, the skater reduces the moment of inertia of her body and her angular velocity increases accordingly.

10.20- The effect also applies to rotating objects undergoing gravitational collapse. At all points in the process, angular momentum must be conserved, both actually and as locally observed. Handbooks provide the moment of inertia of a sphere in terms of its mass, M, and its radius, R, as $0.4*M*R^2$. Since both of these quantities, and the measured rate of the passage of time are subject to the Gravity Transformations of Table 8.13.1, the relative change in the angular velocity of a collapsing object as a function of its locally observed radius is readily determined. As shown in Figure 10.7B, the relationship between the locally observed angular velocity and the locally observed radius of the object during the contraction phase is conventional. Its angular velocity increases in inverse proportion to the square of its radius. During the expansion phase, however, the locally observed angular velocity decreases in inverse proportion to the cube of its locally observed radius. The reason for the difference in behavior between the locally observed expansion and contraction phases is that the locally observed release of gravitational energy increases the object's locally observed mass. At the large locally observed radius existing late in the expansion stage, the internally observed angular velocity is indistinguishable from zero.

10.21- Actually the angular velocity of a collapsing object follows a different path. For the case where no energy is lost by the collapsing object by radiation to space, the mass of the object varies inversely with the square of the velocity of light in accordance with Thomson's $E=M*C^2$ and is proportional to $1/(1-R_H/R)^4$. Figure 10.7A shows that the effect of the mass increase is small for radii which are large compared to the Horizon Radius and the object obeys the conventional Law of Conservation of Angular Momentum as it contracts. When the radius approaches the Horizon Radius, the effects of the increased mass predominate and the rotation of the object comes to a screeching halt. The abruptness of the stoppage is illustrated by Figure 10.8.

10.22- When one recognizes that our Universe is itself a gravitational object of a radius essentially equal to its Horizon Radius, as determined by the gravitational mass of the matter and the radiation it contains, it is obvious that the net angular velocity represented by that matter and radiation must be vanishingly close to zero. The gyroscopically observed zero angular velocity of our Universe is caused by the Law of Conservation of Angular Momentum operating in an 'expanding' Universe. It is not the result of the net angular velocity of all of the matter in the Universe as Drs. Einstein and Mach asserted without having presented any indication as to how such a miraculous effect might come about. Modern General Relativists assert that as a rotating object forms a Black Hole, it drags space along with it and its rotation comes to rest with respect to that space. If any reader believes either of these assertions, the author would like to hear from him. That bridge over the East River is still for sale.

10.23- Gravitational Waves and Gravitons:- Since every accumulation of energy generates a gravitational field, it follows that the energy comprising every object which is undergoing spatial acceleration, such as the Moon in its orbit around the Earth, emits gravitational disturbances. If the acceleration is cyclical, the disturbances are in the form of waves which propagate throughout space and are sufficiently strong to allow their detection, at least in principle, at extreme distances. As is the case for all types of radiation, the wavelength of such a gravitational wave is determined both by the frequency of the cyclical motion generating it and the velocity of the wave's propagation. (If one assumes that the velocity of propagation of a gravitational effect is infinite, gravitational waves can be considered to be a waves of infinite wavelength.) Gravitational

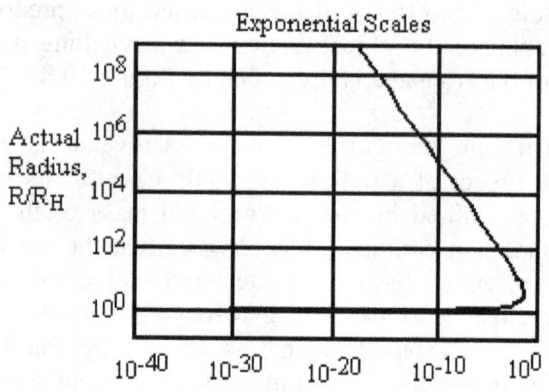

A:- Actual Angular Velocity
Zero Reference Arbitrarily Chosen

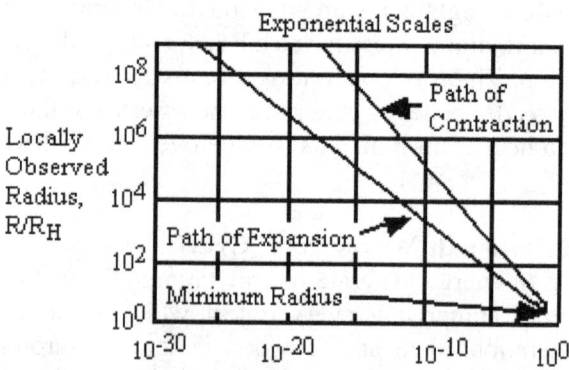

B:- Locally Observed Angular Velocity
Zero Reference Arbitrarily Chosen

Figure 10.7 - Actual and Internally Observed Angular Velocity vs. Radius

waves are implicit in both Newton's and Einstein's concepts of gravitation as well as in Gravity Relativity. Since it has had to be obvious to anyone who has thought about the subject during the intervening years since Newton that gravitational waves propagate throughout the Universe, the practice of crediting Dr. Einstein with predicting their existence seems to be somewhat of a reach.

10.24- A gravity wave is emitted by an accelerated source, such as an object in orbit about another object. Newton's Second Law of Motion requires that interacting objects maintain a common center of gravity which remains stationary. As an example, the Moon does not orbit the center of the Earth. The focus of the orbits of both the Earth and the Moon is at their common center of gravity, about 1000 miles below the Earth's surface on the side nearer to the Moon. As they move in their respective orbits about this focus, both the Earth and the Moon radiate gravitational waves. This radiation can be observed in the near field as separate gravitational waves produced by the orbital motions of the Earth and of the Moon. It cannot be observed in the far field

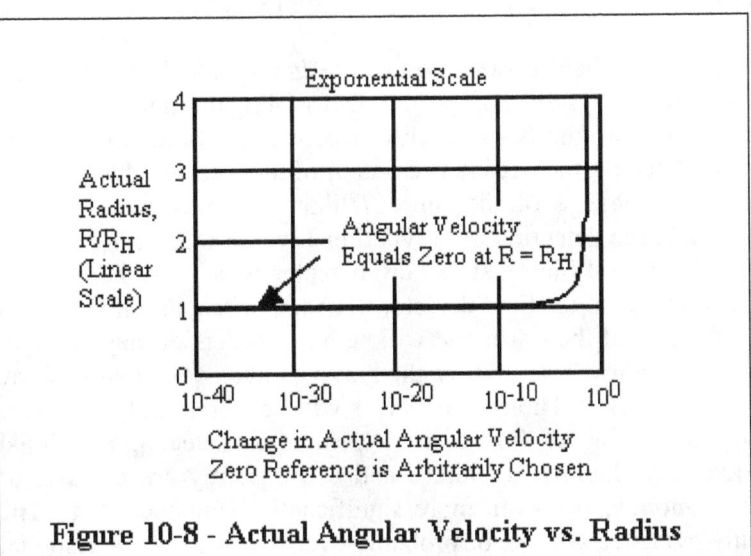

Figure 10-8 - Actual Angular Velocity vs. Radius

because detectors in the far field lack sufficient angular resolution to distinguish the Earth and the Moon as separate sources of gravitational radiation and sense only the net gravitational field from their stationary center of gravity. In the far field, the gravity wave from the Earth is canceled by the gravity wave from the Moon.

10.25- The meaning of near field and far field detection of radiation can be understood by considering the observation of a police car approaching at night along a straight desert road with its headlights flashing alternately. While each headlight is alternately turned on and off, the total light emitted from the car remains constant with its source moving from one side of the car to the other. When the car is 5 miles away, the limited resolution of the observer's eyes prevents him from seeing the headlights as individual light sources, he sees them in the far field as a single source of constant intensity. When the car approaches to within 1 or 2 miles, near field detection begins because the observer's eyes are capable of resolving the two headlights as separate sources. He then observes the cyclical flashing of each headlight, and the light source is seen to move from side to side.

10.26- Near field observation of the gravitational waves of the Earth-Moon system has been a fact of life throughout history. We live in proximity to a highly effective near field gravitational wave detector, the ocean. The output of this near field detector is the daily change of the time of high tide. However, when it comes to the detection of gravitational waves from, for example, closely spaced binary stars, any detector which can be built on Earth will be a far field detector. Its angular resolution is limited by the size of the Earth and will be many orders of magnitude to coarse to allow it to resolve the gravity wave components of any possible source. The gravity waves will be there, but the detector will be unable to find them. Attempts to detect gravitational waves are doomed to failure and consequently are a waste of time, money, and even more significantly, valuable talent. The only evidence we see, or probably ever will see, of gravitational radiation from distant orbiting objects is the decay of their orbits resulting from the loss of energy that such radiation implies. The Universe completely surrounds any source of gravitational radiation and consequently behaves as a near field absorber of its energy.

10.27- There are numerous mentions in the literature of as yet unobserved particles named gravitons as being the carriers of gravitational force. The concept of gravitons arises as an analog to the virtual photons which allegedly produce the electromagnetic forces. Aside from whether or not virtual photons exist, there is no requirement for gravitons to be any more than a figment of the mathematical physicist's imagination and are generated by his need to explain gravity's ability to act at a distance without accepting the existence of the Aether. Since the gravitational field does not contain energy itself, but only modulates the behavior of the energy which is already present, there is no theoretical requirement for it to be quantized. Unless they are observed or are shown to be required, it seems reasonable to take the position that they are a fiction.

Chapter 11 - Gravitational Collapse and the Creation of a Universe

11.1- If one examines the implications of Figure 10.2, it is apparent that every object which undergoes gravitational collapse creates another universe within a sphere having an actual radius slightly in excess of its Horizon Radius. As we have seen, as the object's actual radius contracts towards the Horizon Radius, its locally observed radius initially contracts and then expands. From this point on, gravitational contraction causes the object's locally observed radius to increase and approach infinity. The fact that the Gravity Transformations for Length and for Energy are identical causes both the locally observed energy content of the object and its locally observed radius to approach infinity at the same rate. Since the volume of a sphere varies in proportion to the cube of its radius, once the locally observed expansion phase begins, the locally observed energy density of the object decreases in proportion to the square of its locally observed radius. (The combined effects of the Gravity Transformations on the units of measurement of both length and energy cause the point of maximum locally observed energy density to shift from its nominal value of $4*R_H$ to $4.1*R_H$.)

11.2- As shown in Figure 11.1A, computer simulation reveals that the maximum locally observed energy density of the collapsing object, expressed in terms of nucleon energy per original nucleon, varies inversely with the square of the the actual energy content of the object. At the point of maximum locally observed density, the nucleons of objects smaller than 185 actual solar masses are compressed and energy is stored within them. As the subsequent locally observed expansion phase proceeds, the excess locally observed energy stored within the compressed nucleons is released in a manner which must follow the highest entropy path. As a result, the additional nucleons shown in Figure 11.1B are created while the locally observed temperature of 4.25 trillion Kelvins generated during

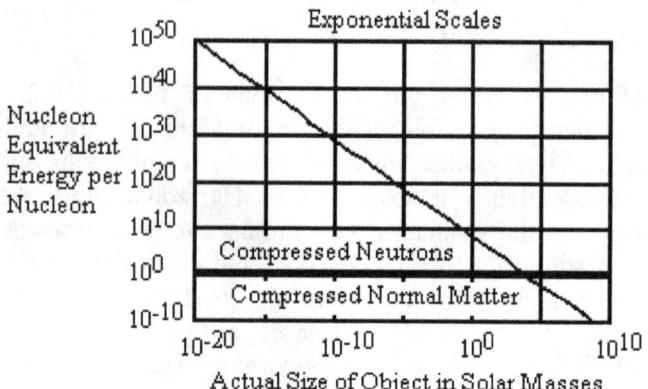

A:- Locally Observed Addition of Energy to Nucleons at Peak of Locally Observed Compression

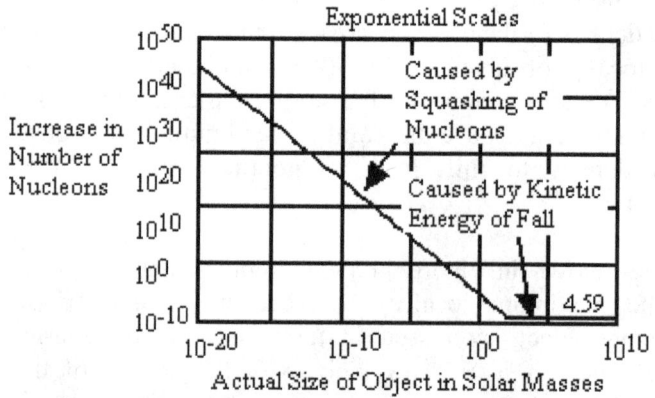

B:- Nucleon Multiplication Resulting From Locally Observed Expansion

Figure 11.1 - Increase in Number of Nucleons Due to Compression and Re-expansion

the contraction phase remains unchanged. Local observations made late in the expansion phase and extrapolated backwards in time to the beginning would lead to the conclusion that matter boiled into being from nothingness and then expanded to become normal matter embedded in a sea of intense radiation. The object, as locally observed, then continues to expand and cool as a gas and, as the locally observed expansion proceeds, additional energy is released from matter to retard but not stop the cooling process. Eventually the object, as locally observed, becomes sparsely populated with matter and the bulk of its energy content is in the form of radiation. The actual contraction and the corresponding locally observed expansion are finally limited by a balance between gravitational and radiational pressure as a function of the factor ς, as shown in Figure 10.6. The collapse of objects more massive than 185 solar masses does not cause compression of nucleons because the volume of such objects at $R=2*R_H$ is sufficient to prevent nucleon compression. For such objects, the energy of fall from an infinite distance causes an increase in the number nucleons by a factor of 4.59.

11.3- Since astrophysicists teach that only objects having an externally observed size larger than 4.5 Solar Masses will undergo gravitational collapse, it is reasonable to use the Gravitational Transformations to examine the interior of such objects after they have undergone that collapse but have not radiated energy to external space. Late in its locally observed expansion stage we would find that:

- The object consists of mostly empty space filled with radiation.

- Nucleons within the object are organized in strings of varying lengths up to a significant portion of the size of the object. If the object is sufficiently large, these strings eventually contain stars and galaxies.

  - Workers in the field of plasma physics have shown that electric and magnetic fields in space are capable of organizing matter in this manner. ("The Big Bang Never Happened" by Eric Lerner, Times Books.)

- The object will be observed to be expanding towards an infinite radius with the fraction of its total energy which is in the form of radiation, ς, approaching unity.

- Extrapolation of observations of the expansion process backwards in time will lead to the conclusion that the object began with an eruption of matter into existence. Simplistically, it would be concluded that the matter was created from nothingness at a single point in space.

- The radius of the object will appear to be increasing at the velocity of light.

- The object will be composed entirely of normal matter or of anti-matter but it will not contain both types.

  - Mutual annihilation will cause matter and anti-matter in the original cloud to annihilate each other until only the predominant type of matter remains.

This description might well apply to our own Universe if it were not for one difficulty. Our Universe contains 10^{80} nucleons. As shown in Figure 11.2A, the nucleon content of a collapsed object is determined by its actually observed size and, for the collapsed object to contain 10^{80} nucleons, the actually observed size of the collapsing object would have to be on the order of $2*10^{12}$ kilograms (approximately that of a rock having a diameter of a kilometer), as shown in Figure 11.2B. Since gravitational collapse does not occur in objects smaller than 4.5 actual solar masses, our Universe could not be formed in this manner unless an additional factor was present.

11.4- While it might be questioned as to whether the process of gravitational collapse qualifies as the creation of a universe, such an objection is groundless. Initially the object is collapsing

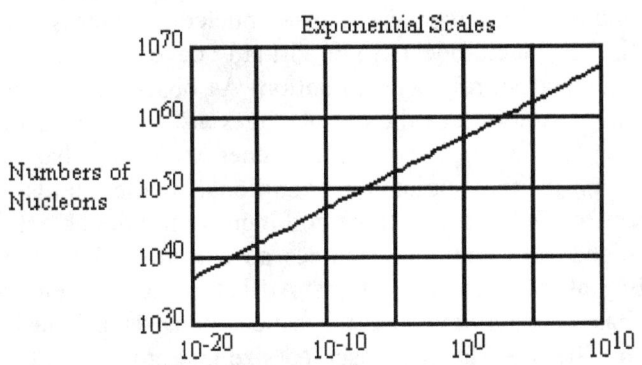

A:- Number of Nucleons Before Contraction

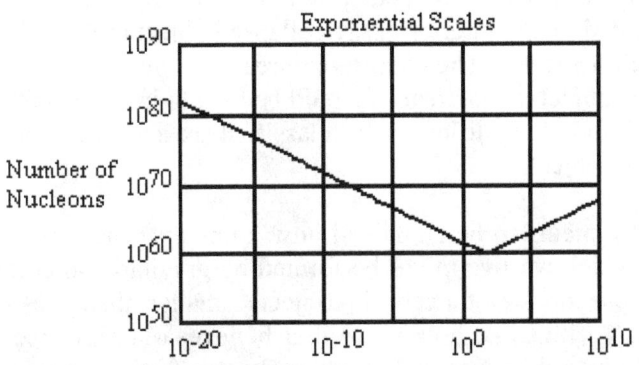

B:- Number of Nucleons After Locally Observed Expansion

Figure 11.2 - Creation of Nucleons by Locally Observed Expansion and Contraction in an existing universe. As observed in that universe, the object collapses to a small ultra dense and ultra hot sphere containing a greatly increased number of collapsed nucleons which shrink in size until the object appears to be virtually devoid of matter and is filled almost entirely with radiation. As observed within the object, the description of the initial stages are virtually identical until after the point of maximum energy density has been reached. From that point on, additional matter boils into existence from what seems to be nothingness and the object then expands and cools until it becomes an extremely large object expanding at the velocity of light. At that point. the density of matter has become low and the object consists primarily of radiation. The large locally observed size and energy content of what is actually an extremely small object containing almost no energy results from the gravitationally induced change in the size of the units of measurement acting in concert with the Principle of Relativity. Even though the boundaries of the object are no longer hot or dense, there is no possible escape from the universe represented by the interior of the collapsed object to the universe in which it resides. The energy required for a nucleon to escape from the object is too great. It would be hard to imagine anything which is more qualified to be classified as a universe as the collapsed object.

11.5- It remains to be explained how a universe the size of the one in which we live might be formed by gravitational collapse if collapse does not occur in objects smaller than 4.5 solar masses. To illustrate how the author believes our Universe was formed, the radii ratios, R/R_H, where the gravitational squashing of nucleons occur, are plotted in Figure 11.3A. In order to magnify the effects of interest, the vertical scale is plotted in terms of (R/R_H-1) rather than R/R_H. (It is the minus sign in the Gravitational Transformation which produces the effect of the field.) As an example, the gravitational collapse of an object of 4.5 solar masses is shown on this plot under conditions where no energy is lost by radiation. (In reality, energy is always radiated.) To show how the loss of energy to space by radiation allows the creation of an extremely large universe, the region of squashed

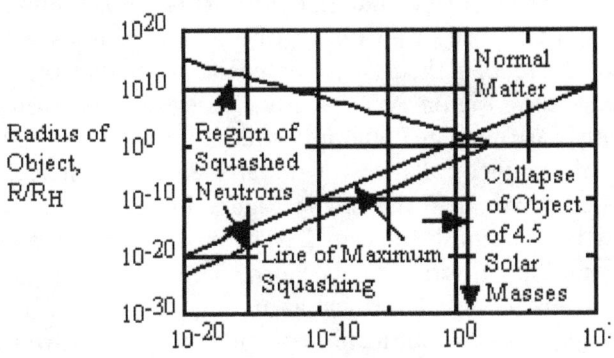

A:- Squashing of Neutrons in Object

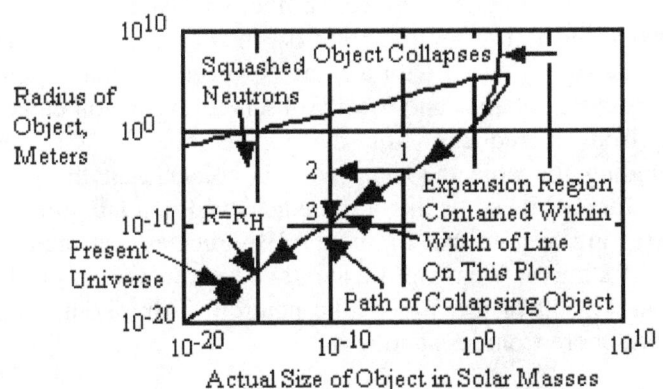

1 - Heat Loss Reduces Size of Object
2 - Object is Now Larger Than Horizon Radius
3 - Object Collapses to Horizon Radius

B:- The Creation of a Universe?

Figure 11.3 - The Compression of Nucleons and the Creation of a Universe

neutrons is replotted in Figure 11.3B with the vertical axis shown in terms of actual meters. This plot shows the gravitational collapse of an object of 4.5 solar masses with the effects of radiational heat loss included. (With the plot scale employed, the region of locally observed expansion is sufficiently narrow to be contained within the width of the lower line.) As the object collapses, it radiates energy to space. That energy loss moves it to the left in Figure 11.3B, as typified by the arrow from point '1' to point '2'. The movement to the left, in turn, means that in terms of its Horizon Radius, it is now too large and must contract further, as typified by the arrow from point '2' to point '3', to again coincide with the lower line, generating more nucleons in the process. The process is continuous and the loss of energy to space by radiation causes the object to move to the left along the lower boundary line. Eventually the object achieves a thermal balance with the external space and, from that point on, the object remains on the lower line of the curve with a ς factor virtually equal to unity. Further actual collapse and locally observed expansion can only result from a reduction of the temperature of the space surrounding the object. (That space is colder than the lowest temperature achieved or likely to be achieved in a laboratory.) It is interesting to note that Figure 11.3B indicates that the actual radius of a universe having the locally observed size of our own is about 1/1000 of the size of the neutron! Talk about making more and more from less and less.

11.6- The examination of gravitational collapse does, however, answer the question as to the final fate of our Universe. In "Gravity", the parameters of a gravitationally collapsed object which had radiated all of its energy to the external space are provided. Such an object has actually reached a zero radius and zero energy content and consequently has ceased to exist. To the local observer, the radius of the object and its Horizon Radius have become infinite. The cosmic crunch which has been postulated as the possible end of our Universe **will not occur**. The Universe, as locally observed, will expand and cool forever as it shrinks to an actual mathematical point.

11.7- The process of gravitational collapse implicit in the Gravity Transformations suggests an interesting scenario. Initially, a single universe contains particles of matter and anti-matter which annihilate each other and release radiation when they meet. In turn, radiation of sufficient energy, after a time, recreates particles of matter to replace the matter which has been annihilated. Eventually the process causes matter and anti-matter to sort themselves into separate groups from which gravitational objects composed of either normal matter or anti-matter can form. Any such objects having a size of at least 4.5 solar masses then contract to form independent universes which may eventually become as large or larger than our own. Within each of these universes, an enormous number of gravitational objects form and undergo their own gravitational collapse to become the next generation of universes. As long as the key parameters which determine the behavior of matter, radiation, and gravitation have Gravitational Transformations equal to unity, the number of generations of universes which can result from this process would seem to be unlimited. The relevant transformations listed in Table 8.13.1 meet this requirement. The Gravity Transformations for Planck's Constant, the Dielectric Constant, the Ergo-gravitational Constant and the Fine Structure Constant are all equal to unity and, if the Principle of Relativity is rigorously valid, then all other fundamental parameters will meet this criteria. There may be, however, an observational disagreement with the concept of an unlimited number of generations of universes. The velocity at which astronomical objects recede from each other due to the "expansion" of our Universe has been reported to vary in steps of about 70 kilometers per second rather than varying uniformly with distance. If this effect actually occurs and if it does not result from an explainable cause, the absolute validity of the Principle of Relativity, and more particularly the knowledge which flows from that principle, would be in question. ***Otherwise, reality may well be the grandfather of all fractals with its basic pattern being that of an individual universe.***

11.8- As will be discussed in the next chapters, the possibility exists that, in the original universe (an ancestor of our own) the velocity of light and the velocity of the quantum field are equal but, while the velocity of light is reduced by the proximity of energy (gravitation), the velocity of quantum field(s) may be affected to a much smaller degree, if at all. If this is the case, the physics of the original universe would differ greatly from the physics of subsequent generations of universes. However, the physics of the later universes would be essentially identical.

Chapter 12 - The Space Time Continuum

12.1- In our conventional experience, an event is normally considered to take place at a location described in terms of three spatial coordinates and one temporal coordinate. For example, if a light bulb is turned on in a room, the source of that light might be described as being four feet from the north wall, three feet from the east wall, and five feet from the floor and the time at which the light was turned on might be described as 2:04 PM on July 3, 1996. One might then ask why Nature employs three spatial dimensions and one time dimension as the basic structure of reality. Why isn't reality built with a different number of spatial dimensions (eg:- one, two, four, or five)? Why does it have a single time dimension along which reality progresses in a single direction from past to future?

12.2- Three Dimensional Euclidian Space:-To examine such questions consider first the possibilities inherent in continuums differing in the number of spatial dimensions they contain. A space consisting of zero dimensions is a mathematical point and events cannot occur. There would not be room. A space consisting of one dimension is a line and, while it can exist, events still cannot occur because points located along that line cannot pass each other to change their sequence. For events to occur, a space of two orthogonal spatial dimensions is required so as to allow points to change their relative locations by going around each other but, for those events to occur, energy must be present to cause them. Energy is a force acting through a distance and, as distance uses up one of the two spatial dimensions, the force must be applied to the remaining spatial dimension. Since that single spatial dimension is a line, it has zero cross-sectional area and the slightest force applied to it will produce an infinite pressure. Energy and the events that energy cause cannot exist in a continuum of two spatial dimensions because infinite pressure is infinitely destructive. Adding a third spatial dimension relieves the difficulty. A force acting along one of the spatial dimensions is then applied to the area represented by the remaining two spatial dimensions and the pressure resulting from the presence of energy remains finite. Except for the conceptual detour introduced by General Relativity, three spatial dimensions would seem to be adequate for Nature's purposes. (Some authors have stated that stable orbits are not possible in a universe having four or more spatial dimensions.) While the author is certain that many physicists will object strongly to the author's conclusions with respect to the dimensional content of space, their disagreement can only considered to have substance if it is accompanied by a demonstration of a small displacement, perhaps a fraction of an inch, in a direction which is orthogonal to our three familiar spatial dimensions. The only justification for a belief that space is represented by a three dimensional non-Euclidian geometry are observations which have been made between reference frames differing in elevation without a correction for the effect of elevation differences on the units of measurement having been made. There is no kind way to say it, those observations represent bad science. When the proper corrections are made, the geometry of space is found to be three dimensional Euclidian.

12.3- Lest there be a misunderstanding, the author has no quarrel with the use of non-Euclidian or multi-dimensional geometry in the mathematical solutions of physical problems. Because the pseudo-dimensions of non-Euclidian and/or multi-dimensional geometry have a property which they share with spatial dimensions, non-interaction between axes, their pseudo-geometry can be quite useful in the solution of physical problems. It is not clear, however, that non-Euclidian geometry can deal correctly with phenomena which involve energy. The author's objection occurs when physicists who have learned their profession by rote assert that the pseudo-geometry of curved space represents reality. Such individuals are apparently satisfied to perform their calculations without understanding the reality they represent and, as a result, can easily be misled when they interpret their findings.

12.4- The Time Dimension:- Nature requires one more dimension in order to function. It requires the dimension of time. The three spatial dimensions provide the room for events to occur and allow the energy required to impel those events to exist. The changes which events represent occur in a sequence which may be considered to be a translation of the three spatial axes along an axis which is orthogonal to them. The effect of that translation of the three spatial axes results in the passage of time in which the present follows the past and the future follows the present. The fourth dimension does not relate to where, it relates to when.

12.5- The passage of time is intuitively obvious. The question which that passage poses is why it progresses from past to future and not from future to past. If the events which are considered are simple enough, there is no way to tell the direction of time. A motion picture showing the impact of two billiard balls appears quite reasonable when it is projected in the forward or in the reverse direction. The situation is different if the motion picture shows the initial break of the balls in a game of pool. If the motion picture is run in the normal direction, the cue ball is observed to strike a triangular cluster of 15 balls. The 16 balls then move in 16 different directions at 16 different velocities and the observer considers the display to be quite reasonable. If the motion picture is run backward, 16 balls approach each other at 16 different velocities from 16 different locations and arrange themselves with 15 of them in a compact stationary triangle and with a single ball departing from the apex of the triangle at a high velocity. An observer viewing such a display will conclude that the motion picture has been run backwards because the probability of the events, as displayed, occurring naturally is vanishingly small.

12.6- There is a fundamental difference between the two examples. In the case of the impact of the two billiard balls, there is no change in the amount of information present before and after their impact. The total relevant information prior to impact is the position and velocity of two billiard balls. The situation is different in the case of the breaking of a rack of pool balls. Prior to impact, of the 16 balls involved, 15 of them are locked into a pattern which has a single position and zero velocity while the 16th ball has both a velocity and position. Only four significant items of information exist. Following the impact of the cue balls, 16 balls have individual positions and velocities and 32 significant items of information exist. (Two items of information are required to represent position and velocity on the surface of the two dimensional table. They are each treated as a single item to avoid complicating the discussion.) One is led to the conclusion that the flow of time from the past to the future is accompanied by an increase in the total amount of information or its equivalent, the total quantity of entropy, in the Universe. Since both gravitational collapse and the universe creation process described earlier involve an enormous increase in total information content, time must progress in the direction of that collapse, from past to present to future. In Chapter 10 it was shown that the energy content of all matter was higher in the past and will be lower in the future. A reversal of the flow of time, even locally, would require an enormous input of energy.

12.7- Surprisingly, there seems to be a school of thought among physicists that Special Relativity allows it to be possible to transfer information from the future to the past if faster than light communication could be achieved. Their concept is based upon the idea that if points A and B were physically separated but moving at the same velocity, an observer at point A could consider that points A and B were moving in a direction from B to A. As a result, a signal sent to point B at a velocity greater than the velocity of light would arrive before it was sent. Similarly, an observer at point B could consider that points A and B were moving in a direction towards point B and that a signal sent to point A at a velocity greater than the velocity of light would also arrive before it had be sent. They then conclude that if a faster than light signal were sent from B to A the instant that a faster than light signal was received from A, the round trip signal from A to B to A would be received before it had been sent and a signal would have been received from the future.

Fortunately for the casino industry, faster than light communication will not achieve such a result. The fallacy is that, while it is legitimate to accept that a faster than light signal sent from A to B can be considered to arrive before it was sent and it is also legitimate to consider that a faster than light signal sent from B to A can be considered to arrive before it was sent, it is not legitimate to hold both viewpoints simultaneously because that view point requires A and B to be moving at two different velocities at the same time. Communication or travel from the future to the past is not possible under any legitimate physical theory. It only appears possible if one's reasoning processes are limited.

12.8- The Barrier at the Past-Future Interface:-The concept of negative time appears many times in mathematical treatments of physical problems ranging from Dr. Feynmann's observation that it is impossible to distinguish between a positive electron traveling forward in time and a negative electron traveling backwards in time to an analysis of AM radio communication. While many individuals consider that because mathematics allows and in many cases requires the concept of negative time, that, in natural processes, time must actually be able to flow in reverse. This presents another situation where one must never ass*u*me.

12.9- A similar effect occurs in suppressed carrier modulation of radio signals, as shown in Figure 12.1. Mathematically, the process consists of the multiplication of the input signal (simplistically shown as a simple sine wave at a frequency of ω_S) by the sinusoidal carrier frequency, ω_C. The result of the multiplication process is two signals, one of which is the sum of the two frequencies and the other is the difference between them, as shown. Passing the composite signal through a low pass filter rejects the higher frequency signal and provides a single signal at the difference frequency which contains all of the information originally present and which occupies half of the bandwidth and requires one quarter of the peak power of a conventional AM radio transmission. (Passing the signal through a high pass filter eliminates the difference frequency signal and transmits the sum frequency signal.) The spectrum of the suppressed carrier modulated signals is shown in Figure 12.2A. With a zero frequency signal, the sum and difference frequency signals superimpose at the carrier frequency. When the signal frequency is equal to half of the carrier frequency, the difference frequency signal also appears at half of the carrier frequency while the sum frequency signal appears at 1.5 times the carrier frequency. The interesting case occurs when one considers the signal frequency to be higher than the carrier frequency. A simple minded mathematical analysis suggests that, for a signal frequency 1.5 times the carrier frequency, the sum frequency will be 2.5 times the carrier frequency (perfectly reasonable) but the difference frequency will have the unreasonable value of minus one half of the carrier frequency. In order for that to occur, time would have to flow in reverse. Nature, however, does not allow this to occur. Instead, the actual physical process relies upon the symmetry of the sinusoidal function to allow the reflection of the difference frequency signal from a "mirror" at the past-present interface and appear as a phase inverted signal at half of the carrier frequency. The reflection process at the past-present interface is analogous to the reflection of an image from the mirror of Figure 12.2B.

12.10- There is an additional factor involved in the passage of time. Information about the past can be remembered in the present while information about the future cannot, at least to a degree which is subject to rigorous verification. This leads to the conclusion that information can only flow from the past to the present, but that conclusion may not be completely true. The Bible recounts the predictions of the Prophets and asserts that many of them have come to pass. It is claimed that psychics can, at times, foresee the future. Individuals have reported experiences which might be interpreted as a memory of a future event and, in the process, some have predicted significant events prior to their occurrence but their reports are invariably attributed to coincidence or outright fakery. Rejection of the possibility by orthodox science may be not be rigorously valid. Information does not necessarily involve energy (e.g.- information represented by the polarization

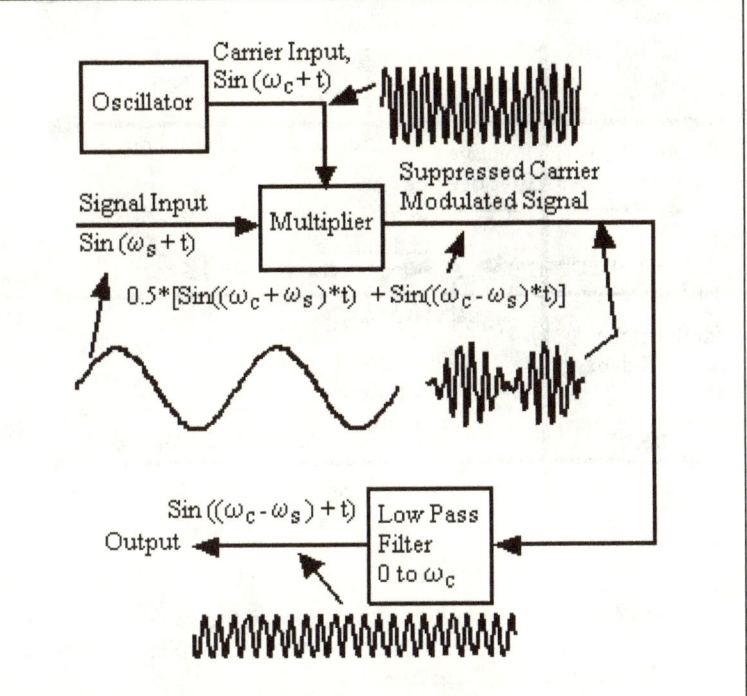

Note:- Low-pass filter rejects upper sideband to produce signal of frequency lower than carrier frequency as shown. Substitution of high-pass filter rejects lower sideband and produces signal higher in frequency than carrier frequency (not shown). Both sidebands contain identical information.

Figure 12.1 - Suppressed Carrier Modulation of Electromagnetic Signals

axis of a photon) and, if the information does not involve energy, its transfer is not necessarily subject to all of the limitations that Nature imposes upon the transfer of information represented by energy. Perhaps there is a backflow of information from the future to the present, but, if that back flow is small compared to the the forward flow of information from the past to the present,

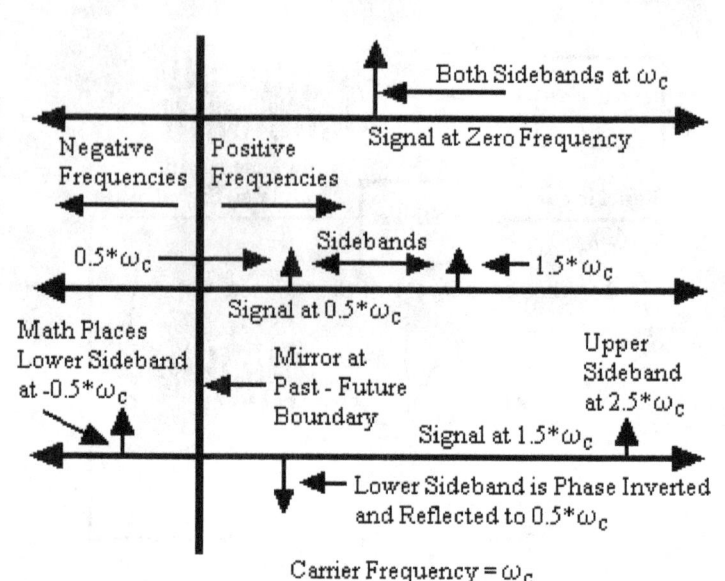

A:- Spectral Components of Suppressed Carrier Modulated Signal Showing Reflection at Past-Future (Zero Frequency) Interface.

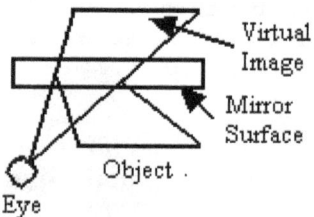

Mathematical Treatment Does Not Recognize Reflection and Predicts Virtual Image Behind Mirror Reality is Affected By Reflection, Actual Object is in Front of Mirror.

B:- Optical Analog of Time Reflection

Figure 12.2 - Reflection at Interface Between Past and Present

experiments will always prove that the reverse information flow path does not exist. Cultural bias acts to prevent the experimental recognition of such an effect, particularly since even the smallest degree of reverse information flow would result in severe philosophical paradoxes.

12.11- Space, Time and Coordinate System- One must recognize that space is a basic characteristic of reality and as such is instinctively understood by all mobile creatures. It is only symbol using mankind that associates a coordinate system with space in order to be able to quantify it. Matter exists in space just as you currently exist in the room in which you are reading this text, but like your room's relationship to you, it continues to exist when you are absent. Space is an unchanging volume (in the "absolute" sense) which may or may not contain energy or matter but which does contain **something** since it has observable properties such as a dielectric constant, permeability, a barrier which prevents energy from exceeding the velocity of light and perhaps other properties which are yet unknown. It is characteristic of space that those properties are affected by the presence of energy and cause matter to reduce its size and to lose energy thereby causing an internal observer to falsely conclude that space has been created by the proximity of energy.

Chapter 13 - The Nature of Particles

13.1- As shown earlier, logical deductions based upon physical observation and experiments by quantum physicists have demonstrated that space must be filled with the Aether of the Lorentz Contraction-Aether Theory rather than the nothingness asserted by adherents of Special Relativity. Furthermore, not only must the Aether be a solid in order for it to propagate the transverse vibrations of electromagnetic radiation, that solid must be absolutely continuous if the marvelous mechanism we experience as reality is to function. (If one observes a conventional automobile traveling down a road, he is safe in concluding that the car has a driveshaft of some type between its transmission and wheels even though he cannot observe that shaft.) The conceptual difficulty that remains is that, while a solid Aether is compatible with the propagation of radiation, solid particles should not be able to move through it. One might conclude that space must both be filled with the Aether and be devoid of the Aether, an obvious absurdity. Nature, however, is a resourceful old bird and has apparently solved the problem in what may be the only way possible. The author believes it configures a wave so as to cause it to behave as if it were a particle. To examine this possibility, it is first necessary to examine the electromagnetic wave and the photons which comprise that wave.

13.2- Consider an electromagnetic wave resulting from the transmission of a vertically polarized signal by a radio transmitter, such as illustrated by Figure 13.1. While it is common practice to consider that such a wave is continuous, progressively reducing its intensity eventually reveals it to be composed of discrete packets of energy called photons. In order for this to be the case, the wave must be composed of synchronized photons which reinforce each other so as to propagate as an apparently continuous wave. Photons must have the following characteristics:

- Separately, the photons must be electromagnetic impulses propagating at the velocity of light.

- The net electric, magnetic, and mechanical effects of each impulse must equal zero when averaged over time.

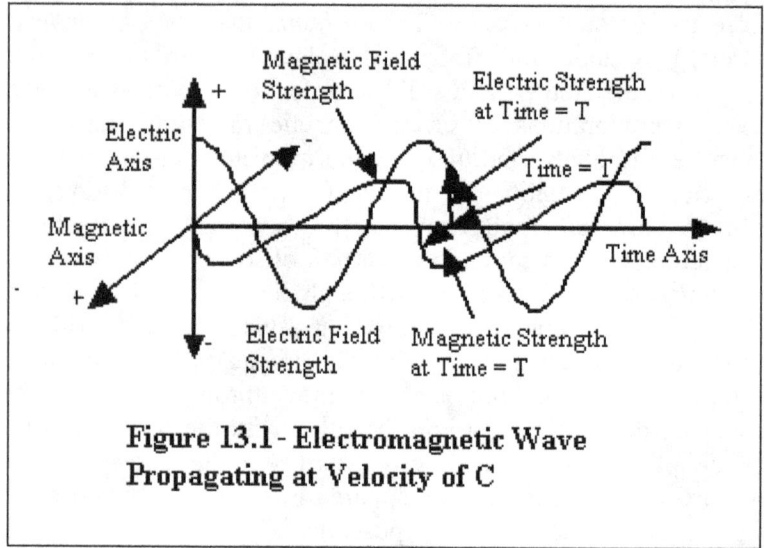

Figure 13.1 - Electromagnetic Wave Propagating at Velocity of C

- When a sufficient number of these impulses are superimposed with the appropriate timing and phasing, they must produce the apparently continuous electromagnetic wave of Figure 13.1.

13.3- To describe how the photon meets these requirements, a mathematical device known as the impulse function, such as shown in Figure 13.2, is required. (There is a class of mathematical functions known as impulse functions. The one shown is the simplest and was therefore selected for the discussion which follows but it must be remembered that Nature may employ a different version in the photon.) Multiplying the electromagnetic wave of Figure 13.1 by the impulse function provides a suitable configuration for the photon, as shown in Figure 13.3. This configuration can travel through the Aether either by itself or, when grouped in a synchronized sequence, as an electromagnetic wave.

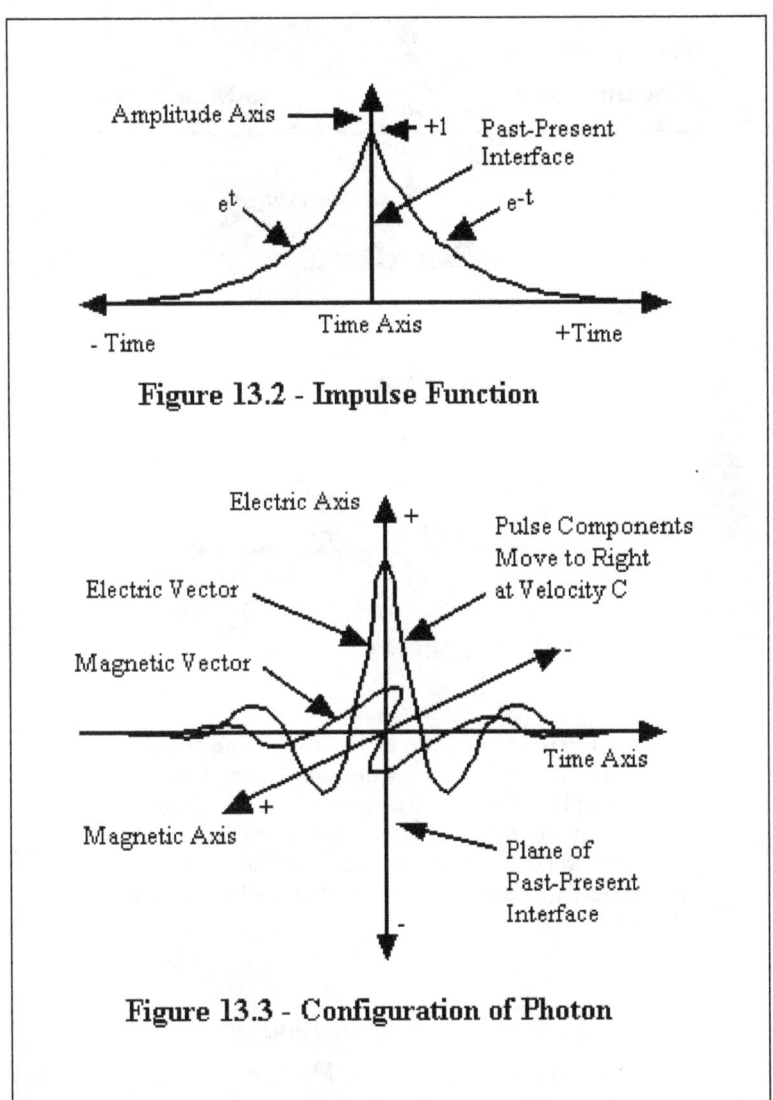

Figure 13.2 - Impulse Function

Figure 13.3 - Configuration of Photon

13.4- The energy content of a photon has been determined to be directly proportional to its frequency (inversely proportional

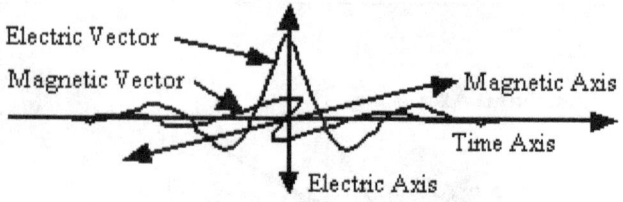

Low Energy Photon

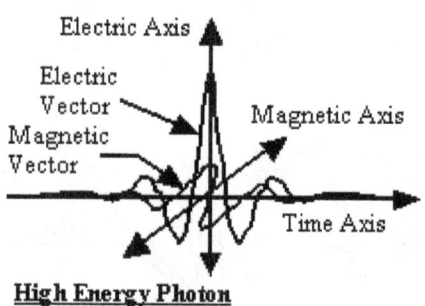

High Energy Photon

High energy photon shown has twice the frequency and energy of the low energy photon shown. The strength of the electric and magnetic vectors of the high energy photon is $2^{0.5}$ times the strength of those vectors in the low energy photon. The time width of the low energy photon is twice that of the high energy photon.

Figure 13.4 - Effect of Energy and/or Frequency on Photon

to its time duration) with the constant of proportionality being equal to Planck's Constant. As a consequence, we may conclude that the amplitude of a photon's electric and magnetic vectors increases in proportion to the square root of its energy, in an manner analogous to the storage of energy in a compressed spring (doubling the distance by which a spring is compressed also doubles the force in the spring and increases the energy stored in the spring by a factor of four). Figure 13.4 illustrates the vectors of two photons which differ in energy by a factor of two. The increased frequency of the higher energy photon corresponds to its having a time duration which is shorter than that of the lower energy photon.

13.5- There is another type of radiation which requires discussion, the radiation represented by neutrinos. Other than the facts that neutrinos lack electromagnetic properties, making them extremely difficult to observe, and they have never been observed to combine into a continuous wave, the neutrino seems to be of the same nature as the photon. It seems reasonable, therefore, to postulate the existence of neutrinic radiation which is akin to electromagnetic radiation. Conceptually, however, propagation of the neutrino presents a difficulty which does not occur with the photon. Propagation of the photon requires the interchange of energy, phased 90 degrees apart, between two mutually perpendicular axes (the electric axis and the magnetic axis). While one might consider the neutrino to result from an impulse of mechanical stress in the Aether along one axis (e.g.- corresponding to the electric stress of the photon), the energy storage mechanism along the other axis (corresponding to the magnetic stress of the photon) is not apparent. However, neutrinos exist and that energy storage mechanism must also exist and may be accounted for by the configuration shown in Figure 6.7. In turn, analogous phenomena to those produced by photons and electromagnetic waves should be possible. The author would hope that the subject will be examined more extensively by others in the future.

13.6- Implicit in the conclusion that solid matter cannot move through a solid Aether is the unproven assumption that solid matter is actually solid. Experiments in particle physics have shown that electrons and nucleons contain no detectable solid structure larger than the observational limit of 10^{-18} meters. This fact and the fact that matter can move effortlessly through a solid Aether require that the assumption that matter is solid be questioned. It is necessary to examine how, what appears to be a solid particle, may actually be a vibration traveling at the velocity of light and thus be capable of moving through the solid Aether as easily as photons or neutrinos. Since particles of matter exist when they are at rest or moving slowly, it would seem that such a possibility must be ruled out, but one must never ass*u*me. Solid matter could move at a low velocity and yet consist of a vibration traveling at the velocity of light if that vibration were traveling in a circle. For this to be the nature of solid matter, single or multiple cycles of radiation would have to be bent into a circular path with the ends joined to form a loop. Providing such loops contain an integral number of cycles, their vibrations will reinforce and there would seem to be no inherent reason why they could not persist indefinitely. Since the loops are vibrations, they are free to travel through the solid Aether at any velocity between the limits of +/-C as easily as photons or neutrinos. (As will be discussed later, these loops imply the existence of enormous local gravitational fields.) In addition, a process which changed the number of cycles contained in such a loop would result in the release or absorption of a photon(s) or a neutrino(s).

13.7- To an observer, such a loop of vibration would appear to be a material particle. If the loop is unstable, it will decay after a time and release some or all of its energy in the form of radiation with the remaining energy in the form of a lower energy particle(s). If the configuration of the loop is stable, barring a severe external disturbance, it will exist forever as a material particle. Of all of the observed solid particles, only electrons and protons and their anti-particles exhibit the stability required for infinite life. Neutrons are slightly unstable and decay into protons and electrons after a few minutes unless they are confined along with protons in the interiors of atomic nuclei. (Perhaps they should be classified, along with all other atomic and sub-atomic particles, as unstable configurations of radiation loops. A possible cause of the neutron's instability will be discussed later.)

13.8- To understand how such a loop of radiation might form a material particle, consider the implications involved in curling one cycle of the continuous electromagnetic wave shown in Figure 13.1 about its magnetic axis to form a loop, as shown in the left side of Figure 13.5. Curling of the single cycle of radiation results in the production of a fixed polarity electric vector rotating at twice the angular rate of the electric vector which formed the original electromagnetic wave. (Mathematically, the process is equivalent to squaring the expression for the electric amplitude vector, $\cos(\omega*t)$, to obtain $0.5+0.5*\cos(2*\omega*t)$. The presence of the term, $2*\omega*t$, indicates that the frequency of the vector rotation has doubled.) As one would expect, the rotation of the electromagnetic wave about its magnetic axis produces a stationary magnetic dipole. It also produces a monopole electric charge because its electric vector is the square of the electric vector of the original electromagnetic wave and, being a squared term, does not change polarity. (As the cycle proceeds, the polarity of both the electric vector and its direction reverse every 180 degrees of the electromagnetic wave and the product of two negative terms is positive.) Positive unipole charges (electrons) would be produced by rotation about the magnetic axis at 0 degrees on the time axis. Negative unipole electric charges (positrons) would be produced by rotation about the magnetic axis at 180 degrees along the time axis.

13.9- It will be noticed that the resultant locus of the electric vectors in the left side of Figure 13.5 is lopsided. It seems likely that electrostatic repulsion would redistribute that locus into a spherical configuration and the magnetic dipole would move to the axis of the sphere, as shown in the right side of Figure 13.5. This configuration seems consistent with the properties of

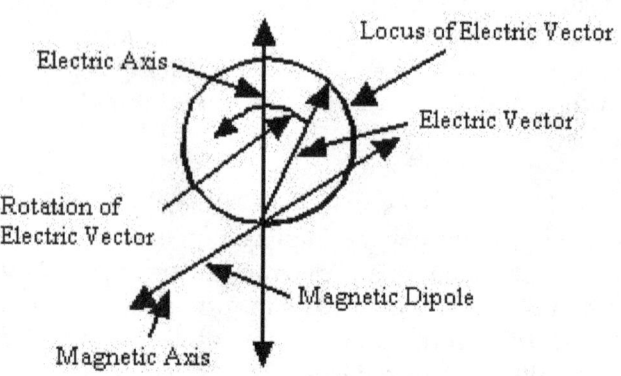

Rotation of One Cycle of Electric Vector About Magnetic Axis to Form Charge Particle With Magnetic Mpment. (Probably Unstable)

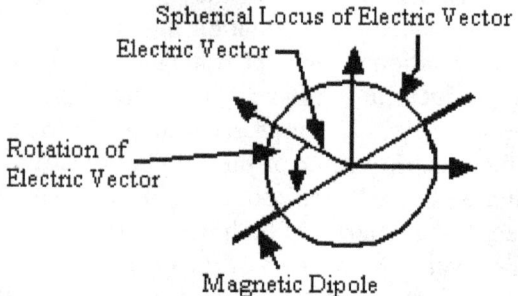

Curling of a single cycle of an electromagnetic wave into a circle about its magnetic axis will produce a charged particle which rotates at twice the rate of the electric vector and which produces a magnetic dipole. It will probably assume the configuration of the lower disgram.

Figure 13.5 - Configuration of Charged Particle

electrons and positrons in that the resultant particle consists of a point charge and a magnetic dipole and which repeats its original state after two rotations about its magnetic axis. The principle objection to the configuration would appear to be the fact that the curling of the electromagnetic wave into a loop requires a variation in the velocity of light along the electric axis such as the radial variation in the velocity of light shown in Figure 13.9. A variation in the velocity of light is not without precedent, the lenses of your eyes do exactly that in order to focus the light reflected from this page as you read it.

13.10- The concept that matter consists of radiation loops can be subjected to a reasonableness test by examining the known size and energy content of the nucleon since the size and energy content of particles formed of radiation loops must be consistent. The radius, r, of a cycle of radiation having an energy of e, in terms of Planck's Constant, h, and the velocity of light, C, is given by $r = h*C/[2*\pi*e]$. Scattering observations have shown that nucleons (protons and neutrons) contain three inseparable internal entities named quarks. Apportioning the energy of the nucleon, about $1.503*10^{-10}$ joules, equally among the three quarks leads one to the conclusion that their radii are $6.315*10^{-16}$ meters, or 63% of the observed radius of the nucleon. This is an interesting percentage because it is equal to 1-1/e", a function which appears frequently in electromagnetic equations. (In this expression, e" represents natural number similar to π having a value of 2.718.) It would appear that the radius of the nucleon is determined by the radius of a loop of radiation having the energy of a single quark. Since the nucleon differs greatly from the electron and it is neutrinos which are emitted and absorbed by processes which are associated with nucleons, one is led to the supposition that the quark is a loop of neutrinic radiation. (Applying this line of reasoning to the electron provides its radius as $3.9*10^{-13}$ meters. This size seems to be consistent with its behavior within atoms.)

13.11- If energy is added to a particle consisting of a loop of radiation, one would expect the wavelength of the radiation composing the particle, and therefore its radius, to reduce because the energy of the particle is Planck's Constant divided by its wavelength, as illustrated by the difference between the low energy particle of Figure 13.6A and the high energy particle of Figure 13.6B. However, if the diameter of a particle is prevented from becoming smaller, it has the alternative of absorbing discrete packets of energy by increasing the number of cycles it contains and/or by absorbing unquantified energy by increasing the velocity of light within the loop (hypothetical possibility for further consideration). These alternatives are illustrated by the difference between the low energy particle of Figure 13.6A and the high energy particle of Figure 13.6C. The significance of these possibilities will become apparent shortly. Recent experimental observations have shown that, like electrons, quarks are also devoid of an internal structure down to the observation limit of 10^{-18} meters. These observations are consistent with the view of the quark described in this Chapter and the requirements of a solid Aether.

13.12- Quarks have an interesting property. They are sociable and cannot be separated. They exist permanently only in groups of three to form the structure of nucleons and they exist temporarily either separately or in pairs (pions and kaons) as the short lived debris products of high energy nuclear collisions. Attempts to separate quarks from the nucleon are extremely difficult because, when quarks are in close proximity the force holding them close together is small but is reported to increase rapidly to a constant level, independent of the separation, as they are moved apart.

13.13- If one compares the size of a quark with the size of a nucleon, it would seem to be impossible that three quarks could fit within it, unless of course, their loops are intertwined. This possibility brings to mind the logo of a popular beer which consists of three intertwined circles (Figure 13.7). If the quarks in a nucleon are intertwined in loops analogous to that logo, the attractive force between them will behave in the manner ascribed to the strong force defined by particle physicists. When the loops in the logo configuration are

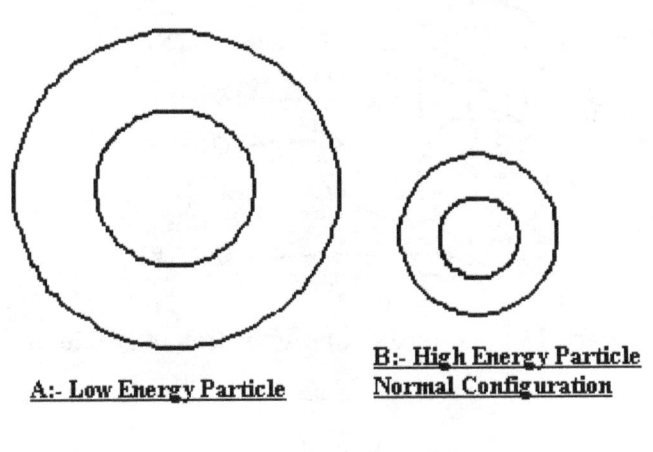

A:- Low Energy Particle

B:- High Energy Particle Normal Configuration

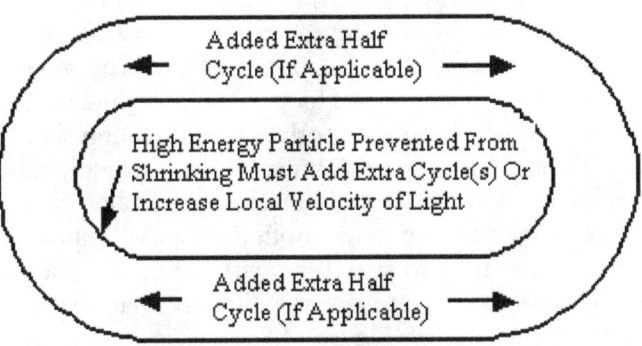

C:- High Energy Particle, Stretched Configuration, Having Same Energy As B Above

Figure 13.6 - Two Methods of Increasing Energy of Loop of Radiation

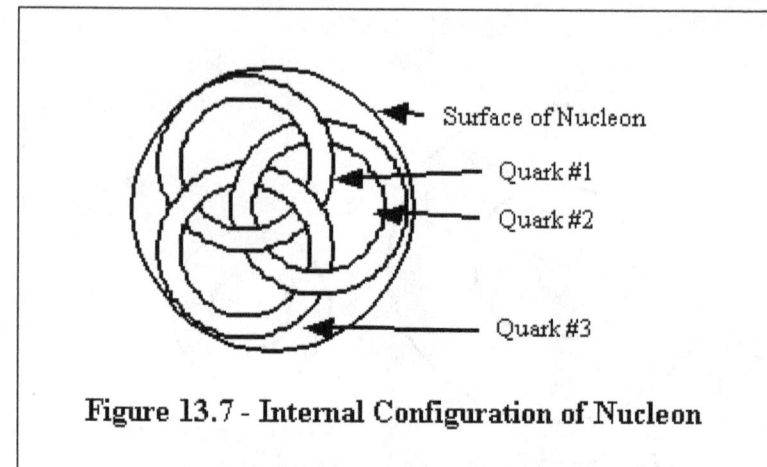

Figure 13.7 - Internal Configuration of Nucleon

nearly centered on each other, virtually no forces are exerted between them. If energy is added to such loops of radiation, one would expect their intertwining to prevent them from absorbing that energy by reducing their size because there is no room in the center to permit shrinkage. The loops composing the quarks in a nucleon would then have to absorb any extra energy in another manner. Their options would seem to be to incorporate the addition of extra cycles and/or increase the velocity of light within one or more of the loops. Both these possibilities suggest that the force required to stretch a quark would be independent of the distance it is stretched and this is what seems to be observed, as illustrated by Figure 13.8.

13.14- Conventional particle physics ascribes the attraction between the three quarks which form nucleons or the two quarks which form pions and kaons to the action of extremely massive and short lived virtual exchange particles named gluons without any explanation as to how such particles can produce attractive forces. With the configuration described above, the forces between quarks in nucleons, pions, and kaons would result from their topological configuration and the strength of their radiation loops. The massive gluons of theoretical particle physics not

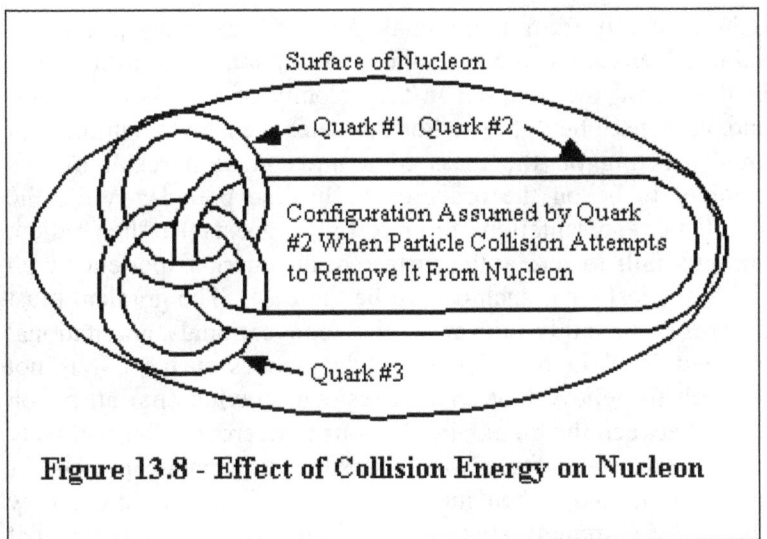

Figure 13.8 - Effect of Collision Energy on Nucleon

only would be unnecessary, they cannot exist within the nucleon without causing the nucleon to be considerably more massive than it is. The only force required to explain the behavior of the intertwined quarks is the tensile strength of the radiation loops of which quarks are composed and the intense gravitational attraction caused by the slowing of the velocity of light within them, as discussed below.

13.15- Protons (or anti-protons) within an atomic nucleus repel each other as a result of electrostatic force acting between like charges. At the distance represented by the size of the nucleon, the level of that force is on the order of 17 pounds. This force is about 10^{40} times as large as the attractive force between the gravitational masses of nucleons as predicted by Newton's Law of Gravitation. Obviously, protons (or anti-protons) in the nucleus of an atom cannot be held together by gravitation. Again it must be repeated, never ass*u*me. As has been shown earlier, the force we experience as gravity results from a gradient in the velocity of light existing as a function of elevation. In order for the quark or the electron to exist as a loop of radiation, it is necessary for the velocity of light within the quark or electron to reduce linearly from its nominal value of C at its periphery to a value of zero at its center, as shown in Figure 13.9 for the quark. In the quark, the gradient in the velocity of light is enormous, and, at its periphery, the author estimates the Newtonian force of gravity acting on the mass of a nucleon as a result of this gradient to be on the order of 11 million pounds. While the simplistic representation of Figure 13.9 suggests that the gradient abruptly falls to zero at the surface of the quark, experience with fields suggests that such cannot be the case. The gradient must decrease smoothly towards the conventional gravitational gradient level in the vicinity of the quark's surface. It is not difficult to believe that enough residual gravitational attraction exists between the quarks in nucleons to overcome the relatively puny electrostatic force which attempts to separate protons or anti-protons and, when nucleons are in close proximity, they should be extremely sticky. (This is the type of question that physicists trained by rote handle extremely well and the author hopes that a few of them will be sufficiently antagonized to examine this conjecture even if their only motivation is to prove it foolish.) With this interpretation of the nature of the nucleon, the electroweak and electrostrong forces postulated by particle physicists are not only unnecessary but are impossible.

13.16- The size of the electron should not be confused with the size of its orbit in an atom. The circumference of that orbit is the wavelength determined by the orbital kinetic energy of that electron. Quantum Mechanics imposes an additional restriction on its size. The circumference of the orbit must contain an integral number of cycles of radiation and the total energy of those cycles must equal the energy of that orbit. As a result, the electrons in an atom are confined to discrete orbits which can change only as a result of a change in the number of cycles contained in them. That change in the number of cycles requires the absorption or emission of a photon of electromagnetic radiation of exactly the correct energy level (wavelength). Diagrammatically, the effect is identical with that shown in the illustrations of Figures 13.6A and 13.6C.

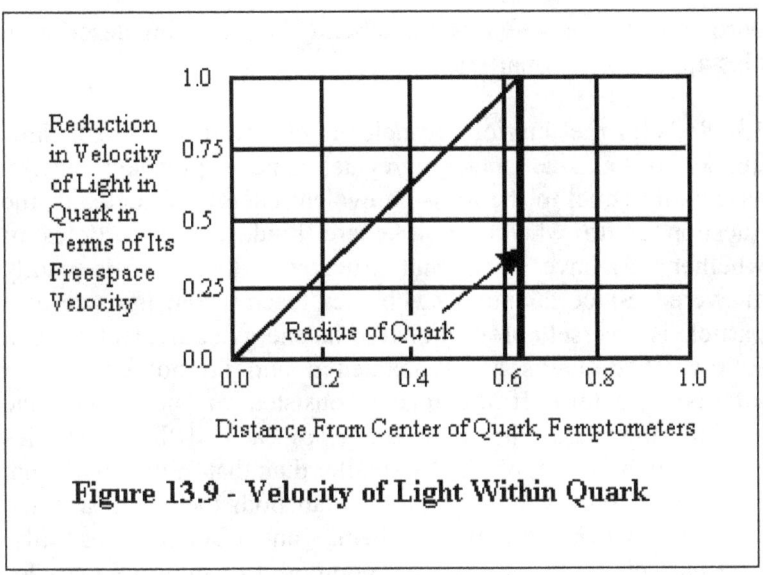

Figure 13.9 - Velocity of Light Within Quark

13.17- It appears that, for the loops of radiation which comprise matter to be stable, the presence of charge of either polarity is required. As mentioned, electrons and protons are stable on their own while neutrons are stable only in the proximity of a proton's charge. This is not an unreasonable result. The electric charge of the proton is repulsive and would act to expand the radius of the quark loops. At a larger radius, these loops contain less energy and a neutron would be able to reduce its internal energy by expelling an electron and becoming a proton. Apparently, this process is restrained in the neutron by the proximity of a proton, and in atomic nuclei, neutrons and protons seem to perform a symbiotic function. The positive charges of the protons prevents neutrons from decaying by emitting electrons to become protons while the extremely strong gravitational attraction between neutrons and protons resulting from the effects implied by Figure 13.9 prevents the mutual repulsion of the proton's charges from destroying atomic nuclei by causing them to fly apart. One might conclude then that the stability of the various atomic nuclei is determined by the parameters of this nuclear symbiosis for each particular atomic weight and number. (The preceding description also applies to anti-particles.)

13.18- With the interior of nucleons observed to contain three quarks of a size and aggregate mass equivalent energy essentially equal to the mass equivalent energy of nucleons, the question as to whether quarks are fundamental particles or whether they have an internal structure of their own is readily answered. Since quantum mechanics teaches that the size of a particle is inversely proportional to its energy equivalent mass, it follows that quarks are fundamental and do not contain an internal structure. If the quark consisted of more than one particle, the mass equivalent energy of each of those particles would either be proportionally smaller than that of the quark and, as a result, they would be larger than both the quark and the nucleon which contained them, an obvious absurdity. Alternatively, if the quark were composed of multiple particles small enough to fit in the space available, the mass equivalent energy of both the quark and the nucleon which contained it would be larger than their observed values, again an obvious absurdity. Even in the microscopic quantum world, blivits are impossible. (A blivit is defined as two pounds of excrement in a one pound bag.) The short lived high energy particles created in nuclear explosions and collision experiments are consistent with the interpretation that they are loops of radiation of high energy content either because they are single cycle loops which have been compressed into a smaller radii and/or have become larger by incorporating additional cycles into their loops or by locally increasing the velocity of light. It would seem reasonable to expect such particles to be unstable and to decay rapidly. ***There is no stable structure in Nature which is smaller or more massive than the quark or the nucleon in which it is contained.***

13.19- The experimental activity in particle physics consists of bombarding particles with other particles at high energies and examining the debris resulting from the collision. This type of experiment is performed, not because of its innate relevance, but because it the only means currently available. In order for such an experiment to provide the desired resolution it is necessary to generate collisions between particles, at least one of which has a kinetic energy which is large compared to the rest mass equivalent energies of the particles. These experiments yield a large number of short lived particles. Many of those particles have mass equivalent energies which are larger (sometimes by a factor of over 500) than the mass equivalent energy of the quark within the nucleon under test. Those particles are conventional particles whose energies have been increased by the process shown in Figures 13.6A and 13.6B or higher energy particles which temporarily form from the excess energy provided by the collision. Such particles apparently cannot be stable and they decay quickly into combinations of the stable particles (electrons, photons, neutrinos, nucleons and their anti-particles). Particle physics experiments involving energies large compared to the mass equivalent energy of quarks may be about as meaningful in determining the nature of matter as an experiment which attempts to determine the nature of fine crystal goblets by shattering them with rifle bullets and studying the fragments while they are in flight. It would appear that what is really under test in high energy particle experiments is the ability of high energy radiation loops to form within the Aether and the stability of those loops.

13.20- The computer program used to evaluate gravitational contraction showed that the maximum internally observed energy density of an object undergoing gravitational collapse is about 10^5 times the energy density of a nucleon. Since this energy density ratio is of the same order as the collision energy to be attained in the now canceled Superconducting Supercollider Program, it might have been hoped that experiments performed on that machine would have revealed information about conditions which existed at our Universe's beginning. But, if the machine had been built, the information it would have provided would probably not be relevant to cosmology. The high energy content imposed on nucleons at the beginning of our Universe resulted from compression while the collision energy of particles produced by the Supercollider produces tension within the particle products. The effects of tension and compression are not the same. The tensile strength of ball bearing steel is on the order of 250,000 pounds per square inch (PSI). The ultimate strength of that steel under the type of compression occurring in bearings is several million PSI! When compression failure (brinelling) does occur in such a bearing, it results from tensile stresses induced by the compression in the region around the area of contact and not from the compression stress itself. Since compression and tension do not produce equivalent results in ordinary mechanics it is hard to justify a belief that they produce equivalent results in particle physics.

Chapter 14 - Adding Quantum Effects to Our Understanding

14.1- The Two Slit Experiment:- There is a classic experiment which leads to one of two interpretations, both of which appear absurd, and which have led to two mutually exclusive schools of thought. Under one interpretation every particle of energy continuously and instantaneous tests all of space to determine the path it should follow and its ultimate destination. Under the other interpretation, every particle in the Universe follows every possible path to every possible destination. In so doing each of these possible paths creates an alternate universe with the Universe we observe as one of a quasi-infinite number of Parallel Universes. Each of those Parallel Universes is as complex as our own, each occupies the same volume of space as our own, and each represents the results of all of the decisions made by particles since time began. While there is a considerable number of physicists who ascribe to the latter viewpoint, it suffers from objections more significant than its quasi-infinite complexity. The creation of each of those alternate Parallel Universes requires the creation, from nothingness, of an amount of energy equal to the energy content of our own Universe and requires them to co-exist in the same volume of space. The trashing, to such a degree, of both the Law of Conservation of Energy and the requirement that no more than one object occupy the same space at the same time suggests a degree of foolishness on the part of those who hold this viewpoint, particular since no justification not involving some form of mysticism has ever been offered. On the other hand, the idea of what might be defined as a quantum stress field continuously testing all of space and controlling the paths of particles does not violate accepted physical laws and allows a reasonable explanation for the results of the Two Slit Experiment and for the behavior of polarized light.

14.2- When a particle (e.g.- photon, nucleon, etc.) passes through an opening, its direction is altered by diffraction through an angle which is inversely proportional to the size of the opening as compared to the wavelength of the particle (Planck's Constant divided by the energy of the particle). When particles pass through a slit, the angle of their diffraction in a direction perpendicular to the slit is large compared to their angle of diffraction in a direction parallel to the slit and its use allows diffraction to be studied as a one dimensional problem. The Two Slit Experiment is illustrated in Figure 14.1, Figure 14.2, and Figure 14.3. In this experiment, a beam of particles, such as electrons or photons, strike a mask which contains two slits. Particles passing through these slits are diffracted and are distributed in a diffraction pattern as they strike the screen. The relative rate at which the particles strike the screen, as a function of position behind a mask with a single slit, is illustrated by the diffraction pattern shown in Figure 14.1. If the mask contains two slits, one would expect the diffraction pattern at the screen to be the sum of the individual diffraction patterns of each slit, as illustrated in Figure 14.2. Instead, the actual diffraction pattern which results is that shown in Figure 14.3. Furthermore, this diffraction pattern persists as the long term distribution of the particles striking the screen even if the the possibility of interaction between particles is precluded by reducing the rate of particle flux to a level where there is not more than one particle in transit at a time.

14.3- The observations described in the preceding paragraph lead to the conclusion, apparently accepted by a significant number of quantum physicists (and the author as well), that somehow, while energy exists in discrete localized packets, the path which those packets follow as they move from point to point is determined by a continuous field propagating throughout all of space at a quasi-infinite velocity. That field, which the author designates for convenience as the Quantum Field, tests the probability of each of the possible paths of each and every particle in the Universe and causes those particles to adjust their paths so that their statistical distribution produces the diffraction patterns characteristic of waves. (The Quantum Field is not directly observable, it is only recognizable by its effects on the behavior of particles.) The author came to this conclusion in the 1980's after reading a book based upon a series of four lectures given by Dr. Richard Feynmann on the subject of Quantum Electrodynamics. His description of the process that determines the path that photons take when they travel from point A, are reflected from a mirror, and arrive at point B, led to what seemed to the author to be an obvious conclusion. For Nature to behave in accordance with the concepts of Quantum Electrodynamics, two requirements must be met. There must be an absolute velocity reference frame (i.e:- the Aether) and the photons must be capable of testing the probability of all of the possible paths between A and B virtually instantaneously and adjusting their paths accordingly while enroute. Since Quantum Electrodynamics has been the

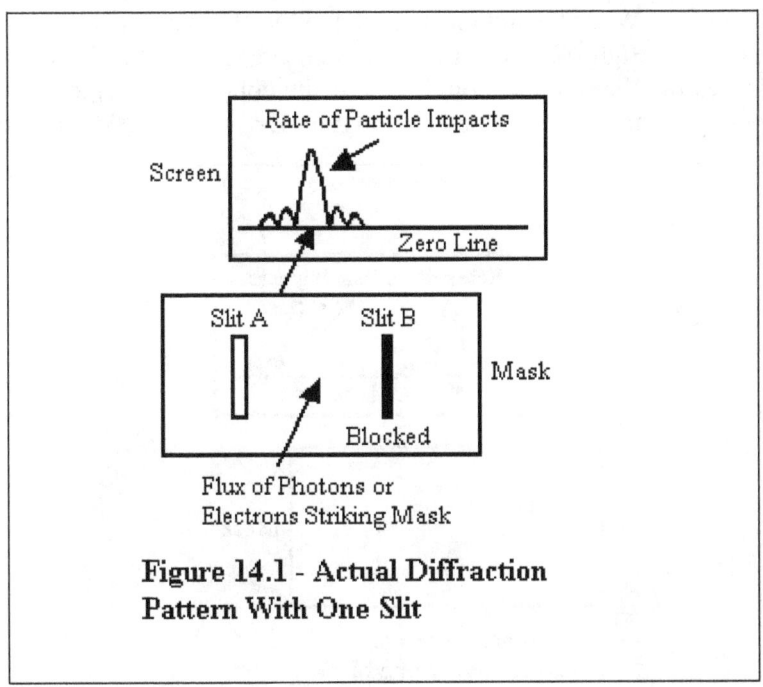

Figure 14.1 - Actual Diffraction Pattern With One Slit

most successful concept ever devised in its ability to accurately predict the results of experiments, it would seem that the concept of a continuous Quantum Field propagating at a quasi-infinite velocity has been observationally verified and the concept of the all pervading Aether has been re-verified.

14.4- It is quite certain that the scientific community will not accept the material in Dr. Feynmann's book as evidence of the existence of the Quantum Field or of the Aether. Their record in recognizing the obvious is rather poor. Acceptance of new knowledge by any group requires that the knowledge be made obvious to the group's least intelligent member in a manner which does not require that member to face up to his inadequacy. One need only note the derision heaped upon Dr. Wigner in the early part of the 20th century for suggesting that the matching continental shapes, geology, flora, and fauna of South America and Africa proved that they were once part of a common continent. When the author first read of Dr. Wigner's conclusions as a pre-High School student in the late 1930's he found it incredible that anyone could look at the continental shapes and not recognize the obvious validity of Dr. Wigner's ideas. It

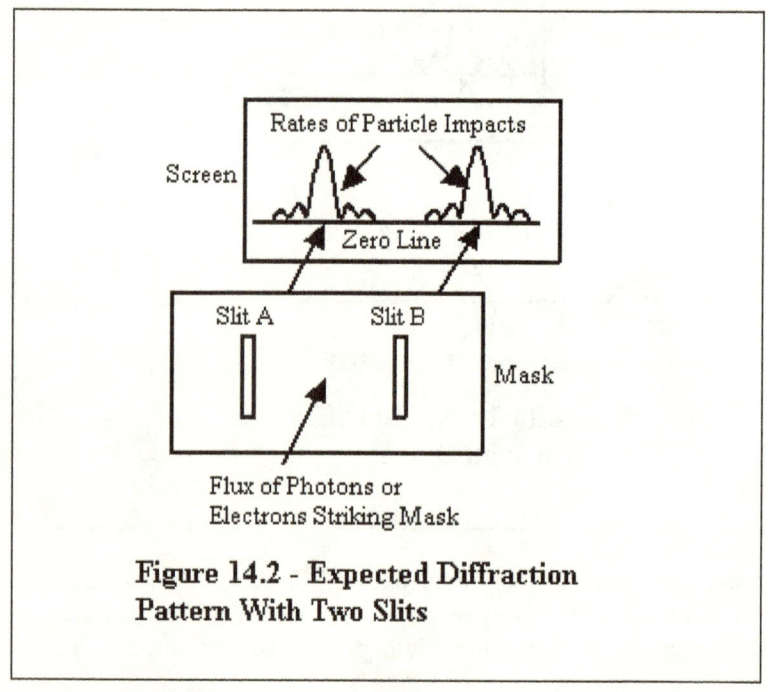

Figure 14.2 - Expected Diffraction Pattern With Two Slits

seemed to him that the only legitimate question remaining was how the continental separation could have come about. Acceptance of Wigner's Continental Drift concept should not have had to await the discovery of the lines of reversed magnetic polarization alongside of a central rift in the ocean's floor. Once Dr. Wigner's ideas had become known, the concept of tectonic plates should have been immediately obvious to any one in the field who was aware of the fact that the core of the Earth was molten and covered with a thin solid crust. It would seem that, in this case, the performance of the academic establishment left something to be desired. The same comment can be made regarding the difficulties that Dr. Shoemaker had in gaining acceptance of the idea that the Earth is periodically bombarded with very large rocks from space. It had already been accepted that rocks of all sizes orbit the Sun and that smaller rocks fall from the sky continuously as meteorites. One would have to be pretty simple minded if he looked at a picture of the Barringer

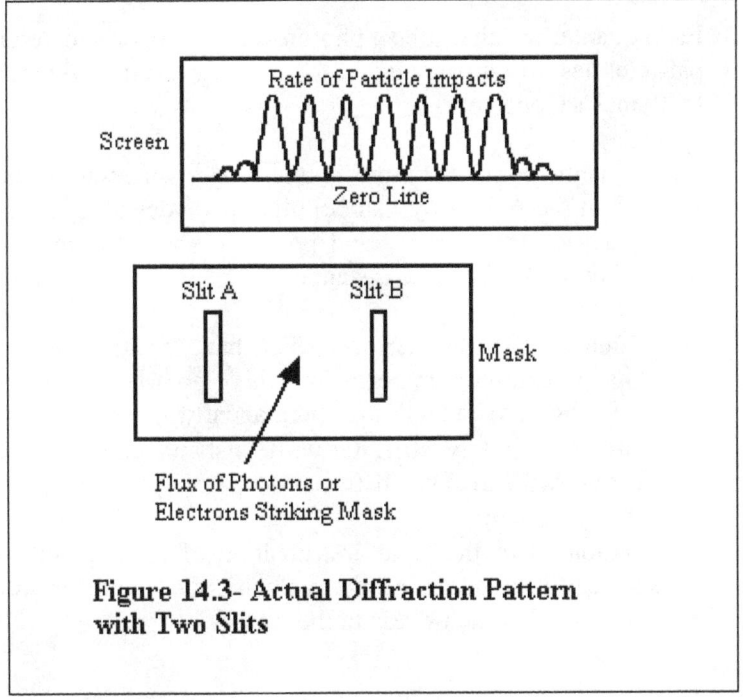

Figure 14.3- Actual Diffraction Pattern with Two Slits

Crater and did not recognize that it could only have been caused by the impact of one of the large space rocks (asteroids or comets) known to exist. The observation of a comet striking Jupiter should have been frosting on the cake but as a verification of cetainty of periodic asteroid strikes on the Earth, it should have been considered trivial.

14.5- The experimentally validated concepts of quantum mechanics yield information which, as did the relativity concepts, modify classical Newtonian Physics. These modifications involve the sub-microscopic world. The modifications of interest to this discussion are provided below:

- In its observable form, energy is not continuous. It appears only in the form of discrete packets which may be photons, neutrinos, protons, neutrons, or their anti-particles, and which are designated as quanta.

- Each quanta, whether it be a photon, a neutrino, or a material particle, has a frequency equal to its energy content divided by Planck's Constant.

- Each quanta couples to the Aether. Any quantum stress induced in the Aether by that coupling pervades all of space at a quasi-infinite velocity and probably obeys the inverse square law with respect to distance.

- Some quanta, such as photons which have been created in pairs, have quantum numbers (the axis of polarization in the case of photons) which are permanently coupled. This coupling is infinitely stiff, virtually instantaneous, and is unaffected by the distance between the paired particles.

- Information as to the state and position of each quanta is transported at a quasi-infinite velocity and with infinite resolution by quantum stress in the Aether.

- The stress pattern in the Quantum Field at any point in space is modified by the configuration of obstacles in the path of the particles producing the stress.

- Particles in motion follow a path which is consistent with the path of the information transported at a quasi-infinite velocity by the stress in the Quantum Field.

- Since the Quantum Field is continuous and the particle is quanticized, the transport of information by the two effects cannot match on an instant by instant basis. The match becomes proportionately more precise when the number of particles observed becomes large enough to allow their positions to be a statistically accurate representation of the information represented by the stress in the Quantum Field. The instantaneous peak to peak error in the information match is equal to the energy of the quanta.

14.6- The Behavior of Polarized Light:- The wave-particle duality of nucleons, photons and neutrinos revealed by the Two Slit Experiment is probably most easily understood by considering the effect of polarizers on a beam of light. As shown in Figure 14.4, a beam of electromagnetic radiation (light) is directed through a pair of polarizers having their polarizing axes at an angle a with respect to each other. The beam of light entering the first polarizer from the left is unpolarized. While the polarization of each photon in the entering beam is at a random angle, providing the entering electromagnetic wave is sufficiently intense, its intensity is independent of direction, as shown by the circular locus diagram in Figure 14.4. When the electromagnetic wave passes though the first polarizer, the randomly polarized input wave is organized into a vertically polarized wave, as shown by the vertical component in the vector diagram of Figure 14.4, with an intensity of $1/2^{0.5}$ times the intensity of the incoming wave. That single vertical polarization vector may be considered to be the vector sum of two orthogonal components. One of those components is parallel

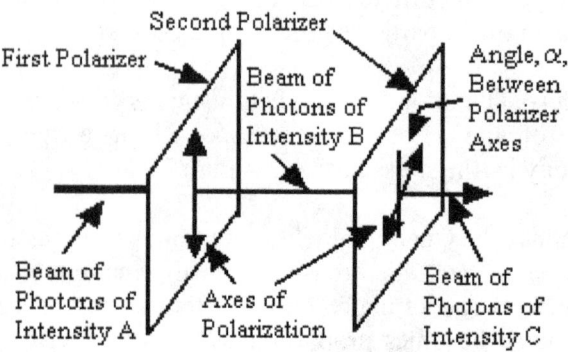

α is Angle Measured With Respect to First Polarizer
Photon Flux is Proportional to Square of Amplitude

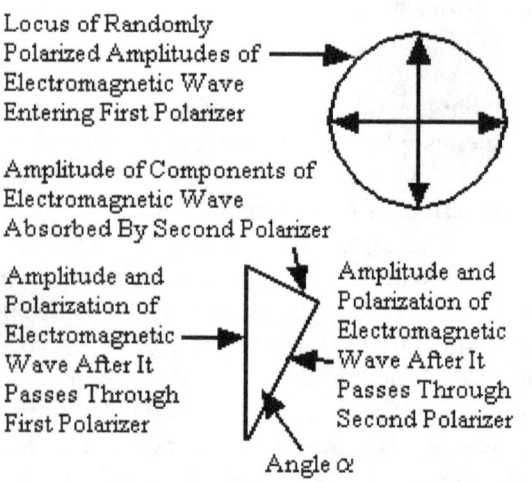

Figure 14.4 - Electromagnetic Wave and/or Beam of Photons Passing Through Polarizers

to the axis of the second polarizer while the other component is perpendicular to that axis. As one would expect, the second polarizer will pass only the parallel component. With the loss of the orthogonal component, the relative intensity between the electromagnetic wave leaving the second polarizer, as compared to the intensity of the wave leaving the first polarizer, is further reduced in proportion to the magnitude of the cosine of the angle between the polarizers. The polarizing process and the preceding analysis may be repeated as often as desired in a series of successive polarizers. A monochromatic electromagnetic wave actually consists of a flow of photons having the same total energy, and the intensity of the wave is proportional to the square root of its energy. As a consequence, the photon flux associated with each of the intensity vectors described varies in proportion to the square of that intensity vector and its polarization is identical to the polarization of the equivalent wave, as shown in Figure 14.5.

14.7- It is both customary and somewhat incorrect to consider that the rate of arrival and the axis of polarization of photons comprising an electromagnetic wave are random. In a monochromatic electromagnetic wave of unchanging intensity, the average rate of arrival of photons is constant. Their instantaneous rate of arrival, however, is a random variation about that average rate. Similarly, if the electromagnetic wave is unpolarized, the average axes of polarization of its photons are uniformly distributed, while their instantaneous angles of polarization are random. If an electromagnetic wave is polarized, both its intensity and the average rate of arrival of its photons vary as a function of their angle with respect to the axis of polarization, as shown in Figure 14.6, and both the instantaneous rate of arrival of the photons and the variation of their axes of polarization are random. In addition, because the intensity of the electromagnetic wave and the distribution of photons are defined by cosine and $cosine^2$ functions respectively, the effect of cascading polarizers may be predicted by multiplication of the effects defined by Figure 14.6 for each polarizer in the chain. [Since the relative randomness of N events varies in proportion

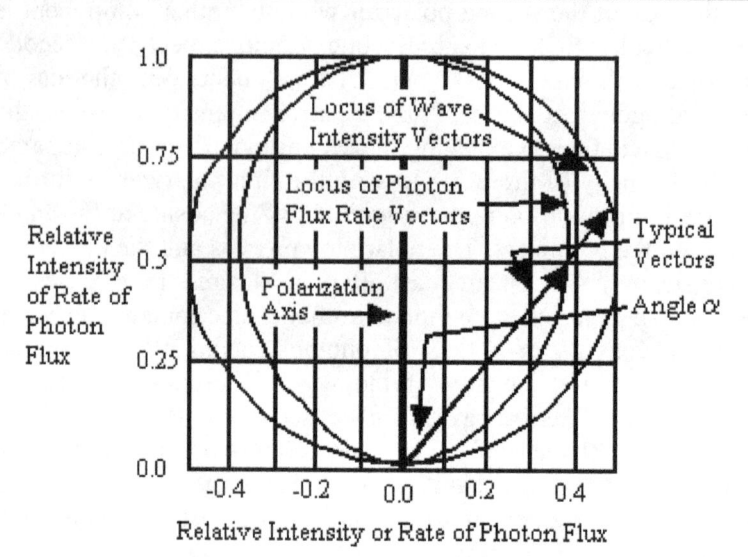

Figure 14.5 - Intensity and Rate of Photon Flux vs. Angle to Polarization Axis

to $1/(2*N)^{0.5}$, the random arrival and polarization of photons in electromagnetic waves is insignificant unless the intensity of the wave is low.]

14.8- Understanding the Quantum Nature of Energy:- The instantaneous time of arrival of photons and their polarizations are individual events which apparently occur at random, yet somehow, in the aggregate, they are organized in an accurate pattern. This is a reasonable result for an unpolarized electromagnetic wave because it can occur by pure chance. This easy answer does not explain the rate of arrival and angle of polarization of photons which comprise a polarized electromagnetic wave. Those photons are organized and that organization requires the action of an organizing mechanism. To understand the characteristics of that mechanism, a brief detour into the operation of a device currently used in digital audio systems is useful.

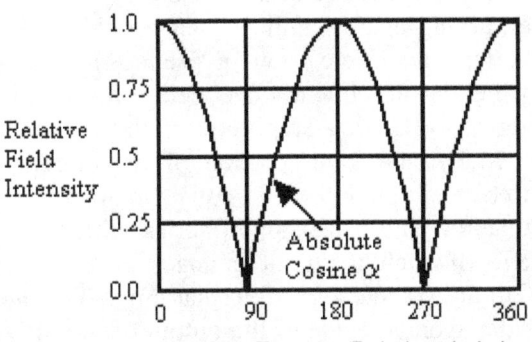

Effect of Polarizers on Electromagnetic Field

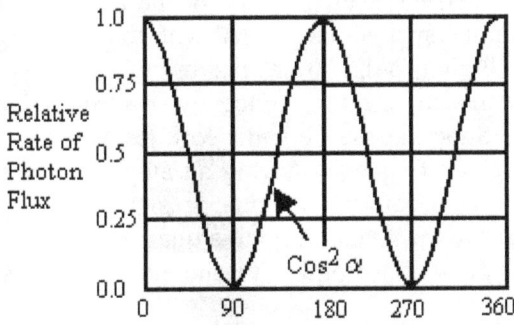

Effect of Polarizers on Rate of Photon Flux

Figure 14.6 - Effect of Pair of Polarizers at Angle α on Field Intensity and Photon Flux From Unpolarized Light Source

14.9- The device of interest is the Sigma-Delta analog to digital converter used for digital signal processing in audio systems. It converts the varying analog audio signal provided by a source (microphone, tape deck, etc.) into a pseudo-random string of digital bits which is subsequently decoded to form digital words corresponding to the analog amplitude of the incoming signal. (See Figure 14.7.) The input portion of the circuitry can be considered to be a simple integrator consisting of an amplifier, a feedback capacitor, and a resistor in series with its input. The output voltage of such a circuit changes at a rate which is proportional to the voltage applied to that input. The input to the integrator is the average value of the output of the SPDT switch subtracted from the input signal. Since the switch is driven by a clocked flip-flop in response to the polarity of the amplifier output and the state of the flip-flop is determined by the output of the amplifier, the average value of the pulse train driving the switch must equal the average value of the analog input signal. By combining the switch drive signal with the clock signal in an and-gate, a digital pulse train results whose average level accurately represents the low frequency spectrum of the analog input signal. Superimposed on this low frequency output is a noise at the clock frequency having an amplitude of one clock pulse. The device converts an analog signal, analogous to a quantum stress in the Aether, into to a quasi-random sequence of standardized pulses, analogous to the quanta represented by discrete particles.

14.10- In order for the Sigma-Delta converter analogy to apply satisfactorily to the Two Slit Experiment, the conditions tabulated below must be met:

- Each quanta must apply a quantum stress to the Aether.

- Quantum stress in the Aether must propagate in all directions at a quasi-infinite velocity.

- Quantum stress in the Aether probably must obey the inverse square law.

- Quantum stress in the Aether must be refracted by the geometrical configuration of matter.

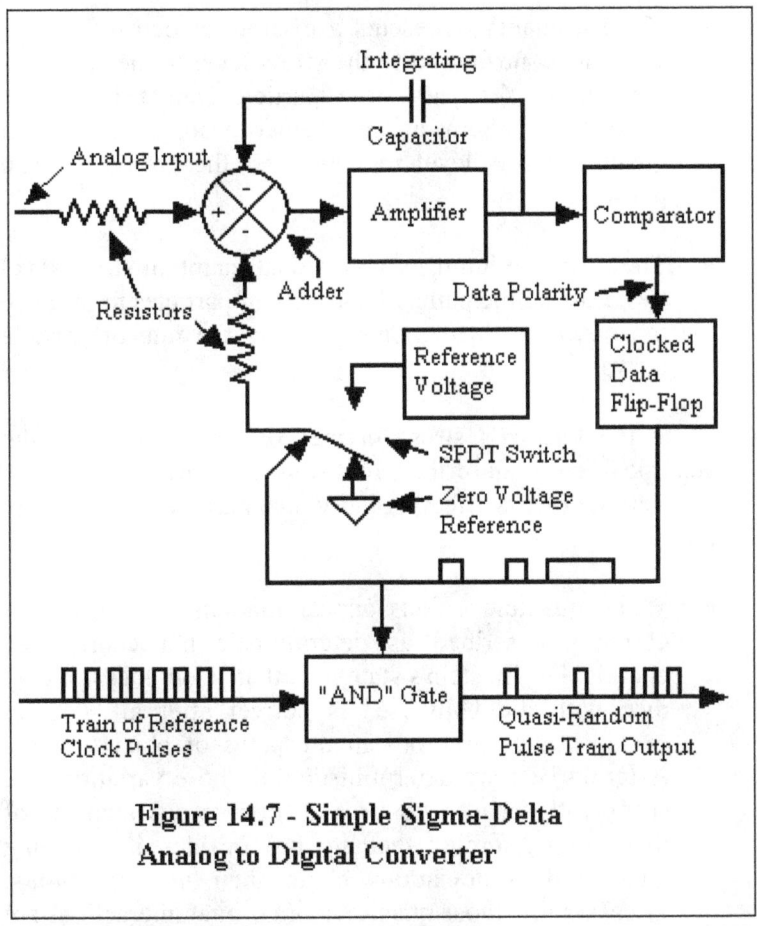

Figure 14.7 - Simple Sigma-Delta Analog to Digital Converter

- The quantum stresses imposed on the Aether by multiple quanta located throughout all of space must superimpose to produce a single quantum stress level at any given point.

- A quanta moving in a free trajectory follows a path consistent with the diffraction pattern of the quantum stress it generates in the Aether.

  - Since a quanta represents a discrete amount of energy while its contribution to the stress level in the Aether is continuous, the path of a particle cannot follow its diffraction pattern in the Aether exactly. A residual quantum stress level remains after the arrival of each particle.

  - The Aether minimizes its residual quantum stress level by adjusting the paths of subsequent particles to produce the observed diffraction pattern in the rate of particle arrival.

- Since the localized stress in the Aether results from the superposition of all of the stress sources in the Universe, the diffraction process affecting individual particles is extremely noisy.

  - While quantum effects appear random, they are more correctly described as deterministic but enormously chaotic. Dr. Einstein's statement that "God does not play dice with the Universe" is correct after all. (As an example, the variations in the paths of objects in the Asteroid Belt are deterministic, but those variations are sufficiently chaotic as to prevent rigorous analysis of their orbits, forcing them to be considered as having quasi-random deviations about their nominal orbits. Occasionally, those quasi-random orbital interactions are sufficient to throw an asteroid out of the Asteroid Belt completely. Occasionally, a radioactive atom throws one or more of its component nucleons from its nucleus.)

  - One would expect that the summation of quantum stresses from all of the events occurring throughout the Universe would occasionally combine to produce extremely large momentary quantum stresses at a given point. The effect would be analogous to the production of an unexpectedly large wave, called a rogue wave, in the ocean by the combination of a large number of different waves of considerably lower amplitude.

14.11- A clear picture of reality emerges when examines the various aspect of quantum theory and the results of experimentation once one's thinking is no longer constrained by the unproven assertion of Special Relativity that information cannot travel faster than light. Velocity Relativity, or Special Relativity if the reader prefers, and Quantum Theory prove only that the propagation of information encoded in the form of energy is limited to the velocity of light and the attendant requirement that such information be quantified. There is no observational or theoretical requirement that subjects information which does not involve energy to those limitations. It is at this point that rote trained physicists can make a contribution. Their skills in the use of mathematics should allow them to apply flesh to our understanding of reality, even though they seem to deny the significance of that reality.

14.12- The Heisenberg Uncertainty Principle and the Wave Equations:- One of the earliest incursions of quantum concepts into classical physics was the Heisenberg Uncertainty Principle. It asserts that it is impossible to determine the product of the position and velocity of a particle to an accuracy smaller than the numerical value of Planck's Constant. This observational limit results from the fact that, in order to determine the position and/or velocity of a particle, whether it be an electron or a baseball, it is necessary to strike the particle to be measured with another particle, such as a photon, and observe the result. Since both of these particles possess their own inertial masses and velocities, they exchange momentum as they interact. To illustrate, if an electron is examined by a photon of much lower energy than its mass equivalent energy, the photon is physically large compared to the electron and cannot measure the electron's position accurately. On the other hand, its inertial mass is low and does not affect the electron's velocity significantly. If the energy of the examining photon is high compared to the mass equivalent energy of the electron, its size is small compared to the electron's and it will observe the electron's position accurately. However, the inertial mass of such a photon is large compared to that of the electron and the interaction changes the electron's velocity considerably. The Heisenberg Uncertainty occurs because, in accordance with Quantum Theory, there can be no particle available to the experimenter which has both negligible mass and negligible size. Other than the fact that Quantum Mechanics predicts the relationship between the physical size and the inertial mass of particles, there does not seem to be a connection between the Heisenberg Uncertainty and either the Schroedinger Wave Equation or other quantum effects.

14.13- The Schroedinger Wave Equations were derived to deal with the observed impossibility of determining the location of electrons in their orbits as one would expect if they were distinct objects. Their locations were found to be fuzzy and it was necessary to describe their locations in terms of probability, as defined by the Schroedinger Wave Equations, rather than in terms of Newtonian Mechanics. Once it is recognized that the particles under observation, and the particles which are used to examine them, are themselves composed of waves, it is not unreasonable to expect imprecision in the observation of position and velocity. The observation of one wave with another is an imprecise process. While the probability waves described by the Schroedinger Wave Equations include this effect, they also result from a more subtle effect. Earlier, the storage of kinetic energy in a particle was shown to occur as a change in the energy of the particle itself and in the stressing of a disk like region of space around it. The storage of that energy probably requires time, and for periods of time which are shorter than the time required for that storage, the velocity restrictions imposed by Velocity Relativity would not apply and the particle would be free to bounce around within a small region of space at velocities which are greater than the velocity of light. To an external observer whose ability to observe is limited to the velocity of light, however, the particle would appear to exist simultaneously at more than one point within that region. As a result, the location of the particle can only described by the Schroedinger Wave Equations in terms of its probable position within a fuzzy volume of space.

Chapter 15 - Changing the Paradigms

15.1- The Chimera of a Unified Field Theory:- The last decades of Dr. Einstein's career are reported as having been spent in an attempt to develop a Unified Field Theory which would reduce all of physics to a single set of equations. While he was not successful himself, continuation of his efforts was at least successful in encompassing all forces but gravity into such a set of equations. The need for such a theory resulted from the fact that both General Relativity and the Quantum theory were accepted as correct and yet were incompatible with each other. That incompatibility reveals the weakness of limiting one's understanding to mathematical approaches. With such a methodology, a single unrecognized error can destroy the validity of enormous amounts of subsequent work. In this case, the unrecognized mathematical error in the derivation of General Relativity was the culprit. While the descriptions of curved space, Black Holes, Singularities, Worm Holes, etc. should have provided ample warning of that error, theoretical physicists trained to ignore common sense doubts relied on their mathematics and did not find it. (One is reminded of the medieval theologians who allegedly engaged in debate over the number of angels who could dance on the head of a pin.)

15.2- Eventually, the Standard Model and its proposed successor, the Superstring Theory and its permutations evolved. In that approach, it was found that both a 10 dimensional and a 26 dimensional matrix of equations were capable of defining the observed gravitational, electromagnetic, the strong and weak nuclear forces, and the characteristics of quarks and leptons. When those matrix sizes were employed, all of the troublesome cross-products which occurred with other matrix sizes canceled and tractable solutions were obtained. (The arrangement of the ten dimensional set of equations is provided in Figure 15.1. In their writings on the subject, authors have raised the question as to what special properties 10 dimensional and 26 dimensional matrices have which allows them to work when other matrix sizes fail. The answer to that question should be apparent to anyone who is old enough to have operated a television set which had a horizontal hold control. If the horizontal hold control was not properly set, the horizontal scan lines would not line up vertically and an unviewable picture would result. Similarly, if an unsuitable number of columns is used in the matrix, the data in subsequent rows will not align properly in

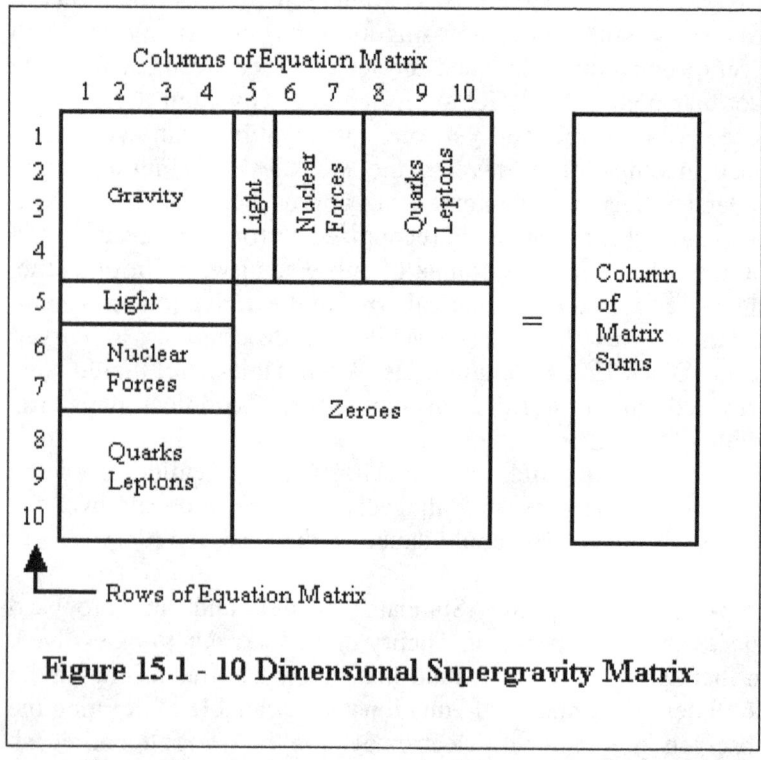

Figure 15.1 - 10 Dimensional Supergravity Matrix

columns, cross products will result, and the equations will be unsolvable. In order to accommodate those ten or twenty six dimensions in the four dimensional space of General Relativity, however, theoreticians found it necessary to consider dimensions beyond the fourth to be curled into tiny circles having a diameter more than twenty orders of magnitude smaller than the nucleon. (Somehow, accepting such entities as dimensions strains the author's credulity. Even though they can be intermixed with the conventional dimensions in the solution of mathematical equations, they should be thought of differently and be given a different designation.) It would appear that these dimensions were necessary for the mathematical solution but it was necessary to consider them to be in that peculiar form and of that size in order to prevent them from being experimentally observable as new particles.

15.3- Examination of the matrix illustrated in Figure 15.1 reveals that all of the terms referring to each item of interest (gravity, electromagnetics, strong and weak forces, quarks and leptons) are grouped in such a manner as to be independent of each other. None of the groups contain cross-products which tie them to the other groups. ***The matrix is not a single set of equations defining everything. It is a grouping of four independent sets of equations which describe unrelated effects occurring in the same volume of space.*** As such, they hardly represent a unification of physical theory. Descriptions of the mathematical efforts involved in the development of such unified theories assert that difficulties experienced by the theoretician are reduced as dimensions are added. This is to be expected, the addition of superfluous degrees of freedom reduces the constraints on the theoretician's work. What separates the men from the boys is the capability of dealing with physical problems while retaining the number of degrees of freedom actually required by Nature.

15.4- The concept that matter consists of loops of radiation yields a simpler matrix of equations, as shown by Figure 15.2. Gravity is described by a 3x3 matrix of equations. Added to that matrix are a single row and column representing photonic effects and another row and column representing neutrinic effects. The strong and weak forces of the Standard Model vanish and electrons and quarks revert to permutations of electromagnetic and neutrinic radiation.

15.5- The Role of Paradox:- The recognition of a paradox always raises a flag. It tells us that two or more concepts which have been accepted as true are in conflict and it reveals, as did the Right Angle Lever Paradox described earlier, that an error exists in one or more of the concepts and it is necessary to rethink the ideas involved. In the case of the Right Angle Lever Paradox, the rethinking process resulted in a revision of the Lorentz Transformation for Transverse Force. Surprisingly, the community of physical scientists does not seem to make a significant use of the opportunities that paradoxes present. For illustration, two well known paradoxes, the Paradox of Zeno and Schroedinger's Cat Paradox are described.

15.6- The Paradox of Zeno is 2000 years old and its apparent ability to prove that all motion is impossible was not resolved until the mathematical techniques of Calculus became available, even though that technique is not required. One form of the paradox describes the flight of an arrow which has been shot at a target. The arrow is shot at a constant velocity, V, to a target at a

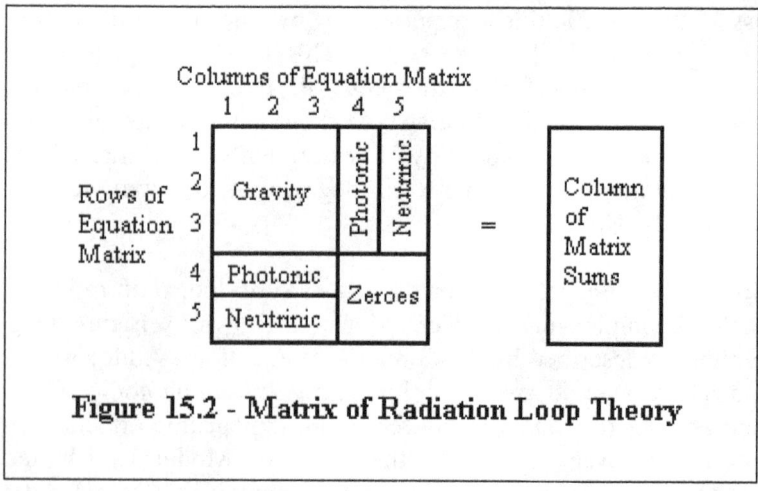

Figure 15.2 - Matrix of Radiation Loop Theory

distance, L, and the time of flight is divided into intervals. In the first interval, the arrow covers half of the distance to the target and, in each succeeding interval of time, it covers half of the remaining distance. Under the line of reasoning presented, the arrow never reaches the target because, after each successive interval of time, one half of the distance to the target that existed at the beginning of the interval remains. The author finds it incredible that this paradox has been taken seriously by intelligent men for over two millennia and has not been recognized as a form of trickery. If one accepts that in each successive interval of time the arrow traveled half of the remaining distance to the target, he must also accept that each of those successive intervals of time is half of the duration of the interval which preceded it. As a result, under the Paradox of Zeno, not only does the arrow never reach the target, the elapsed time of its flight never reaches the time, T, when the arrow would reach the target. There is no paradox in the Paradox of the Zeno. As long as one allows a cheap trick to fool him into considering only those events which occur prior to the arrival of the arrow at the target, he most certainly will believe that the arrow never reaches the target. The reality is that the passage of time does not slow as the arrow approaches the target and the arrow reaches the target on schedule.

15.7- The Schroedinger Cat Paradox is more esoteric. In one form of this paradox, a closed box contains a source of cyanide gas, a means of releasing that gas when a signal from a radiation detector senses an alpha particle (helium nucleus), and a radioactive element which releases alpha particles at a rate determined by its half life. A cat is placed in the box for a period of time equal to the probable time of the next release of an alpha particle. If the alpha particle is emitted before the cat is removed from the box, the cyanide is released and the cat dies. If the alpha particle is emitted after the cat is removed, it lives. According to quantum mechanics, the decay of the atom which results in the emission of the alpha particle is determined entirely by chance and, under the interpretations of quantum physics, the atom simultaneously exists in the both the decayed and undecayed states and only reverts to one of those states when it is observed. Schroedinger's Cat Paradox asserts that since there is no way to observe whether the cat is dead or alive while the chamber remains closed, quantum mechanics requires that the cat simultaneously exists in two states, both alive and dead, until its state is actually observed when the box is opened.

15.8- To the layman, Schroedinger's Cat Paradox seems absurd. How could it be possible for the cat to be both alive and dead at the same time? The layman is correct, the cat does not exist in both states simultaneously and quantum mechanics does not require that it do so. The significant observation is not the observation made by the experimenter when he opens the box, the observation which counts is the decay of the radioactive atom. Even the detection of its decay by the radiation detector is not a quantum mechanical consideration. Asserting that the cat exists in a dual alive-dead state until it is observed by the experimenter is about as sensible as asserting that the verdict in the O.J. Simpson trial was simultaneously 'guilty' and 'not guilty' as it lay overnight in a sealed envelope prior to being read. The significant observation in Quantum Mechanics is one which affects a physical particle. Whether an intelligent observer is aware of that observation isn't significant. The Schroedinger Cat Paradox is not a paradox as normally stated. It is an absurdity similar to the absurdity of the question as to whether a tree makes a sound when it falls in a forest with no one present to hear its fall. In the physical sciences, sound is a physical phenomena, not a psychological phenomena. It does not require the presence of an intelligent observer to exist.

15.9- The error flag presented by Schroedinger's Cat Paradox does reveal that our thought processes about quantum physics are inadequate. They are based upon the rigid acceptance of the idea that no entity can travel faster than light. Recognition of the fact that time must be required for the kinetic energy level of a particle to change demonstrates that the idea that the velocity of light represents an absolute speed limit is not quite true. That time delay involved in acquiring(losing) kinetic energy allows a particle to travel a small distances at a velocity greater than that of light. To an observer, such a particle will falsely appear to be in two places simultaneously. When one adds the fact that stress in the quantum field propagates at a quasi-infinite velocity, the paradoxical effects attributed to the Schroedinger Wave Equations will probably vanish. ***There can be no paradoxes in Nature.***

15.10- Can the Second Law of Thermodynamics be circumvented?- The validity of the First and Second Laws of Thermodynamics seems to be beyond question. Under the first law, the total energy content of a closed system must remain constant. Under the second law, the availability of energy for useful purposes must always decrease or remain constant within that closed system. In effect, the First Law of Thermodynamics states that you can't win and the Second Law states that, furthermore, you can't break even. The First Law of Thermodynamics is unquestionably true, energy can neither be created or destroyed, General Relativity not withstanding. The Second Law of Thermodynamics stands on less firm ground because it is a law based upon statistics. As such, it is in the same category as an actuarial table. An insurance company can predict quite accurately how many people will die in a given year. They cannot predict who those people will be. Statistical laws are valid for large numbers of events, they become less significant as the number of events is reduced. As an example, if one patronizes a casino, he might initially win a large sum of money playing a slot machine, but if he continues to play he not only will give all his winnings back to the casino, he will sustain a significant loss. The question then arises as to whether it is possible to by-pass the Second Law of Thermodynamics though the use of nanomachines. (A nanomachine is a mechanism whose significant dimensions are measured in nanometers, the size scale of atoms.)

15.11- One who observes Brownian motion in a microscope might reasonably conclude that, in principle at least, a nanomachine could be built which would bypass the Second Law of Thermodynamics. When a liquid containing microscopic particles is observed, the particles are seen to be in continuous (Brownian) motion. That motion is caused by random thermal impacts between the molecules of the liquid and the particles. If the thermal motion of water molecules can produce a visibly observable motion in particles which are at least 10^{15} times as massive, it certainly not unreasonable to believe that suitable nanomachines could organize the effect to produce a useful mechanical output. The postulated nanomachines would then be able to export energy to the outside environment which it obtained by reducing the temperature of the liquid. The exported energy would be converted to heat and raise the temperature of the external environment as the output performed useful work. The resultant temperature difference between the environment and the liquid will then cause the energy which had done useful work to flow back into the liquid to return it to its original temperature and allow the process to continue indefinitely.

15.12- James Clark Maxwell proposed a hypothetical perpetual motion machine, known as Maxwell's Demon, which was not proven to be theoretically unworkable for 75 years. In that machine, Maxwell imagined that a demon controlled a microscopic gate between two gas filled chambers. Making use of the fact that, in a gas, the velocity of the molecules is random and that the temperature of the gas is determined by the mean square velocity of those molecules, Maxwell proposed the concept that, if an appropriate demon existed, he could sense the speed of molecules approaching the gate and open the gate only when a fast molecule approached it from one side or when a slow molecule approached it from the other side. By operating the gate in this manner, the demon would sort the molecules so that one chamber contained fast molecules and the other chamber contained slow molecules. Since the temperature of a gas is determined by the mean square velocity of its molecules, such a process will maintain a temperature difference between the chambers which can be exploited to produce useful work in a direct violation of the Second Law of Thermodynamics. It took 75 years before a rigorous proof was found which was able to show that the energy required for Maxwell's Demon to identify the fast and slow molecules and allow the gate to operate was at least as great as the energy which could be released and Maxwell's Demon was shown to be an unworkable concept.

15.13- There is a modification to the concept of Maxwell's Demon for which there is, at least as yet, no valid theoretical objection. Suppose that the two chambers of the Maxwell's Demon example no longer rely on a demon but are separated by a diffusion membrane having a permeability from side A to side B which is higher than the permeability from side B to side A, as shown in Figure 15.3. The energy required to allow the membrane to make the decisions it needs to make in order for it to function in this manner is available in the kinetic energy of the gas molecules passing through it. In diffusing through the membrane, molecules can provide the energy needed by being slowed from their average room temperature velocity of about 1300 feet per second to a much lower exit velocity. The lower velocity of the gas leaving the membrane means that side B is colder than the ambient temperature. The loss of kinetic energy by the molecules as they pass though membrane provides the energy required to operate the differential diffusion mechanism in the membrane pores, and the membrane becomes warmer than the ambient temperature. If the surface areas are sufficiently large, the temperature of the gas on both sides of the membrane and of the membrane itself must remain close to the temperature of the environment. As a result, the pressure in chamber B will be higher than the pressure in chamber A. That difference in pressure can be used to operate a turbine and provide useful output power. As the output power is produced by the gas flowing through the turbine, the chambers are cooled below the ambient temperature and energy flows from the environment to the chambers to replace the energy delivered by the turbine. The arrangement would extract useful energy from its environment in direct contradiction to the Second Law of Thermodynamics.

15.14- Conceptually, the membrane might be constructed with pores which were covered by spring loaded trapdoors, as shown in Figure 15.4. In this illustration, a molecule represented by a ball would approach the right side of the membrane at a velocity which was appropriate to its temperature, knock the trapdoor open, and pass through it. A similar molecule approaching the trapdoor from the left side would bounce back and not pass through to the right side. When the molecule on the right passed through the trapdoor, it would lose most of its kinetic energy to the trapdoor and exit at a low velocity. As a result, the trapdoor and the membrane would be heated and the molecule which passed though it would be cooled. The process would generate a

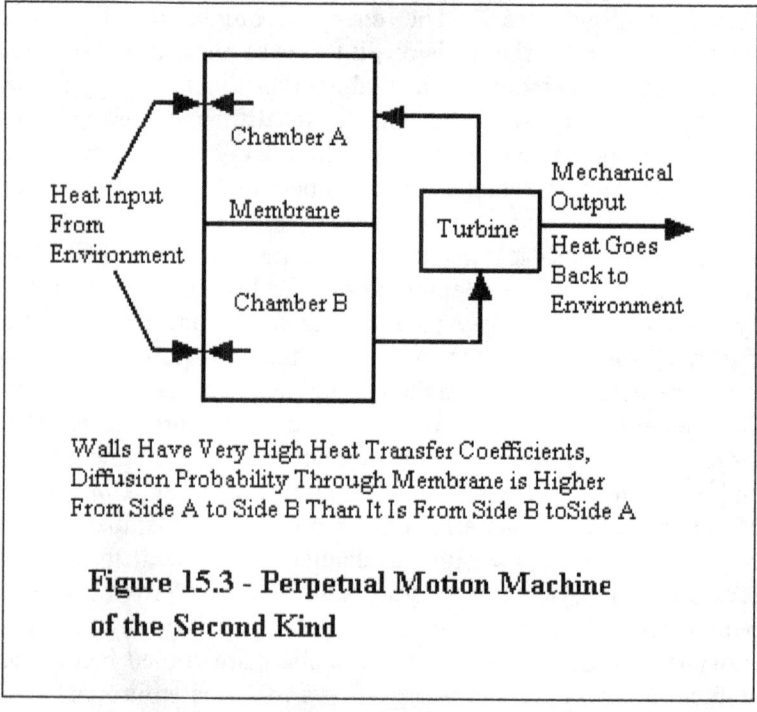

Walls Have Very High Heat Transfer Coefficients, Diffusion Probability Through Membrane is Higher From Side A to Side B Than It Is From Side B toSide A

Figure 15.3 - Perpetual Motion Machine of the Second Kind

local temperature difference which would quickly be equalized by any reasonable level of heat transfer.

15.15- The first theoretical objection to this type of perpetual motion machine that the author has found in literature is that it cannot work because it violates the Second Law of Thermodynamics. This is hardly a valid objection since the arrangement is specifically designed to bypass the limitations of that law. The Second Law of Thermodynamics is a statistical law and it is not binding on nanomachinery since such mechanisms deal with molecules on an individual basis and the pores of the diffusion membrane certainly qualify as a nanomechanism. The only other theoretical objection that the author has found was provided in another book by Dr. Feynmann in which he described a nanomechanism consisting of a riverboat type of paddle wheel mounted on a shaft inside a cylinder containing a fluid. The paddle wheel was bombarded by the random motion of the molecules of the fluid and caused the shaft to undergo a random rotary oscillation. To convert this motion to a useful output, an external one-way ratchet was attached to the shaft. Dr.

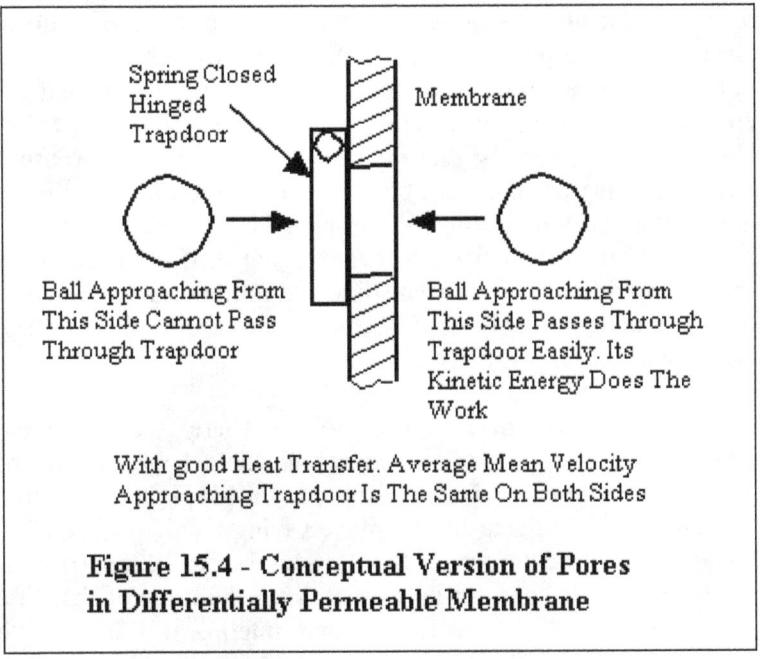

Figure 15.4 - Conceptual Version of Pores in Differentially Permeable Membrane

Feynmann then demonstrated that the device would not work because the motion of the ratchet pawl would generate enough heat so that the resultant molecular motion would make the pawl bounce sufficiently to render the one way mechanism inoperative.

15.16- From the description provided, it is obvious that, while Dr. Feynmann is an excellent theoretical physicist, he is not as effective as a design engineer. Relocating the ratchet mechanism to the interior of the fluid chamber cools it and dampens its bounce, causing Dr. Feynmann's objections to vanish. When the model is modified, Dr. Feynmann probably would be forced to agree, that unless he could devise another objection, the concept should represent a physically realizable device which would by-pass the Second Law of Thermodynamics. A theoretical demonstration that the mechanism suggested by the author cannot work requires a proof that the permeability of all possible diffusion membranes must be the same in both directions. Deriving such a proof may be particularly difficult because the pores of the required membrane are are permitted to extract energy from the molecules which pass though them. Considerable effort is justified in developing such a proof because, if that proof cannot be found, the possibility of building a perpetual motion machine of the second kind is not foolish and an all out effort is justified to demonstrate it. If it can be built, society would have an inexhaustible and pollution free source of energy which probably could be sized for use in the smallest homes and the largest factories.

15.17- The Uniqueness of the Earth:- There is considerable interest in the possibility that there may be a large number of planets in our galaxy which are suitable for life. In the hope that there may be intelligent life on planets lying within a reasonable distance, a project named SETI (Search for Extraterrestrial Intelligence) has been set up to search for evidence of that life. The idea behind the project is that intelligent life may be generating signals which can be received on Earth that are either a by-product of their civilization (such as our own radio broadcasts) or a deliberate attempt to communicate. Unfortunately, the probability of success of those programs is far lower than currently believed. If an Earth sized planet existed

93,000,000 miles from a star which was virtually identical to the Sun, it is extremely unlikely that it would be capable of supporting life. To see why this should be so, an examination of our own Solar System is order.

15.18- With the exception of Mercury, the Earth, Mars, and Pluto, all of the planets have enormous atmospheres (relative to the Earth). One can draw no conclusions about the the original conditions on Mercury or Pluto. Mercury is too small and too close to the Sun to have prevented its atmosphere, regardless of its original quantity, from boiling away to space. (There may be a remnant of an atmosphere frozen at the poles.) At the other extreme, due to its distance from the Sun, any atmosphere that Pluto may have had at its beginning and which has not been lost by evaporation to space is frozen solid and is therefore unobservable. Observations have shown that Mars once had a significant atmosphere which supported running water (and, by implication, oceans) but has lost both. Apparently, its low gravitational mass has made it too easy for the Sun's radiation to cause Mar's atmosphere to evaporate to space. Of all the planets, it is Earth that is the anomaly.

15.19- Due to its location, Venus receives about twice the heat input from the Sun as does the Earth. Its gravitational mass is slightly less than that of the Earth and yet it has an atmosphere about 70 times as dense as the Earth. It seems reasonable to conclude that the early history of the Solar System probably went through the following stages:

- The planets were formed by the collision of smaller objects circling the Sun in eccentric orbits. The collision process continued until the Solar System was virtually cleared of objects in non-circular orbits.

- During the planetary formation stage, the planets could not acquire atmospheres because the bombardment which was forming them made their surfaces extremely hot. Any atmospheric gases which impacted the planet from interplanetary space would boil away immediately.

- Once the rate of bombardment forming the planets reduced to the point where the planets could cool sufficiently, they proceeded to collect atmospheres from the gases which remained in the Solar System.

  - For Venus to have its present atmospheric density, all of the planets, including the Earth, must have acquired enormous (by Earth standards) atmospheres.

    - A planet gains atmosphere by sweeping up gases from interplanetary space. It losses atmosphere through evaporation of those gases to that same space from its uppermost layer.

- In order for a molecule of gas to be lost to the planet, it must acquire a thermal velocity greater than the planet's escape velocity. This must occur at an altitude at which the atmosphere is sufficiently thin so that it does not strike other molecules while escaping.

  - The rate at which atmospheric gases are lost to space is determined almost entirely by the rate of energy input from the Sun and by the escape velocity of the planet at the top of its atmosphere The rate of atmosphere loss is virtually independent of the amount of atmosphere the planet owns.

15.20- The Earth-Moon system has two characteristics which are anomalous compared to the other planets. The first is that it has far too much angular momentum (orbital angular momentum, rotational angular momentum of the Earth and the Moon, and orbital angular momentum of the Moon around the Earth). As pointed out in a text by Dr. Urey, an exponential plot of angular momentum vs total mass for all of the other planets yields a straight line. The total angular momentum of the Earth-Moon system lies far above that line. The second anomaly is that it contains far too little atmosphere and, unlike Mars, the density of that atmosphere has remained almost unchanged. A satisfactory explanation for both of these anomalies seems to have been provided about a decade ago by a computer simulation of a glancing impact on the Earth by an object having a mass about one sixth of its mass. The simulation predicted the formation of a binary system with a Moon sized object orbiting the Earth an altitude of about 12,000 miles, with the Earth having a 4 hour day, and with the Earth having captured the iron cores of both objects. Since the length of the Earth's day was, is, and will remain less than the Moon's orbital period until the Sun enters its red giant stage, tidal effects on the Earth will perpetually transfer angular momentum from the Earth to the Moon. This transfer has lengthened the Earth's day to 24 hours and has caused the Moon's orbit to increase to 238,000 miles. More important, such an impact would have blasted away most if not all of the Earth's atmosphere and, if the collision occurred late enough in the formation period of the Solar System, most of the interplanetary gases would have been absorbed by the other planets and/or lost to interstellar space. This scenario could easily allow the Earth to have the comparatively puny but stable atmosphere required to support the evolution of intelligent life.

15.21- In order for a planet to support life, not only must it be in the "life zone" about a suitable star, it must possess an atmosphere of a suitable density for a sufficient period of time for life to evolve. On the Earth, life does not seem to prosper above an altitude where the density is half an atmosphere. At the other end of the scale, the atmosphere must not be too thick or the wavelengths of radiation needed for photosynthesis not only will not reach the atmosphere-water interface where life begins, that interface is likely to be too hot. Making the optimistic assumption that four and a half atmospheres is the highest suitable atmospheric pressure requires that a life supporting planet not lose more than four atmospheres of density in the period required for intelligent life to evolve. For a planet starting with the atmospheric density of Venus to lose 60 PSI of surface atmospheric pressure in 3 billion years (the time required for intelligent life to have evolved on Earth), the existence of such life would require an age of 50 billion years for the planetary system. Such a conclusion presents problems. A star similar to our Sun will become a red giant about 10 billion years after its formation and the apparent age of the Universe is only 15 billion years. On the other hand, if a planet such as Mars lost its atmosphere at a sufficient rate to reach compatibility with the requirements of life before its star became a red giant, it would pass though the "life range" so quickly that intelligent life would probably not have had time to evolve. It is the author's belief that, without the addition of the 'wild card' implicit in the postulated Earth-Moon collision, a planet capable of supporting life cannot exist. (It is hoped that this question would be examined further.) It is the author's belief that intelligent life is much rarer in the Universe than Dr. Sagan suggested.

15.22- A Comment on Meteorites and the Asteroid Belt:- The asteroid belt exists as a ring of stony and iron rocks in orbit about the Sun between the orbits of Mars and Jupiter. The radius of that orbit coincides with the anticipated location of a planet's orbit under the conventional theory of planetary formation. If one examines the objects in the asteroid belt, the moons of Mars, and the meteorites which strike the Earth, one finds that, unlike comets, they are densely composed of stone or of iron. Unlike the flimsy comets, such objects cannot form by accretion, they can only be formed within a planet sized object. One must conclude, therefore, that initially a planet did form at the radius of the asteroid belt and was later shattered by a collision. Such a collision would drive away most of the planetary material and leave a residue of rocks from the planets upper layers and iron objects from the planet's core. It seems probable that the object which impacted the early Earth to form the Earth-Moon system, the meteorites which strike the Earth, and the moons of Mars result from that collision.

Chapter 16 - What Can We Conclude?

16.1- Prior to the early years of the 20th Century, the assertion (which the author's memory suggests originated with Lord Raleigh) that there was no phenomena in Nature for which a model could not be built was an accepted philosophy. The building of such conceptual models was considered to be a goal of the physical sciences. Obviously, since we exist, Nature has already built that model and it is our task to understand it. The task of the physical scientist was deemed to be one of postulating mechanisms by which nature might function in the observed manner, mathematically analyzing the behavior of those mechanisms, and finally devising and performing experiments to determine the validity of the envisioned mechanization. It was expected, of course, that such a process would be iterative. One could never hope to get it right the first time, but each iteration would hopefully be closer to representing reality until any discrepancy which remained was sufficiently small as to be unobservable. With this philosophy, our methodology in understanding Nature was akin to a three legged stool with one leg representing the conceptual model, the second leg representing the mathematical analysis of that model, and the third leg representing the experimental verification of the validity of the first two legs. Just as a stool with less that three legs is at best unreliable, one would expect that a science missing one of these legs to also be unreliable and, as we have seen, this is the case.

- Without a conceptual model we cannot know to whether our mathematical analyses are relevant and what information our experiments have provided. (As an example, consider the experimental measurement of the velocity of light, as discussed in Chapter 6. The experiment measures something but it most certainly does not measure the velocity of light.)

- Without a valid mathematical analysis of the model, it cannot be known whether the model will function in the anticipated manner and what experimental results are required to validate the model and its mathematical analysis.

- Without experimental verification, one cannot be sure that the model and its mathematical analysis represents reality or delusion.

Unless all three legs of the stool are used in the acquisition of knowledge understanding is likely to be faulty and, as this text shows, is likely to produce weird and untenable conclusions when used as the basis for further work.

16.2- Mathematics is an extremely powerful tool, and properly applied it yields perfect results. Like all powerful tools, it is dangerous. Mathematics relies upon a sequence of logical steps and, since the steps are in sequence, the probability of an error having been made increases in proportion to the number of steps involved. In addition, the mathematical model provides no assurance that it is consistent with the reality it is attempting to describe. On the other hand, the building of a conceptual model involves the generation of a pattern whose agreement with reality is readily determinable. (It is impossible to complete a jigsaw puzzle incorrectly.) The probability of an error in a conceptual model decreases as the number of its components increase regardless of the level of the individual errors, but the model does not have the rigor of the mathematical approach. (The error reducing capability of the conceptual model approach is analogous to the error reducing capability of the process, known as Kalman Filtering, used in advanced navigational systems. Kalman Filtering combines the weighted "best guess" accuracy estimates of many lower accuracy inputs to produce a conclusion of higher accuracy.)

16.3- The conceptual model has an advantage over the mathematical approach in solving physical problems. It makes use of the most powerful capability of the mind, its pattern recognizing ability. The power of the mind's pattern recognizing capability was dramatically brought home to the author at the start of the missile era when he had the responsibility for the development of a star tracker capable of operating in full daylight using a television camera tube. When an image of a simulated star and sky was viewed on a television screen, the image of the star abruptly became recognizable to the eye when the peak star signal to RMS noise ratio of the display exceeded 1:1. The peak star signal to RMS noise ratio required for recognition by logic circuitry (analogous to a mathematical approach of solving problems) was, and still is, 6:1. A reasonable man might well conclude that the brain's inherent pattern recognizing capability is much more powerful than its capability of performing logical reasoning.

16.4- Relying upon the pattern recognizing capability of the brain to solve physical problems is difficult for an organized structure of knowledge. Its successful use depends upon the inborn talent of the individual involved, the proper training of that talent, and the protection provided to that talent from the strongly motivated efforts of untalented teachers to force him to conform to their own limitations. The untalented majority will invariably complain that talented individuals are intellectual mavericks whose reasoning is impossible to understand even though it seems to lead to valid results. Their complaints are understandable, but the problem lies in the fact that the majority is attempting to engage in activities for which it does not have the required aptitude.

16.5- Science and technology are the only areas in human endeavor where the requirement for talent is not considered to be primary. The need for talent in sports or the arts is unquestionable. No one would attempt to assemble a major league ball team by training a highly motivated group of randomly selected individuals. If those individuals were not innately talented, they could not be trained to a level which would permit success and the team would be a dismal failure. Talent cannot be taught. Michael Jordan could not teach untalented individuals to reliably sink a basket from midcourt nor could Hank Aaron teach them to swing a bat so as to drive a baseball into the centerfield stands. When the occasion arises, talented individuals don't consciously think about how they are going to perform their task, they just do it.

16.6- Up until the 20th century, innate talent was necessary for leadership in the physical sciences but, with the introduction of Dr. Einstein's work, it became possible for the untalented majority to gain ascendancy. The same intellectual takeover occurred after World War II in Engineering and the costs and failure rates of sophisticated projects greatly increased. (It is fortunate for the USA that this effect did not take place earlier or the war is likely to have been lost.) Admittedly, engineering has accomplished a great deal since that time but at an excessive cost in money and time. It is no accident that, when a project such as the development of the U-2 aircraft is required, the work is performed in a skunk works, it is not performed in a mainstream organization. (The one statement that the author can quote verbatim from a college instructor after almost 50 years is that "an idiot can do anything that an engineer can, the difference is that it will take three time as long and cost five times as much".) In the basic sciences the situation is much worse. While anyone can tell whether an airplane flies, they cannot tell whether there is a wormhole in space or whether the Law of Conservation of Energy is consistent with the gravitational field. The end result is that the rules of a religion have been imposed on the physical sciences and those sciences have been degraded to a level analogous to that of a medieval theology which, as mentioned, debated the number of angels who could dance on the head of a pin. It is no accident that the PC industry was started by two men working in a garage rather than by a huge corporation.

16.7- The selection and training of talented individuals is a difficult process. They can only be identified by their ability to arrive at valid results by unanticipated, and perhaps incomprehensible routes. The conventional educational approach in the sciences and the technologies is to attempt break those individuals away from their unorthodox thought processes and force them to conform to accepted practices and ideas rather than to recognize their talent and assist them in its development. Talent in both the physical sciences and in technology operates by making use of the pattern recognition capability of the mind and, because it operates in the background, the individual possessing the talent is normally unaware of the thought processes involved and ascribes their results to intuition. Depending upon the lifelong personality of the individual, the innate level of that talent and the discipline that he has exerted in training it, the reliability of an individual's intuition varies from zero to perfection. As is the case with a computer, GIGO applies. In order for the pattern recognizing capability of his brain to be effective in the solution of physical problems it is necessary for an individual to have accepted a certain set of learning rules for himself. These rules are:

- Everything he accepts as true must agree with everything else that he accepts as true.

  - In the event of contradiction, the validity of the relevant information, both old and new, must be considered to be tentative.

  - If the information is more than trivially important, the necessary effort must be made to resolve any inconsistencies by correcting the new and/or the old information.

  - The more items of information processed in this manner, the closer one's store of knowledge will approach absolute validity. (It will never be perfect, no one lives that long.)

- One must never assume that his source of information, no matter how revered, is correct.

  - Where necessary in an academic or business situation, one must keep two sets of books. One set of books must contain the information expected by the individuals possessing power over your future and the other set must contain the information you have determined to be true.

- One must never allow himself to be brainwashed or bullied into not following the preceding rules.

If an individual follows these rules from an early age, he will find that his intuition, also known as common sense, will be extremely reliable and will provide him with information long before he understands it consciously. As a bonus, the information will be correctly labeled as to its reliability. Intuition only yields faulty results if an individual has failed to adequately follow the rules described above.

16.8- Is the information presented in this book more nearly correct than presently accepted concepts? The author believes that it is for the simple reason that he can find no contradiction between the various items of information involved. On the other hand, the presently accepted concepts are severely inconsistent. They are as full of holes as lacy swiss cheese. Will the material that the author has provided require improvement and correction? Certainly it will, no one can cover this much material without error. As to where our understanding of reality will lead, only time will tell. Unfortunately the author's age insures that he will not be around to see the results.

16.9- *Common Sense is another name for intelligence. While mathematics and experimentation require intelligence for their development, their usage does not directly involve intelligence. In their application, they are merely tools which must be used as an adjunct to intelligence.*

Appendix 1 - "Gravity"

Original Copyright 1987

Note:- The original version of "Gravity" was written in the 1960's but until wordprocessing PC's became available, it was not practical to convert it into a text document suitable for the copyright process. It is included here to provide a rigorous backup for "The Einstein Hoax".

Index to "Gravity"

Part	Subject	Page
1	Introduction to Discussion of Gravity	227
2	Laying the Groundwork	241
3	Evaluating the Gravitational Converse Factors	265
4	Comparison with the 'Real World'	283
5	The Complete Gravitational Field	351

Part 1 - Introduction to Discussion of Gravity

List of Topics

Sections	Section Titles	Page
1.0	Introductory Comments	227
1.1	A Tabular Comparison with General Relativity	230
1.2	Concerns with the Derivation of General Relativity	234
1.3	The Derivation of a Gravitational Theory	236
1.4	References	238
1.5	Author's Notes	239

Section 1.0 - Introductory Comments

1.0.1 - In high school, the author asked the question: "Where is the energy stored when a weight is raised?" The answer which was received, that it was stored as potential energy, was frustrating. It seemed, at the time, to be a face saving means of saying that the answer was unknown. The author's interest in gravitation was rekindled in the early 1960's by an article in 'Scientific American'. In the hope that General Relativity might provide an answer to that old high school question, the author began to study the subject. The effort ranged from reading popular books by reputable workers in the field to laboriously studying undergraduate and postgraduate level text. As the activity proceeded, the authors initial open acceptance changed to bewilderment, then to disbelief and then to an overwhelming sense of disappointment as internal and external contradictions associated with General Relativity became more and more apparent.

- General Relativity provides an expression for time dilation which is additively rather than multiplicatively commutative. As a result, it cannot yield correct results in strong gravitational fields (e.g.- neutron stars, gravitational collapse, etc.) even though its conclusions are verified to the limits of observational accuracy within the weak fields of the Solar System.

- The time dilation provided by General Relativity implies the existence of equivalent dilations for other basic parameters such as length and force. The possible existence of such dilations is not only ignored, but seems to be denied.

- General Relativity teaches that space in a gravitational field is curved (non-Euclidian). Older texts teach that the straight line of that curved space is the "null geodesic", which is the path of a ray of light in the gravitational field and represents the least time path between two points. Recent texts recognize that an ideal massless string stretched between points in a gravitational field would follow a path having half the curvature of the null geodesic. In keeping with the accepted definition of a straight line, that path is the straight line of the non-Euclidian geometry of the curved space since it is the shortest distance between two points. This viewpoint apportions the observed curvature of a ray of light in the gravitational field as resulting from equal parts of conventional refraction and spatial curvature. However, it is easily shown that, if space is non-Euclidian in the gravitational field, a closed cycle perpetual motion machine is possible in principle.

Some texts teach that, under General Relativity, the force which we experience as gravity results from the curvature of space. Curved space, of itself, would seem to be incapable of producing an observable force either on a moving or stationary object.

Where the object is moving within the space, the inertia forces resulting from motion within the curved geometry would occur in a direction orthogonal to that geometry and could not be observed within it. Other texts have taken a different view and state that General Relativity has shown that the force (energy) of gravity has been shown not to exist! This seems to contradict the most rudimentary experimental data, such as is obtained when one slips on the ice.

- Under both Newtonian Theory and General Relativity, the gravitational field is capable of creating energy. Indeed, in an early text, the statement is made that "the presence of mass calls into being additional mass" (in the form of gravitational energy), but this does not constitute a violation of the Law of Conservation of Energy because, if that energy were to escape the field, the work required for that escape would reduce the energy which would reach the external universe at a level which was no greater than the original mass energy which entered the field." Some individuals may find such a statement satisfactory, to the author it appears to be a rationalization.

1.0.2 - As a result of these and other considerations, the author decided to facilitate his understanding by attempting to generate an energy balance for the process of lowering a weight from the ceiling to the floor. At first glance, this appears to be simple. At the ceiling, the weight has a mass energy equal to its mass times the square of the velocity of light. At the floor, it has the same mass energy and has released an energy of fall equivalent to its mass energy times the gravitational potential between the ceiling and the floor. However, in providing a time dilation as a function of gravitational potential difference, General Relativity opened a door. Time dilation, which has been experimentally verified, may be equivalent to a change in the size of the units of measurement (Section 3.2 shows that this is actually the case). With this door opened, the possibility must be considered that other quantities are subject to a similar dilation. Thus, performing the desired energy balance requires that the relationship between the size of the unit of measurement for energy at the ceiling and the size of the unit of measurement for energy at the floor be established by a method which does not involve circular reasoning. It is not valid to assume that they are equal.

1.0.3 - The result of this effort was more fruitful than could have been hoped. Not only did the source of gravitational energy become apparent, it was seen to be released in complete compatibility with the Laws of Conservation of Energy and Momentum, to occur in three dimensional Euclidian space, and to be compatible with the second order gravitational effects which are cited as proofs of the validity of General Relativity. At the low levels of gravitational potential observable within the Solar System (10^{-6}), the difference between the conclusions provided herein and those of General Relativity are, and may always remain, undetectable. At much higher levels, such as in the vicinity of a neutron star, the differences are quite pronounced and lead to an unexpected bonus. They demonstrate that, contrary to the predictions of General Relativity, gravitational collapse has an end limit and does not proceed to a singularity. Near this end limit, the conditions which would be observed within a collapsed object bear a startling resemblance to present descriptions of our external universe!

Section 1.1 - A Tabular Comparison with General Relativity

1.1.1 - A tabular comparison of the compatibility of General Relativity and of the results of this discussion with our observed and/or currently accepted conceptual external reality is provided in Table 1.1.1:

Table 1.1.1 - A Comparison with General Relativity

Item	General Relativity	Discussion
Consistent with Principle of Relativity	No	Yes
Consistent with Unnormalized Principle of Equivalence	Yes	No
Consistent with Renormalized Principle of Equivalence	No	Yes
Locally Consistent with Law of Conservation of Energy	Yes	Yes
Globally Consistent with Law of Conservation of Energy	No	Yes
Locally Consistent with Law of Conservation of Momentum	Yes	Yes
Globally Consistent with Law of Conservation of Momentum	No	Yes
Consistent with Eotvos Experiment within Experimental Limits	Yes	Yes
Consistent with Observed Time Dilation within Experimental Limits	Yes	Yes

Table 1.1.1 - A Comparison with General Relativity - (Continued)

Item	General Relativity	Discussion
Time Dilation Multiplicatively Commutative	No	Yes
Other Dilations Multiplicatively Commutative	Unstated	Yes
Consistent with Orbital Precession Data within Experimental Limits	Yes	Yes
Consistent with Refraction of Starlight within Experimental Limits	Yes	Yes
Reveal Source of Gravitational Force/Energy	No	Yes
Valid for Intense Fields	No	Yes
Singularity at Final State	Yes	No
Gravitational Collapse Self Limiting	No	Yes
Leads Directly to Observed Cosmology	No	Yes

1.1.2 - A comparison of the most significant dilations, hereafter designated as gravitational transformations, as provided by General Relativity and by this discussion, is made in Table 1.1.2

Table 1.1.2 - A Comparison of Gravitational Transformation Predictions

Transformation	General Relativity	Discussion
Time	$1/(1+\Theta)$	$(1-\Theta)$
Length	1 or $1/(1+\Theta)$	$1/(1-\Theta)$
Force	1 or $(1+\Theta)$	1
Energy (Force x Length)	1	$1/(1-\Theta)$
Space	$(1+\Theta)$ or $(1+\Theta)^2$	1

Notes on Table 1.1.2:

- A force-length-time system of units is employed rather than the conventional mass-length-time system. The reasons for this choice are outlined in Section 2.3.

- The symbol 'Θ' is employed to represent the gravitational potential between elevations. This symbol is defined in Section 3.1 and is identical with the equivalent function employed by General Relativity.

- Two values are provided for the transformations for length, force, and space under General Relativity. The ambiguity results from a conflict implicit in the use of Tensor Calculus in its derivation.

- Time dilation, refraction of starlight, and orbital precession differ between the two approaches in proportion to Θ^2. The difference is unobservable in the weak fields of the Solar System but are very pronounced in the region around a neutron star.

- Energy differs in proportion to Θ. The difference is grossly observable in the weak fields in the Solar System and manifests itself as the energy of fall. It allows the discussion to pinpoint the source of gravitational energy.

Section 1.2 - Concerns with the Derivation of General Relativity

1.2.1 - In examining the derivation and the usage of General Relativity, the author has been left with a feeling of uneasiness by certain questions which present themselves and which, when they have been addressed, seem to have been answered ineffectively.

1.2.2 - To be useful in generating a gravitational theory, the Principle of Equivalence must have more significance than the fact that all objects (particles) gravitate equally. That requirement need only mean that things (inertial mass, gravitational mass) proportional to the same thing are proportional to each other, and is, of itself, rather trite. To be significant, it is necessary that the Principle of Equivalence means that the effects of an 'inertial field' are identical to the effects of a gravitational field. This interpretation, as it is employed in the generation of General Relativity, seems to present difficulties which can be described in words by reference to Einstein's elevator model. That model asserts that observations made within a closed room on the Earth's surface are indistinguishable from the observations which would be made in an elevator being accelerated 'upwards' in free space. It also assumes that the room is sufficiently small so that the effects of the Earth's curvature are undetectable.

- The first difficulty occurs from the fact that, with the elevator being accelerated upwards, an external force must be applied to the elevator. This force acts through the distance that the elevator moves under the applied inertial acceleration. Since a force applied to a moving body results in a transfer of energy to or from that body, the elevator model implies that the elevator is receiving a steady input of energy from a source external to the model. In model analysis, any relevant quantity which crosses the boundary of the model must be taken into account as an input or an output. For the treatment to be correct, an accounting must be made of the energy transfer to or from the elevator. The derivation of General Relativity does not take this energy transfer into account, and, by failing to do so, produces internal contradictions which, in turn requires a non-Euclidian geometry for space for their resolution.

- Such a geometry can be shown to violate the Law of Conservation of Momentum as applied to a closed system (Section 4.2).

- The second difficulty with the usage of the Principle of Equivalence results from the assertion that the effects of curvature can be reduced to insignificance by considering a sufficiently small region of space. While Cartesian coordinates tend to conceal its nature, polar coordinates clearly show curvature to be a first derivative. Its effects are therefore scale independent and this assertion cannot be true. In a field of a given curvature, the same results are obtained, in proportion to the gravitational potential difference, regardless of the size of the region of space considered.

- A final difficulty is that if the Principle of Equivalence (i.e.- that a gravitational field is equivalent to an 'inertial field') is validly applied in General Relativity, it should be possible to describe a single inertial reference frame which accounts for the gravitational and inertial accelerations observed on opposite sides of the Earth. Such a reference frame can be described in mathematical terms (Section 4.9), but it is rather bizarre.

1.2.3 - General Relativity employs Tensor Calculus in its derivation. In so doing, it utilizes the mathematical process of integration. When one remembers that, in performing a mathematical integration, it is not valid to assume that the integral of K times δX is equal to K times the integral of δX without first determining that K is independent of X, the use of Tensor Calculus becomes highly suspect. By using the Principle of Relativity as a postulate, the derivation of General Relativity admits of the possibility that the size of the units of measurement for various quantities are a function of the gravitational potential difference between reference frames. (Many descriptions of Special Relativity refer to the shrinkage of metersticks and the slowing of clocks. Section 3.2 shows that the gravitational time dilation can only be interpreted as a change in the size of the units of measurement for time.) In order to validly use Tensor Calculus in the gravitational field then, it is necessary to establish the effects, if any, of the field on the units of measurement for the relevant quantities. Once that effect is determined, Tensor Calculus is no longer required except as convenient computational tool. The manner in which Tensor Calculus was employed in the derivation of General Relativity leads to the dual tabulations in Table 1.1.2.

Section 1.3 - The Derivation of a Gravitational Theory

1.3.1 - In introducing the concept of 'invariance' between reference frames, the Principle of Relativity introduced the concept of 'constancy' between reference frames as a corollary. To be rigorous, therefore, it is not adequate to just consider whether relationships or quantities change between reference frames in terms of units of measurements as they exist within each reference frame. It is also necessary to consider whether or not those relationships or quantities change between reference frames after a compensation has been made for any change in the size of the units of measurement which may occur as a result of the change in reference frame. The Principle of Relativity implies four, not two possibilities. They are 'invariant' vs. 'non-invariant' and, independently, 'constant' vs. 'not-constant'. Inclusion of both of these concepts may lead to unnecessary complexity where both are not required but will produce no error. Omission of one of these concepts where it is required will force an incorrect solution to the problem under consideration. The importance of the concept of 'constancy' will become apparent as the discussion proceeds. The force and energy of gravitation are shown (without circular reasoning) to result from the fact that total energy is 'constant' rather than 'invariant' in the gravitational field.

1.3.2 - It is normally considered that the time dilation predicted by General Relativity results from the application of the Principle of Equivalence for its derivation. In this discussion, the necessary derivations will be performed without reliance on that principle. It will be shown, after the derivations have been completed and verified by comparison with external observations, that the results are consistent with the Principle of Equivalence after it has been renormalized to account for the energy input to the accelerated system as described in Section 1.2.

1.3.3 - In the discussion which follows, the relationship between the gravitational time dilation, an equivalent energy dilation, and the gravitational potential are derived based upon the following precepts:

- The Principle of Relativity is valid.

- The energy of a photon is proportional to its frequency.

- Energy which is capable of existence 'at rest' gravitates.

- A perpetual motion machine of the first kind is impossible in principle.

The equation which results provides a family of solutions which include both the time dilation expression of General Relativity and the time dilation expression provided by the present discussion. Adding the requirement that the time dilation be multiplicatively commutative parses the solution into separate expressions for the time and for the energy dilations as a function of the gravitational potential. It is at this point where the conclusions of this discussion diverge from those of General Relativity.

1.3.4 - A complete description of the gravitational field requires a third dilation expression. This is obtained by factoring the dilation expression for energy into expressions for force and length dilations. Symmetry considerations suggest that a 'Law of Conservation of Existence' should hold in addition to the conservation laws for energy, momentum, and angular momentum. Such a law is therefore postulated and employed to permit the required factoring. With the three dilations determined, the gravitational field is defined completely.

1.3.5 - In Part 4, the predictions which these dilations make of the gravitational field are compared with experimental results. They are shown to be in complete agreement with observation to the limits of accuracy of measurement possible within the Solar System, to be compatible with three dimensional Euclidian Space, and to be consistent with the absolute validity of the Law of Conservation of Energy. The proposed 'Law of Conservation of Existence' is therefore considered to be verified. In Part 5, the gravitational dilations which were developed are employed to describe the complete field. Among what is shown is that gravitational collapse does not proceed to a cataclysmic singularity within a 'black hole'. Instead, it is a self limiting process which proceeds to a state which, when observed internally, bears a striking resemblance to our own universe.

Section 1.4 - References

1.4.1 - No references are provided for the material presented since the development of the arguments which follow requires only those relationships which are accepted as general knowledge at the undergraduate level in the physical sciences and in engineering. The treatment of the material which follows is based solely on the requirement that these relationships obey the Principle of Relativity, that all material will be both internally and externally consistent, and that all currently accepted physical laws will be followed. (It shall be impossible in principle, for example, to construct a perpetual motion machine of the first kind.) The relationships which are employed are chosen so as to be verifiable within a given reference frame and may be tested experimentally if in doubt.

Section 1.5 - Author's Notes:-

1.5.1 - The statement that curvature is a first derivative has raised a question among some readers who, interestingly enough, possessed Ph.D. degrees. These individuals cited the handbook expressions for curvature. When curvature is expressed in terms of polar or Cartesian coordinates, second derivatives are present, however, in terms of Cartesian coordinates, second derivatives must appear. This coordinate system is incapable of dealing with curvature without introducing second derivatives even if they are not related to the nature of that curvature. In the polar coordinate system, second derivatives appear as a function of the distance between the instantaneous center of curvature and the origin of the polar coordinate system. If the coordinate system is translated to the instantaneous center of curvature, those second derivatives vanish. Obviously, if translating the coordinate system causes the second derivatives to vanish, they are artifacts introduced by the coordinate system rather than properties of the curvature itself. This conclusion is verified by the fact that handbooks also provide curvature as the rate of change of direction with respect to distance. Curvature, by its nature, is a first derivative (1999 comment).

1.5.2 - Some question has been raised as to whether the need for multiplicative commutivity has been proven. A brief reflection should demonstrate that multiplicative commutivity is a requirement of the Principle of Relativity and it is the fact that the Lorentz Transformations are multiplicative commutive that allows Special Relativity to work. The product of the time dilations between levels A and B (in terms of the gravitational potential between those levels measured in terms of level A units of measurement) times the time dilation between levels B and C (in terms of the gravitational potential between those levels measured in terms of level B units of measurement) must equal the time dilation between levels A and C (in terms of the gravitational potential between those levels measured in terms of level A units of measurement). General Relativity does not meet this requirement and must therefore be in conflict with one of its postulates, the Principle of Relativity. General Relativity's lack of a length dilation corresponding the Lorentz Transformation for Length of Special Relativity means that General Relativity does not satisfy the Principle of Equivalence either. It would seem that General Relativity is a theory which contradicts its own postulates (1999 comment).

1.5.3 - The copyrighted text of "Gravity" was sent in 1988 both to individuals and publications identified as having a reputation in the field of gravitation.

Part 2 - Laying the Groundwork

List of Topics

Sections	Section Titles	Page
2.0	Introduction	241
2.1	Basic Postulates	242
2.2	Invariance vs. Constancy	243
2.3	The Concept of 'Dimensional Entities'	246
2.4	The Rules of Dimensional Analysis	250
2.5	Dimensional Analysis Applied to Different Reference Frames	251
2.6	The Notation System	253
2.7	Applying Dimensional Analysis to the Gravitational Field	255
2.8	The Isotropic Nature of the Gravitational Field	259

Section 2.0 - Introduction

2.0.1 - In this Part, the basic precepts employed in the remainder of this discussion are introduced. The concept of 'invariance' vs. 'constancy' is provided and its relevance is illustrated by an example which is analogous to the situation which occurs when observations are made between reference frames in which the basic units of measurement may differ.

2.0.2 - In order to evaluate the effects of possible changes in the magnitudes of the units of measurement occurring between reference frames, the rules of Dimensional Analysis are employed. The pertinent rules of Dimensional Analysis are defined, and the Laws and Constants of the Science of Physics which will be employed in later Parts are described. (References are not provided since both Dimensional Analysis and the pertinent Laws and Constants of the Science of Physics are amply described in undergraduate level texts.)

2.0.3 - Tables are provided of various physical quantities and of the gravitational conversion factors for these quantities, the values of which are evaluated in Part 3. This Part closes with a Section which demonstrates that the gravitational field is isotropic with respect to these conversion factors, allowing the field to be treated as a scalar with respect to elevation.

Section 2.1 - Basic Postulates

2.1.1 - The basic postulates employed in this discussion are:

- The Laws of the Science of Physics are invariant but not necessarily constant between reference frames.

- 'Entities' which are subject to the macroscopic conservation laws, and which are not vector quantities, are constant but not necessarily invariant between reference frames.

- The rules of Dimensional Analysis are valid and must be used for organizing concepts, experiments, and the results of experiments between reference frames.

2.1.2 - The first postulate is true by definition. A Law or Constant of the Science of Physics must have the property of invariance between reference frames in order to qualify for that status. In addition, the property of invariance is readily verified and may be tested at any time that an assumed invariance is in question.

2.1.3 - The second postulate seems self evident, if an 'entity' which is not a vector quantity is conserved, the amount of that 'entity' which is present remains unchanged, regardless of any change in the units of measurement. When the effects of any change in the units of measurement which occur as a result of a change in reference frame are compensated, the amount of the conserved quantity must be found to be unchanged between reference frames. When a conservation law applies to a vector sum of an 'entity', however, it is not necessary that the amount of that entity remain fixed, and the second postulate does not apply. (Section 2.2 provides a discussion of invariance and constancy.)

2.1.4 - Finally, the rules of Dimensional Analysis have been found to be empirically valid for all physical phenomena to the point where no equation would be considered valid if it were not 'dimensionally correct'.

2.1.5 - The reference to the 'Science of Physics' is to that conceptual structure of laws and constants which man has devised to enable him to deal objectively with reality. It must be distinguished from the possibly unknowable underlying structure of Nature which produces our objective reality.

Section 2.2 - Invariance vs. Constancy

2.2.1 - The Science of Physics is based upon the property of invariance between reference frames. If a quantity or a relationship is to qualify as a Constant or a Law of the Science of Physics, that quantity or relationship must be the same regardless of where or when (i.e.- in which reference frame) it is measured. It is necessary to recognize, however, that it is possible for quantities or relationships to possess the property of constancy between reference frames in addition to, or in lieu of, the property of invariance. It is also necessary to recognize that invariance and constancy are independent properties which must be determined separately.

2.2.2 - While for most purposes, it is desirable to consider only the property of invariance so as to remove all considerations of reference frame from physical measurements, it must also be remembered that ignoring the property of constancy may result in the destruction of information which would otherwise be available. Since it is the purpose of this discussion to extract such information, it is important to define carefully what is meant by both of these terms.

- A quantity is *invariant* between reference frames when that quantity, as measured with units of measurement native to each reference frame, is found to have the same numerical value in both reference frames.

- A quantity is *constant* between reference frames when that quantity, as measured with units of measurement which have been corrected for any change in their magnitude occurring as a result of the change in reference frames, is found to have the same numerical value in both reference frames.

2.2.3 - It should be noted that, for the purposes of this discussion, ideal instruments which are perfectly calibrated to each other are assumed. The effects to be considered result from possible changes in the very size of the units of measurement which the instruments employ (e.g.- the relativistic shrinkage of metersticks, the relativistic slowing of clocks, etc.), and not from any deficiencies in the instruments themselves. The idea that there may be changes in the basic units of measurement between reference frames may seem strange to some readers, but, in reality, it is a valid, if unconventional, way of considering the significance of the Lorentz Transformations of Special Relativity. Conflict between this interpretation and the conventional interpretation of Special Relativity exists only at the metaphysical level since both viewpoints lead to identical conclusions when properly applied.

2.2.4 - Needless to say, it is more difficult to examine the constancy of a quantity between reference frames than to determine its invariance. Invariance may be determined by a simple measurement. Constancy is determined by adjusting the results of measurement for the effects of a change in reference frame on the units of measurement which were employed in making the measurement. Constancy can only be determined where it is possible to evaluate unambiguously the effects of a change in reference frame on the units of measurement involved. Constancy and invariance are independent properties which must be determined separately.

2.2.5 - It may be helpful to illustrate the above definitions. Consider the possibilities inherent in the price of gasoline between New York City and Toronto, Canada. The imperial gallon in use in Canada is 25% larger than the gallon in use in the USA. In Canada, gasoline is purchased with Canadian dollars which do not have the same value as US dollars. The first possibility is the case where the price of gasoline is the same in New York City and in Toronto, using local units of currency and fluid measure. To employ the terminology of the physicist, the price of gasoline is invariant between the reference frames (New York City and Toronto).

2.2.6 - It is not possible, from the above description, to state whether the price of gasoline is constant between these cities. In order to make that determination, it is necessary to correct the purchase prices for the effects of the difference in size of the US and imperial gallons and the values of the US and Canadian dollars. A purchase of one gallon of gasoline in Toronto provides the purchaser with the equivalent of 1.25 US gallons. In order for the price of gasoline to be both constant and invariant between New York City and Toronto, it is necessary for the Canadian dollar to be worth 1.25 times the value of the US dollar. For any other rate of exchange, the price of gasoline will not be constant (providing that it is invariant) between those reference frames.

2.2.7 - The final possibility is illustrated by the case where the Canadian and the US dollar have the same value and where a gallon of gasoline costs $1.25 in Toronto and $1.00 in New York City. The price of gasoline is not invariant between these reference frames, but it is constant. At both locations, the equivalent of $1.00 of US money will buy the equivalent of one US gallon.

2.2.8 - While dealing with the concept of invariance is conventional in the Science of Physics and presents no difficulty, dealing with the concept of constancy requires more care. It is necessary to have a means of determining the change in size of the units of measurement occurring as result of a change in reference frame. It is then necessary to be able to employ those changes in size of the units of measurement so that the results of measurements made in one reference frame may be compared with the results of measurements made in the second reference frame in terms of units of measurement which have been compensated so as to be the same for both reference frames.

2.2.9 - To allow the required correction factors to be determined and to provide the rules for their use, the procedures of Dimensional Analysis are required. Dimensional analysis is not rigorous, it is empirical. It does have one thing in common with the Science of Physics, its rules agree with the results of observation.

Section 2.3 - The Concept of 'Dimensional Entities'

2.3.1 - The Laws of the Science of Physics are normally stated in the form of equations. Implicit in these equations are fundamental 'entities' called 'dimensions' which define the quantities related by the equations. The inclusion of these 'dimensional entities' is vital if the equations are to have meaning. It would be meaningless to state, for example, that the acceleration of gravity on Earth is 32.2. To provide meaning, it is necessary to include the 'dimensional entities' of length and time and state that the acceleration of gravity on Earth is 32.2 feet per second per second.

2.3.2 - It has been determined by observation (a rigorous proof does not seem to exist) that three 'dimensional entities' are required to describe the laws of our macroscopic physical universe. A system of 'dimensional entities' greater than three in number is always found to be reducible, without loss of information, to three basic 'entities' by application of the appropriate equation(s) provided by the Science of Physics. On the other hand, all attempts to reduce the required number of 'entities' to less than three have required the substitution of one or more universal constants (e.g.- the velocity of light, the gravitational constant, the permeability of space, etc.) for one or more of the 'dimensional entities'. It is then found that these constants must themselves be treated as if they were 'dimensional entities' if information is not to be lost, and, in effect, the required number of 'dimensional entities' remains at three. (While many tables of 'dimensional entities' include more than three components, careful examination reveals a sufficient interrelationship between them to permit a reduction in their number to three. For example, the commonly employed 'dimensional entity' of temperature may be defined as energy per degree of freedom. The 'dimensional entity' content of temperature is thus seen to be identical with the 'dimensional entity' content of energy.)

2.3.3 - It does not seem to have been possible to establish that any of the possible choices of 'dimensional entities' is more basic than any of the others. Any group of three may be chosen providing that there is some degree of independence between them. It is conventional, however, to employ the 'entities' of mass (M), length (L), and time (T) in the physical sciences. For the purpose of these discussions, a change is made in the selection. The 'entity' of force (F) is substituted for the 'entity' of mass (M). The reason for this change is threefold:

- Unlike force, mass is a derived property which is not directly observable. A measurement of the mass of an object requires a measurement of the gravitational force applied to it by another object of a known mass (weighing), by the inertial force applied to it in response to a spatial acceleration (shaking), or by the energy released when the matter composing the mass is annihilated.

- The 'entities' of force, length, and time provide the simplest representation of Planck's Constant, the 'dimensional content' of which becomes F*L*T.

- The partial products of these 'entities' represent directly the physical quantities which are subject to the macroscopic conservation laws, namely energy (F*L), momentum (F*T), and angular momentum (F*L*T). As suggested by D. L. Shapiro, considerations of symmetry would indicate that an unrecognized 'entity' which shall be designated as 'existence', and having the 'dimensional entity' content of L*T, should also obey a conservation law. (It should be noted that, in the absolute sense, 'existence' is conserved under both Special Relativity and the gravitational field as shown in the discussion which follows. The application of the "Law of Conservation of Existence" to Special Relativity, however, is valid only for directions parallel to the relative velocity vector.)

2.3.4 - With the selection of force, length, and time as the basic 'dimensional entities' to be employed. It is possible to provide the 'dimensional entity' content of various physical quantities by the application of the rules of Dimensional Analysis, as described in Section 2.4, to the appropriate equations represented by the Laws of the Science of Physics. The 'dimensional entity' content of various physical quantities of interest are tabulated below:

Table 2.3.1 - Dimensional Entity Content of Physical Quantities

Quantity	Symbol	Dimension
Length	L	L
Time	T	T
Force	F	F
Charge	Q	L
Energy	E	F*L
Angular Momentum	J	F*L*T
Velocity	V	L/T
Acceleration	A	L/T^2
Mass	M	F*T^2/L
Gravitational Constant	G	L^4/(F*T^4)
Dielectric Constant of Space	ε	1
Permeability of Space	μ	F*T^2/L^2
Existence	B	L*T
Ergo-Gravitational Constant	D	1/F

2.3.5 - An unfamiliar quantity, the 'Ergo-Gravitational Constant', is provided in the above table. This constant is determined by dividing the conventional gravitational constant by the fourth power of the velocity of light. It relates the gravitational force to the energy equivalents of the gravitating masses.

- The 'dimensional entity' content for charge, the dielectric constant of space, and the permeability of space are found by combining the expression for the electrostatic force between charges, the expression for the magnetic force between moving charges, and the expression for the velocity of light as a function of the dielectric constant and the permeability of space. The result becomes unambiguous when it is recognized that the magnetic force is velocity dependent while the electric force is between stationary charges. These quantities are presented for reference and, since they are not employed in the discussions which follow, further justification is not provided.

Section 2.4 - The Rules of Dimensional Analysis

2.4.1 - In order to apply Dimensional Analysis in this text, it is necessary first to provide the rules which have evolved by this study that are pertinent to the discussion. These rules, which are provided in recognized texts on the subject, are provided below:

- The net exponent assigned to each dimensional entity in a term of a physical equation is equal to the algebraic sum of the exponents occurring for each appearance of that entity in that term of the equation.

- The net exponent assigned to each dimensional entity in a term of an equation must be identical to the net exponent assigned to the same dimensional entity in every other term of the equation.

- When the units of measurement which define the magnitude of a dimensional entity are altered by a ratio of $1/K^N$, the numerical magnitude of the term in a physical equation is altered by a ratio of K^N, where 'N' is the net exponent assigned to that dimensional entity.

2.4.2 - There are, of course, other rules of Dimensional Analysis, such as those provided by Buckingham's Pi Theorem, but they are not listed since they are not required for further development of this discussion.

Section 2.5 - Dimensional Analysis Applied to Different Reference Frames

2.5.1 - Consider two frames of reference, which for convenience of notation, are designated as the upper and lower reference frames. Consider further that the units of measurement are not necessarily identical in both frames of reference despite the fact that they have identical designations. (This result is obtained, for example, in the Special Theory of Relativity which advises that yardsticks change in length and clocks change in rate as their velocity is altered.) To keep track of possible changes in units of measurement in this discussion, consider that the units of measurement for the 'dimensional entities' of length, time, and force are 'X', 'Y', and 'Z' times larger respectively than the same units of measurement in the lower reference frame. For the above considerations, the following equations may be written:

Eq. 2.5.1 $X = l/L$

Eq. 2.5.2 $Y = t/T$

Eq. 2.5.3 $Z = f/F$

Where:

- l is a length measured with the units of measurement of the lower reference frame.

- L is the same length measured with the units of measurement of the upper reference frame.

- t is a duration of time measured with the units of measurement of the lower reference frame.

- T is the same duration of time measured with the units of measurement of the upper reference frame.

- f is a force measured with the units of measurement of the lower reference frame.

- F is the same force measured with the units of measurement of the upper reference frame.

2.5.2 - It must be noted that the above equations are provided for the purpose of defining the conversion factors for length, time, and force between reference frames. They do not imply that it is necessarily possible to determine whether a length, a duration of time, or a force are identical between reference frames by observational means.

2.5.3 - With the conversion equations for the fundamental 'dimensional entities' established, the conversion equations for the other quantities of interest may be provided in accordance with the 'dimensional entity' content of these quantities as shown in Table 2.3.1:

Table 2.5.1 - Conversion Factors for Physical Quantities

Quantity	Conversion Factor
Length	$l = (X)*L$
Time	$t = (Y)*T$
Force	$f = (Z)*F$
Charge	$q = (X)*Q$
Energy	$e = (X*Z)*E$
Angular Momentum	$j = (X*Y*Z)*J$
Velocity	$v = (X/Y)*V$
Acceleration	$a = (X/Y^2)*A$
Mass	$m = (Y^2*Z/X)*M$
Gravitational Constant	$g = (X^4/[Z*Y^4])*G$
Dielectric Constant of Space	$\varepsilon = (Z)*E'$
Permeability of Space	$\mu = (Y^2*Z/X^2)*M'$
Existence	$b = (X*Y)*B$
Ergo-Gravitational Constant	$d = (1/Z)*D$

2.5.4 - It should be noted that the use of factors such as X, Y, and Z as conversion factors for 'dimensional entities' between reference frames is not new to the Science of Physics. The Lorentz Transformations provided by Special Relativity are the equivalent transformations for reference frames having a relative velocity but based upon the 'dimensional entities' of mass, length, and time.

- The quantities of velocity, acceleration, and mass must be treated carefully when considering reference frames having relative velocity due to the effects introduced by the finite velocity of light as evidenced by the correction factor, $(1+V_1*V_2/C^2)$ in the denominator of Special Relativity's expression for the addition of velocities. The transformations for velocity and for mass are only rigorously valid when the product of the velocities, V_1 and V_2, is sufficiently small compared to the square of the velocity of light. The expression for acceleration is valid for all levels of acceleration provided that the duration of that acceleration is sufficiently small so as to result in a small velocity change. This subject is discussed completely in Appendix 2 entitled "Corrections to Residual Errors in the Special Theory of Relativity".

Section 2.6 - The Notation System

2.6.1 - The discussion which follows make use of what may be described as 'dimensional equations'. These equations will refer to measurements made in different reference frames, either with the units of measurement native to each reference frame or with the units of measurement of a different reference frame. In addition, use will be made of the numerical values of various physical constants which will be postulated to be invariant between reference frames. In order to unambiguously denote these various elements of the discussion, the following notation system will be employed:

- The numerical value of a physical constant will be denoted as the upper case symbol for that constant primed. The numerical value for the velocity of light, for example, would be denoted as C'.

- A measurement made with a unit of measurement of the lower reference frame will be denoted by a lower case symbol (e.g.- time=t).

- A measurement made with a unit of measurement of the upper reference frame will be denoted by an upper case symbol (e.g.- time=T)

- Where convenient, the reference frame in which the measurement is made is denoted by the use of the appropriate subscript:

    - 'U' refers to the upper reference frame.

    - 'L' refers to the lower reference frame.

    - 'M' refers to a reference frame between the upper and lower reference frames.

    - 'V' refers to reference frames having a relative velocity.

    - 'P' refers to a direction parallel to that relative velocity.

    - 'T' refers to a direction transverse to that relative velocity.

- Where convenient, a quantity which is pertinent only to a single Section of this document will be valid only for that Section and will be defined within that Section.

2.6.2 - Thus, a measurement of time in the upper reference frame using the units of measurement of the lower reference frame will be denoted as t_U, while the same measurement made with the units of measurement of the upper reference frame will be denoted as T_U.

Section 2.7 - Applying Dimensional Analysis to the Gravitational Field

2.7.1 - In order to draw meaningful conclusions as to the nature and behavior of the gravitational field, it is necessary to be able to evaluate the effects of changes in elevation (reference frame) within the field upon the various physical quantities. Since, in apparent agreement with the results of observation, quantities and relationships accepted as laws and constants of the Science of Physics are invariant between reference frames, the results of measurements of these quantities and relationships will be identical at all points within the field. Any changes which actually occur as a result of a change in elevation must therefore be concealed by suitable changes in the units of measurement between these elevations. (The laws and constants of the Science of Physics are invariant between reference frames, in accordance with the Principle of Relativity.)

2.7.2 - Changes in physical quantities or constants which occur as a result of changes in elevation will be revealed only when compensation has been made for the effects of any changes in the magnitudes of the units of measurement which result from the change in elevation. Such a compensation permits the results of measurement to be recorded in terms of units of measurement which are unaffected by the change in elevation. The required compensation of the results of experiment may be accomplished between any two elevations providing the factors X, Y, and Z can be evaluated unambiguously between these elevations. It is toward the evalutation of these factors that Part 3 of this discussion is directed. The upper elevation will be designated as the upper reference frame and the lower elevation will be designated as the lower reference frame.

2.7.3 - Much of the discussion which follows is based upon 'ideal thought experiments' which are assumed to employ ideal error free instruments with the requirement that their results must be consistent with the laws and constants already accepted by the Science of Physics. The justification for the use of these laws and constants is that, if their validity is questioned, they may be verified by observation using local units of measurement. These 'thought experiments' will be considered to take place under the following conditions:

- The determinations are made on an ideal, non-rotating planet located sufficiently remote from all other gravitating objects so that their effects may be ignored.

- The determinations are made between two elevations, a fixed distance apart, which are separated by an unchanging gravitational potential which is sufficiently small as compared to unity so that the gravitational field may be considered to be ideally linear.

2.7.4 - In addition to the postulates regarding *invariance*, *constancy*, and *dimensional analysis* provided in the earlier Sections of this Part, the following statements are considered to be true, subject, of course, to re-verification by physical observation in a single reference frame with local units of measurement:

- The Law of Conservation of Energy (including the energy represented by 'rest mass') is valid for closed systems. If this were not true, it would be possible in principle for a closed system to continuously export energy without depleting its internal resources.

- The Law of Conservation of Momentum is also valid for closed systems. If this were not true, it would be possible, in principle, to construct a machine which would violate the Law of Conservation of Energy in contradiction to the above.

- The energy of the photon, as measured with local units of measurement, is equal to Planck's Constant times its frequency. The inertial mass of the photon, as measured with local units of measurement, is equal to its energy divided by the square of the velocity of light. This inertial mass is evidenced by the radiation pressure observed when light is reflected or absorbed.

- The energy stored in material objects is invariant between reference frames. For example, one might consider, as a thought experiment, that the energy was stored in material particles by breaking up a helium atom into four hydrogen atoms and was recovered by recombining the hydrogen atoms into a helium atom. The energy represented by the mass difference between the hydrogen atoms and the helium atom would then be considered to be 'stored energy'.

- The gravitational transformations, such as represented by Table 2.5.1, must be multiplicatively commutative. If this were not true, one would obtain the absurdity of a different result occurring between the first floor and the third floor of a building depending upon whether the elevator happened to stop at the second floor. (It should also be noted that, unless the gravitational transformations meet this requirement, the resultant conclusions will not be consistent with the Principle of Relativity [1999 comment])

- 'Existence' is conserved between reference frames. This postulate is employed to allow a prediction of the results of astronomical observation to be made based upon basic principles. If one wished, the procedure could be inverted, and the results of astronomical observation could be used to verify the 'Law of Conservation of Existence'. (It should also be noted that, under Special Relativity, 'existence' is conserved in a direction parallel to the relative velocity vector. If 'existence' is not conserved in the gravitational field, then the Principle of Equivalence would not be valid [1999 comment]).

- The gravitational field is isotropic. This assumption provides a considerable simplification of the discussion since it eliminates the need to consider horizontal and vertical transformations separately. In order to provide rigor, Section 2.8 which follows discusses the isotropicity of space in the gravitational field and provides a description of the observable effect which would occur if space were not isotropic. An experimental verification may be made if desired, otherwise Section 2.8 may be omitted.

- The derivations will be made without consideration of the 'Principle of Equivalence' which forms the basis of General Relativity. After it is shown that the conclusions which are derived are in complete agreement with the results of observation, the conflict between the results which are obtained herein and those of General Relativity will be shown to result from the failure of General Relativity to perform a required renormalization in applying that principle.

- Where it is desired to duplicate a thought experiment as a real experiment performed on the Earth, compensations can be made for the effects of the Earth's motion through space and for the effects of other astronomical bodies through the laws of celestial mechanics and through the use of the Special Theory of Relativity.

Note:- A sufficiently large number of particles are assumed so as to permit quantum uncertainties to be ignored.

Section 2.8 - The Isotropic Nature of the Gravitational Field

2.8.1 - It is conventional to assume that space is isotropic since, in making physical measurements, the orientation of the measuring apparatus in space produces no detectable effect on the result of measurement. While this insures that space is isotropic in terms of 'invariant' units of measurement, the possibility exists that space may not be isotropic in terms of 'constant' units of measurement in the presence of the gravitational field. Changes in the units of measurement between orientations may occur which serve to conceal a lack of isotropicity. It is the purpose of this Section to examine the isotropicity of space in the gravitational field in terms of units of measurement which are corrected for the effects of changes in orientation. This examination must be performed in order to permit the factors X, Y, and Z to be evaluated in the Sections which follow. Should observation show that space is not isotropic in the gravitational field, this evaluation is probably still possible, but will be much more complicated.

2.8.2 - Conditions of symmetry dictate that there will be no difference in the characteristics of space in the horizontal directions. A difference may exist, however, between the horizontal and vertical orientations. Further, since length and force are vector quantities while time is a scalar quantity, orientation in the gravitational field is significant only in its effect on the units of measurement of length and force. Evaluation of the isotropicity of space will be based upon determining the change in the units of measurement for length and force between horizontal and vertical orientations, as denoted by the conversion factors for isotropicity, X_I and Z_I, as defined below:

Eq. 2.8.1 $X_I = l_I/L_I$

Eq. 2.8.2 $Z_I = f_I/F_I$

Where:

- 'l_I' is a length measured with horizontal units of length.

- 'L_I' is the same length measured with vertical units of length.

- 'f_I' is a force measured with horizontal units of force.

- 'F_I' is the same force measured with vertical units of force.

Additional subscripts, 'V' and 'H' will be employed to denoted measurements of length and force made in the vertical and horizontal directions respectively.

2.8.3 - Consider a simple bell crank having both arms of equal length, L, as measured in a horizontal plane. This bell crank is mounted with one arm horizontal and one arm vertical. A horizontal force, f_{IH}, is applied to the vertical arm and is balanced by a vertical force, F_{IV} applied to the horizontal arm such that the bell crank does not rotate on its ideal frictionless pivot Figure 2.8.1.

2.8.4 - The application of the vertical force, F_{IV}, to the horizontal arm produces a clockwise torque on the bell crank equal to $F_{IV}*l_{IH}$, while the application of the horizontal force, f_{IH}, to the vertical arm produces a counterclockwise torque on the bell crank equal to $f_{IH}*L_{IV}$. The net torque on the bell crank, as measured with uncorrected units of measurement, is the difference between these torques. Since the bell crank is observed not to rotate, the net torque must be zero and we may write:

Eq. 2.8.3 $F_{IV}*l_{IH} = f_{IH}*L_{IV}$

In terms of 'constant' units of measurement, the net torque applied to the bell crank must also equal zero, therefore:

Eq. 2.8.4 $F_{IV}*L_{IH} = F_{IH}*L_{IV}$

Converting the horizontal units of measurement in Equation 2.8.3 to vertical units of measurement by the use of the conversion factors X_I and Z_I from Equations 2.8.1 and 2.8.2 provides:

Eq. 2.8.5 $X_I F_{IV} * L_{IH} = Z_I * F_{IH} * L_{IV}$

Combining Equations 2.8.4 and 2.8.5 provides:

Eq. 2.8.6 $X_I = Z_I$

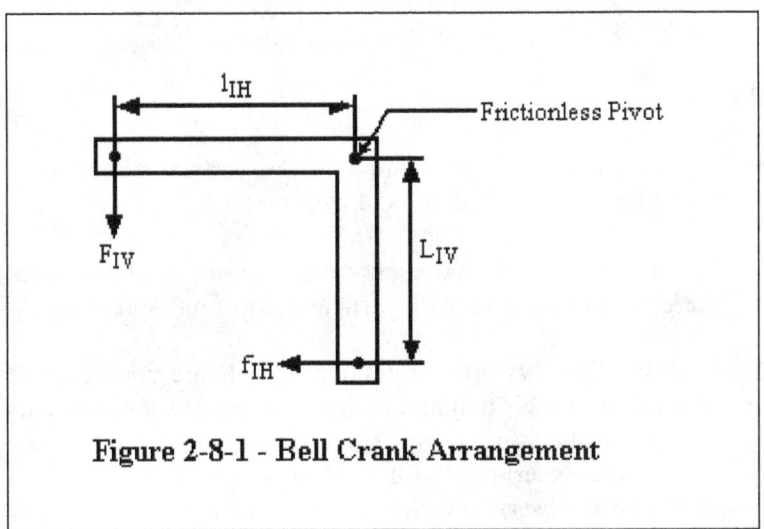

Figure 2-8-1 - Bell Crank Arrangement

2.8.5 - Consider next the case of an ideal coil spring which is oriented with its axis in a horizontal plane, compressed, and tied. Since the ideal spring is presumed to obey Hooke's Law, the energy stored in the spring, e_S, is provided by:

Eq. 2.8.7 $e_S = 0.5 * f_{IH} * l_{IH}$

Where:

- 'f_{IH}' is the force in the spring in the horizontal direction as measured with horizontal units of measurement.

- 'l_{IH}' is the change in length of the spring in the horizontal direction as measured with horizontal units of measurement.

The spring is then rotated so as to have its axis vertical and the tie is released. The spring then returns an amount of energy equal to E_S as provided by:

Eq. 2.8.8 $\quad E_S = 0.5 * F_{IV} * L_{IV}$

Where:

- 'F_{IV}' is the force in the spring in the vertical direction as measured with vertical units of measurement.

- 'L_{IV}' is the change in length of the spring in the vertical direction as measured with vertical units of measurement.

2.8.6 - Providing that gravity gradient effects are compensated by knowledge of their magnitude, by insuring that the moments of inertia of the spring are equal in all axes, and/or by performing the experiment in a field of infinite radius, any net torque which is observed which tends to align the spring either vertically or horizontally must produce a change in the energy stored in the spring. The existence of such a torque is subject to experimental verification and, to the author's knowledge, has never been reported. We may write, therefore:

Eq. 2.8.9 $\quad E_S = e_S$

Combining Equations 2.8.1, 2.8.2, 2.8.7, 2.8.8, and 2.8.9 provides:

Eq. 2.8.10 $\quad X_I * Z_I = 1$

And combining Equations 2.8.6 and 2.8.10 provides:

Eq. 2.8.11 $X_I = 1$

Eq. 2.8.12 $Z_I = 1$

Since the factors X_I and Z_I are equal to unity, they may be ignored.

2.8.7 - In the absence of an observed tendency for an object in which energy is stored anisotropically to align itself either vertically or horizontally in the gravitational field after the gravity gradient effects on anisotropic moments of inertia have been compensted, space must be considered to be isotropic in the gravitational field both in terms of 'invariant' and of 'constant' units of measurement if the Law of Conservation of Energy is to be satisfied. Since no such tendency has been reported, its seems permissible to proceed with the determination of the factors X, Y, and Z in the following Sections without regard to orientation in the gravitational field.

Part 3 - Evaluating the Gravitational Conversion Factors

List of Topics

Sections	Section Titles	Page
3.0	Introduction	265
3.1	The Gravitational Potential	266
3.2	The Relationship Between Y, X*Z, and Θ	268
3.3	The Evaluation of the X*Z Product and of Y in Terms of Θ	274
3.4	The Evaluation of X and Z in Terms of Θ	277
3.5	The Evaluated Gravitational Conversion Factors	281

Section 3.0 - Introduction

3.0.1 - In this Part, the definition of the gravitational potential, Θ, is provided and the gravitational conversion factors for time (Y), and for energy (X*Z) are determined using 'ideal thought experiments' and the requirement that the gravitational transformation for time be multiplicatively commutative. A postulated 'Law of Conservation of Existence' is then employed to separate the conversion factor for energy into the component factors for force (Z) and for length (X). (As shall be seen in Part 4, the use of the postulated 'Law of Conservation of Existence' yields results which are in agreement with observations and which agree with the results which are obtained when the Principle of Equivalence is correctly applied.) The gravitational conversion factors listed in Table 2.5.1 are evaluated and are tabulated in Table 3.5.1.

Section 3.1 - The Gravitational Potential

3.1.1 - The behavior of the gravitational field between elevations is best described in terms of the gravitational potential, denoted by the symbol Θ, existing between those elevations. It is necessary, therefore, to define the meaning of the gravitational potential as used in this discussion.

3.1.2 - The gravitational potential, Θ, between elevations (reference frames) is a dimensionless quantity which is defined as the energy, as measured with upper elevation units of measurement, released by a material test object in falling from the upper reference frame to the lower reference frame divided by the upper reference frame energy equivalent of the mass of that object ($E=M*C^2$). The upper reference frame is chosen as a basis for the definition to facilitate the treatment of the gravitational field as a whole. The choice permits the conditions which exist at an infinite distance to be employed where convenient to the discussion.

3.1.3 - The definition of gravitational potential allows its value between elevations to be determined in terms of the gravitational parameters which exist at the upper elevation and the distance between the upper and lower elevations. For small values of gravitational potential, second order effects may be ignored, and in terms of upper elevation units of measurement, Newton's Law of Gravitation provides:

Eq. 3.1.1 $F_T = G*M*M_T/R^2$

Where:

- 'G' is the gravitational constant.

- 'M' is the mass of the central attracting object.

- 'M_T' is the mass of the test object.

- 'R' is the distance between the centers of the test object and the central attracting object.

- 'F_T' is the gravitational force on the test object.

3.1.4 - Multiplying both sides of Equation 3.1.1 by the vertical distance, L, as measured with upper elevation units of measurement, and specifying the mass of the test object in terms of energy equivalence provides:

Eq. 3.1.2 $\quad F_T*L/E_T = G*M*L/(R^2*C^2)$

But, F_T*L is the energy of fall of the test object, then:

Eq. 3.1.3 $\quad E_F/E_T = G*M*L/(R^2*C^2)$

And since, by definition, E_F/E_T is equal to Θ then:

Eq. 3.1.4 $\quad \Theta = G*M*L/(R^2*C^2)$

3.1.3 - It should be noted that the preceding equations are approximations which are valid only where the gravitational potential is sufficiently low to permit higher order effects to be ignored. However, since the definition of Θ is based upon the upper reference frame units of measurement for both the energy of fall and for the rest mass equivalent energy of the mass which is falling, Θ is exact for all fields. (It should be noted that the factor 'Θ' is identical to the factor incorporated by General Relativity in its expression for time dilation).

Section 3.2 - The Relationship Between Y, X*Z, and Θ

3.2.1 - In this Section, the definition of the gravitational potential, the law of the Science of Physics which provides the energy of the photon in terms of its locally measured frequency, and the Law of Conservation of Energy as applied to a closed system will be employed to establish the relationship between the factor Y, the product of the factors X and Z and the gravitational potential, Θ. An ideal thought experiment will be described which employs a sufficient number of photons to reduce the uncertainty resulting from quantum effects to insignificance compared to the gravitational potential involved. The diagram for the experiment is provided in Figure 3.2.1.

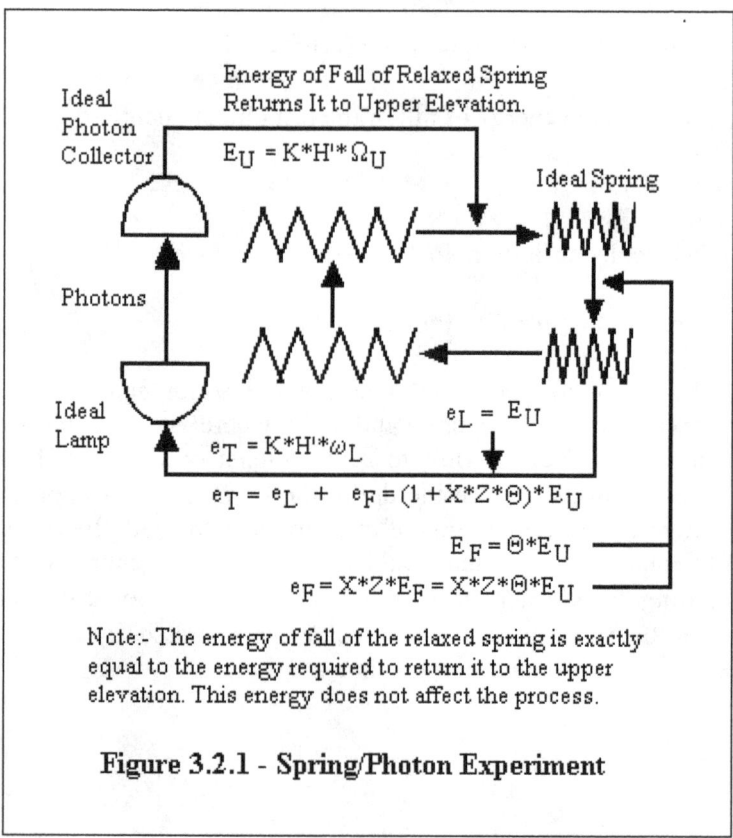

Figure 3.2.1 - Spring/Photon Experiment

3.2.2 - At the upper elevation, a spring is compressed and tied, thereby storing an amount of energy E_U at the upper elevation. Lowering the spring to the lower elevation releases an amount of gravitational energy due to the reduction of elevation of the energy of compression. The amount of energy released in this manner is equal to $E_U*\Theta$ since the energy equivalent of the mass of the stored energy is identically equal to that energy. As measured with the units of measurement of the upper elevation, the energy released by the fall, E_U, is provided by:

Eq. 3.2.1 $E_F = E_U*\Theta$

Converting E_F to the units of measurement of the lower elevation by the use of the conversion factors from Table 2.5.1 provides:

Eq. 3.2.2 $e_F = X*Z*E_F$

And, by combining with the previous expression:

Eq. 3.2.3 $e_F = X*Z*\Theta*E_U$

3.2.3 - At the lower elevation, the spring is untied and its stored energy is released. The release of the compressional energy of the spring provides an amount of energy at the lower elevation, as measured with lower elevation units of measurement, of e_L. Since the energy of compression of the ideal spring obeys the laws of the Science of Physics, the energy released at the lower elevation will be numerically equal to the energy installed in the spring at the upper elevation, E_U. (In place of the energy stored in the spring, the energy could be stored, in principle, by separating a helium atom into four hydrogen atoms. It would then be released at the lower elevation by re-combining the hydrogen atoms into a helium atom. The stored energy would then be equal to the energy equivalent of the mass difference between an atom of helium and the four hydrogen atoms. This mass [energy] difference is a constant of the Science of Physics and is therefore invariant between reference frames. Quantum effects may be ignored in this example if a sufficiently large number of atoms are considered.) Then:

Eq. 3.2.4 $e_L = E_U$

The total energy received at the lower elevation as a result of the lowering of the compressional energy of the spring, as measured with local units of measurement, is equal to the sum of these energies:

Eq. 3.2.5 $e_T = e_L + e_F$

And, by combining Equations 3.2.3, 3.2.4, and 3.2.5:

Eq. 3.2.6 $e_T = (1+X*Z*\Theta)*E_U$

3.2.4 - The released spring is then returned to the upper elevation, the energy of fall of the spring itself is exactly equal to the energy required to raise the relaxed spring to the upper elevation. The spring acts as a 'passive working fluid' and it may be ignored.

3.2.5 - The net energy received at the lower elevation, e_T, is converted to a quantity of photons numerically equal to K, and of locally measured frequency ω_L. It follows from the law of the Science of Physics which provides the energy of a photon in terms of its frequency:

Eq. 3.2.7 $e_T = K*H'*\omega_L$

Where:

- 'H' is the numerical value of Planck's Constant as measured with local units of measurement. As a constant of the Science of Physics, Planck's Constant is invariant between reference frames.

Combining Equations 3.2.6 and 3.2.7 provides:

Eq. 3.2.8 $K*H'*\omega_L = (1+X*Z*\Theta)*E_U$

3.2.6 - At the upper elevation, the frequency of the photons, as measured with local units of measurement, is Ω_U. (A sufficiently large number of photons are assumed so as to allow quantum effects to be ignored.) The energy of these photons, if the Law of Conservation of Energy is be valid over the closed cycle, must be equal to the energy originally installed in the spring:

Eq. 3.2.9 $E_U = K*H'*\Omega_U$

Combining Equations 3.2.8 and 3.2.9 provides:

Eq. 3.2.10 $\omega_L = (1 + X*Z*\Theta)*\Omega_U$

3.2.7 - The symbol ω_L represents the frequency of the photons as measured at the lower elevation with local units of measurement while the symbol Ω_U represents the frequency of the same photons as measured at the upper elevation with local units of measurement. The value of ω_L may differ from the value of Ω_U as a result of either or both of two effects. The change in elevation may cause a change in the frequency of the photons as measured with constant units of time and/or the units of time which are used to measure frequency may change between elevations. To dispose of the first possibility, let us consider another (physically realizable) thought experiment.

3.2.8 - A radio wave may be considered to be a large group of photons of identical frequency and phase traveling together as a group. The effect on frequency experienced by a radio wave as a result of a change in elevation is therefore identical to the effect of such a change in elevation on the frequency of the photons which compose that wave. Unlike individual photons, however, the radio wave may be observed continuously enroute. Each cycle of that wave may, in principle and in practice, be observed as it travels from transmitter to receiver.

3.2.9 - Consider a radio transmitter at the bottom of a vertical shaft. Connected to the transmitter is a counter which counts the number of cycles of the radio wave which have been transmitted. At the top of the shaft is a receiver to which is connected another counter which counts the number of cycles which have been received. Initially, both counters are set to zero. The transmitter is then turned on and its counter counts the number of cycles which have been transmitted. When the radio wave reaches the receiver, the upper counter begins to count the number of cycles which have been received. The transmitter is operated for an extended period of time and is then turned off, stopping its counter. When the last cycle of the transmitted wave is received, the receiver counter also stops. See Figure 3.2.2.

3.2.10 - The maximum simultaneity error possible in this experiment is twice the time for the radio wave to travel between the transmitter and the receiver, and, as a result, the duration of the experiment may be made identical for both elevations, to any desired accuracy, by continuing the experiment for a sufficient period of time. (Since the change in the velocity vector of the shaft in space between the beginning and the end of the experiment is known from astronomical data, simultaneity effects can also be calculated and compensated using Special Relativity.) The frequency of the radio wave at each elevation, as measured with 'constant' units of time, will then be proportional to the counts on each counter. The change in frequency of the photons which make up the radio wave, in terms of 'constant' units of time, is then provided by the ratio of the counter readings.

3.2.11 - In order for the outputs of the counters to differ, cycles of the radio wave would have to be created or destroyed enroute. Such an occurrence would involve the appearance or disappearance of discrete observable entities (cycles) without any means to cause such a creation or destruction, rather than just a change in the size of the units of measurement. As a result,

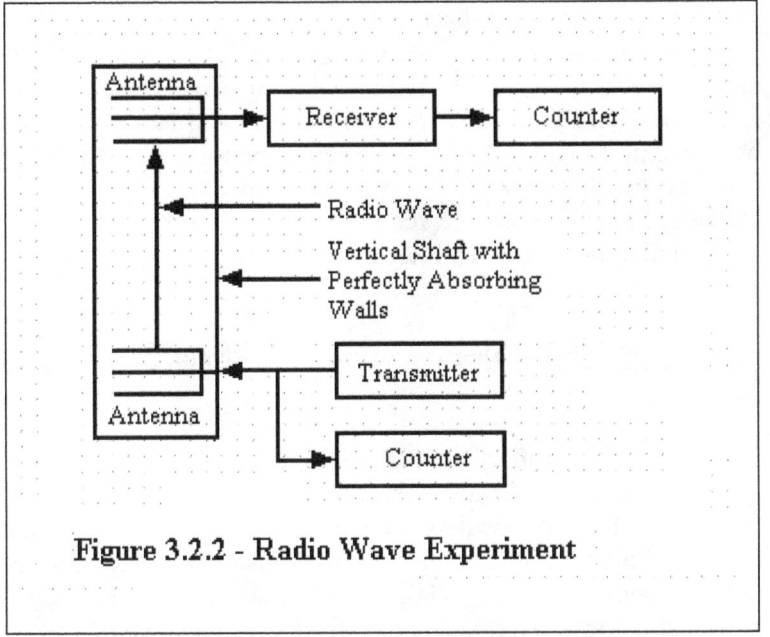

Figure 3.2.2 - Radio Wave Experiment

it must be concluded that the counts on the counter will be identical since any other conclusion is absurd. We may conclude, therefore, that any observed change in the frequency of the photon between elevations results in a change in the size of the units of measurement for time. Since frequency is defined as cycles (dimensionless) per unit time, Equation 3.2.10 may be written as:

Eq. 3.2.11 $\quad t_L = T_U/(1+X*Z*\Theta)$

And from the definition of the factor Y:

Eq. 3.2.12 $\quad Y = 1/(1+X*Z*\Theta)$

Note:- The conclusions to this point are identical to those of General Relativity. General Relativity asserts, through the manner in which it uses the Principle of Equivalence, that no transformation akin to the transformation for time (time dilation) exists for energy. This means that the X*Z product would be unity in Equation 3.2.12. Substituting unity for the X*Z product and substituting the value of Θ provided by Equation 3.1.4 into Equation 3.2.12 yields the expression for time dilation provided by General Relativity. The divergence between the approach taken in this discussion and that of General Relativity begins in the next Section when the requirement that the expression for time dilation be multiplicatively commutative is added.

Section 3.3 - The Evaluation of the X*Z Product and of Y in Terms of Θ

3.3.1 - It is a corollary of the Principle of Relativity that the gravitational transformations represented by the factors X, Y, and Z must be multiplicatively commutative. In other words, the results of observations at elevation #1 made with the units of measurement of elevation #1, converted to the results of obtained when using the measurements of elevation #2 and then converted to the results obtained when using the units of measurement of elevation #3 must be the same as the results which are obtained when the results which are obtained when the results of these observations are converted directly to the results obtained when using the units of measurement of elevation #3. (Colloquially, when going from the first floor to the third floor, it must not make a difference if the elevator stops at the second floor.) It is the fact that the Lorentz Transformations associated with Special Relativity are multiplicatively commutative in combination with effects resulting from the finite velocity of light which prevent the measurement of an absolute velocity through space. The requirement that the time dilation be multiplicatively commutative is not met by General Relativity and from this point on, General Relativity and this discussion diverge. See Figure 3.3.1.

3.3.2 - Consider an ideal thought experiment between three elevations, the upper (U), the middle (M), and the lower (L), which are spaced vertically in a gravitational field. The

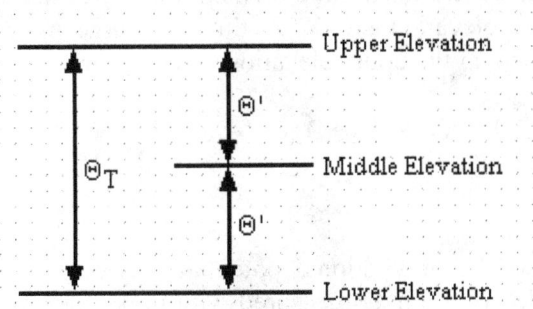

Figure 3.3.1 - Multiplicatively Commutative Field

elevations are chosen so that both the gravitational potential between the upper and middle elevation, as measured with the units of measurement of the upper elevation, and the gravitational potential between the middle and lower elevation, as measured with the units of measurement of the middle elevation are numerically equal to Θ'. The gravitational potential between the upper and lower elevations, as measured with the units of measurement of the upper elevation is designated as Θ_T. (The gravitational potential is assumed to be sufficiently small compared to unity between elevations so that higher order effects may be ignored.

3.3.3 - Since the locally measured gravitational potential represented by the level changes (upper to middle, middle to lower are both equal to Θ', the locally measured (upper to middle, middle to lower) changes in the units of measurement for time and for energy associated with the two level changes are identical and may be denoted as Y' and $X'*Z'$ respectively. (The relationship between these quantities and the gravitational potential is considered to be a law of the Science of Physics and therefore to be invariant between reference frames in accordance with the Principle of Relativity.)

3.3.4 - Since the units of measurement for energy at the energy at the middle elevation are $X'*Z'$ times as large as the units of measurement at the upper elevation, we may write:

Eq. 3.3.1 $\Theta_T = \Theta_{1U} + \Theta_{2U}$

Where:

- 'Θ_{1U}' is the gravitational potential between the upper and middle elevation as measured with upper elevation units of measurement.

- 'Θ_{2U}' is the gravitational potential between the middle and the lower elevation as measured with upper elevation units of measurement.

But:

Eq. 3.3.2 $\Theta_{1U} = \Theta'$

Eq. 3.3.3 $\Theta_{2U} = \Theta'/(X'*Z')$

Then:

Eq. 3.3.4 $\Theta_T = (1+1/[X'*Z'])*\Theta'$

3.3.5 - In order for the conversion factors for time and for energy to be multiplicatively commutative, it is necessary that:

Eq. 3.3.5 $Y_T = Y'^2$

Eq. 3.3.6 $X_T*Y_T = (X'*Z')^2$

From Equation 3.2.12:

Eq. 3.3.7 $Y = 1/(1+X*Z*\Theta)$

Then:

Eq. 3.3.8 $\quad Y_T = 1/(1+X'*Z'*\Theta)^2$

And:

Eq. 3.3.9 $\quad Y_T = 1/(1+X_T*Z_T*\Theta_T)$

Combining Equations 3.3.4, 3.3.6, 3.3.8, and 3.3.9 provides:

Eq. 3.3.10 $\quad X'*Z' = 1/(1-\Theta')$

Or, for a single elevation change:

Eq. 3.3.11 $\quad X*Z = 1/(1-\Theta)$

Combining Equations 3.3.7 and 3.3.11 provides:

Eq. 3.3.12 $\quad Y = (1-\Theta)$

3.3.6 - It will be noted that the requirement for multiplicative commutivity requires an energy dilation which is the inverse of the time dilation associated with the gravitational field. It will be further noted that the time dilation differs from the time dilation of General Relativity by Θ^2, an amount which is currently undetectable. More significantly, the energy dilation differs from that provided by General Relativity (i.e.- unity) by Θ, an amount which is highly detectable, and will be the subject of further discussion in Section 4.7.

Section 3.4 - The Evaluation of X and Z in Terms of Θ

3.4.1 - In the preceding Sections, the value of Y and the combined factor X*Z have been evaluated in terms of the gravitational potential, Θ. This determination is basic and requires only that the Principle of Relativity and locally testable laws of the Science of Physics be valid. It is now necessary to separately determine the factors X and Z. With the correct form for the transformations known it would be possible to apply the results of astronomical observation and the proof of the Euclidean nature of space in the gravitational field, which will be provided in Section 4.2, to make the necessary evaluation. It is more satisfying however, to perform the evaluation from basic principles.

3.4.2 - Separation of the factor X*Z into its components without relying upon the Principle of Equivalence, requires the application of a separate conservation law. This law must be independent of the Law of Conservation of Energy since that law has already been used to separate the energy and non-energy effects of the gravitational field. Of the commonly accepted macroscopic conservation laws, there remains only the Law of Conservation of Momentum and the Law of Conservation of Angular Momentum. (The Law of Conservation of Mass is not suitable since it is another form of the Law of Conservation of Energy.) The Law of Conservation of Angular Momentum is unsuitable. Its conversion factor is X*Y*Z and, since the X*Z product is the reciprocal of Y, the Law of Conservation of Angular Momentum is satisfied for all possible values of X and Z. It might be hoped that the Law of Conservation of Momentum would be useful, but it yields indeterminate results since it may be derived from the Law of Conservation of Energy. Since momentum is conserved as a vector quantity and not as a magnitude, it is not required to be constant between reference frames. (The magnitude of the momentum present in a system is not conserved, only its vector sum, and, as a result, it is not a true conservation law. Colloquially, one cannot satisfy hunger by creating a positive apple and a negative apple, discarding the negative apple, and eating the positive apple.)

3.4.3 - In Section 2.3, it was suggested that conditions of symmetry required that an entity having the dimensional content of a length-time product (L*T) should also obey a conservation law. To facilitate the discussion, the proposed entity has be designated as 'existence'. It is not a vector quantity and should therefore, like energy, be conserved in the absolute rather than the vector sense. (It should be noted that 'existence' is conserved under under Special Relativity - 1999 comment.) Unlike energy, 'existence' is not subject to partition as a result of a change in elevation since one of its components is known. (The conservation of energy in the gravitational field is discussed further in Section 4.7.)

3.4.4 - On might consider, for example, that an unstable atomic particle, such as a neutron, possesses 'existence' which might be defined as the product of its diameter and its half-life. The 'existence' of the neutron would be a constant of the Science of Physics and would therefore be invariant between reference frame. (As previously, it is assumed that a sufficient number of particles are involved to permit quantum effects to be ignored.) If the 'existence' of the neutron is to be conserved between the reference frames which represent the different elevations, then its 'existence' must also be constant between those reference frames. For 'existence' to be both constant and invariant in the gravitational field, the gravitational conversion factor for 'existence', X*Y, must be equal to unity. Therefore:

Eq 3.4.1 $X*Y = 1$

Combining Equation 3.3.12 and 3.4.1 provides:

Eq. 3.4.2 $X = 1/(1-\Theta)$

And combining Equations 3.3.11 and 3.4.2 provides:

Eq.3.4.3 $Z=1$

3.4.5 - Confidence in the proposed 'Law of Conservation of Existence' is provided by examining the 'Law' as applied to reference frames having a relative velocity. In the direction of the relative velocity vector, the equivalent conversion factors are the Lorentz Transformation for Length and for time associated with the Special Theory of Relativity are:

Eq. 3.4.4 $X_V = 1/(1-V^2/C^2)^{0.5}$

Eq. 3.4.5 $Y_V = (1-V^2/C^2)^{0.5}$

Where:

- 'X_V' is the Lorentz Transformation for Length.

- 'Y_V' is the Lorentz Transformation for Time.

- 'V' is the relative velocity.

- 'C' is the velocity of light.

The Lorentz Transformation for 'existence' (in the direction of relative motion) is the product of the Lorentz Transformations for length and for time and is equal to unity. The proposed "Law of Conservation of Existence is therefore valid for reference frames having relative velocity. If the Principle of Equivalence is to be valid, the 'Law of Conservation of Existence' must also be valid for reference frames which differ in elevation. Final verification will occur when it is shown in Part 4 that the values of X, Y, and Z which result are in agreement with the results of observation.

Note:- Some readers may experience misgivings at this point. It will be noted that the reciprocal identity between the factors X and Y, and the Lorentz Transformations for length and time mean that the conversion factor for velocity, X/Y, cannot be equal to unity unless both X and Y are also equal to unity. (This is certainly not the case.) This means that the velocity of light cannot be both constant and invariant between reference frames having relative velocity. The velocity of light is invariant between these reference frames. Special Relativity then must be then be telling us that it is *not* constant between moving reference frames. The assumption that the velocity of light is both *constant* and *invariant* between elevations in the gravitational field is exactly that, an unproven assumption, which if early writings are to be believed, was made because "we have to keep something constant".

Section 3.5 - The Evaluated Gravitational Conversion Factors

3.5.1 - Table 3.5.1 provides the evaluated gravitational conversion factors for the quantities tabulated in Table 2.5.1.

Table 3.5.1 - Evaluated Conversion Factors for Physical Quantities

Quantity	Conversion Factor
Length	$l = L/(1-\Theta)$
Time	$t = T*(1-\Theta)$
Force	$f = F$
Charge	$q = Q/(1-\Theta)$
Energy	$e = E/(1-\Theta)$
Momentum	$u = U*(1-\Theta)$
Angular Momentum	$j = J$
Planck's Constant	$h = H$
Velocity	$v = V/(1-\Theta)^2$
Acceleration	$a = A/(1-\Theta)^3$
Mass	$m = M*(1-\Theta)^3$
Gravitational Constant	$g = G/(1-\Theta)^8$
Dielectric Constant of Space	$\varepsilon = E'$
Permeability of Space	$\mu = M'*(1-\Theta)^4$
Existence	$b = B$
Ergo-Gravitational Constant	$d = D$

Part 4 - Comparison with the 'Real World'

List of Topics

Sections	Section Titles	Page
4.0	Introduction	284
4.1	The Observational Verification of Y	286
4.2	The Euclidian Nature of Space in the Gravitational Field	287
4.3	The Gravitational Acceleration of the Photon	297
4.4	The Observational Verification of X by the Bending of the Path of Starlight	303
4.5	The Gravitational Effects on the Velocity Vector and the Precession of Mercury's Orbit	304
4.6	Gravitational Acceleration of the Confined Photon and the Equivalence of Inertial and Gravitational Mass	309
4.7	Conservation of Energy in the Gravitational Field	315
4.8	Conservation of Momentum in the Gravitation Field	325
4.9	Gravitation and the Principle of Equivalence	330
4.10	A Speculation on the Gravitation of and Nature of Rest Mass	346

Section 4.0 - Introduction

4.0.1 - To be considered valid, a physical theory must be both internally and externally consistent. It must agree internally with itself and it must be consistent with *all* external realities. The discussion up to this point meets the first requirement, it is internally consistent. The purpose of this Part is to show that the conclusions presented are consistent with the results of observations made in the external universe, to highlight the points where General Relativity is in conflict with the reality represented by that universe, and to show where errors in General Relativity arise.

4.0.2 - Part 4 opens with a demonstration that the time dilation factor predicted in this discussion and the same prediction by General Relativity differ in a second order term which is too small to be detected by present day technology within the confines of the Solar System. The observed time dilation is thus consistent with both concepts. It is next shown that space, in the presence of the gravitational field, must be Euclidian if the Laws of Conservation of Energy and of Momentum are to be valid for a closed system. It follows, therefore, that any observed bending of the path of a ray of light or precession of orbits due to the gravitational field must result from conventional refraction rather than from a 'curvature of space'.

4.0.3 - It is then shown that unconfined photons experience a gravitational acceleration equal to twice that experienced by a material particle and that this acceleration is the correct acceleration to produce the observed refraction of light by the gravitational field. It is also shown that the velocity vector of a moving object is refracted by the gravitational field to the same degree as is the path of a ray of light and that the degree of refraction is the amount necessary to cause the observed anomalous precession of Mercury's orbit.

4.0.4 - It is next demonstrated that, while both free electromagnetic and free kinetic energy experience a gravitational acceleration twice that which is experienced by material particles, when such energy is confined by matter, lowering the elevation causes the confined energy to do work on the confining matter equal in magnitude to that of conventional gravitational acceleration but of reversed direction. The net effect is to cause confined electromagnetic and/or kinetic energy, in combination with the matter which confines it, to experience a net gravitational acceleration equal to that experienced by rest mass equivalent energy. As a result, all matter gravitates equally, regardless of the fraction of its total inertial mass that results from electromagnetic or kinetic energy associated with matter.

4.0.5 - The primary deficiency in both Einsteinian and Newtonian gravitational theory is their inability to deal with gravitational energy. Under both of these concepts, the gravitational field is ultimately capable of *creating* an infinite amount of energy, in flagrant violation of the Law of Conservation of Energy. It is shown that such a creation of energy does not occur, but rather, that energy which is released in falling is provided by a reduction in the energy content of the falling object due to the combination of the Principle of Relativity and the reduction in the size of the unit of measurement for length. Energy is conserved absolutely in the gravitational field, as is the momentum-velocity of light product.

4.0.6 - It is disturbing that the combination of the Principle of Relativity and the Principle of Equivalence, as applied in the derivation of General Relativity, did not yield correct results for the gravitational field. Accordingly, the Principle of Equivalence is examined, and it is shown that the incorrect results are caused by an improper application of the concept. When the Principle of Equivalence is renormalized to take into account the work done on the 'accelerating' reference frame, it, in combination with the Principle of Relativity, yields results which are identical to those derived independently in this discussion.

Section 4.1 - The Observational Verification of Y

4.1.1 - The factor Y, is, by definition, the 'time dilation' in the gravitational field. General Relativity predicted that that 'time dilation' would occur and provided an expression, as reported in reliable texts, for the 'time dilation' as a function of the field:

Eq. 4.1.1 $Y_{GR} = 1/(1+G*M*L/[R*C^2])$

Where:

- 'Y_{GR}' is the gravitational transformation for time (time dilation) under General Relativity.

- 'G' is the gravitational constant.

- 'M' is the mass of the attracting body.

- 'R' is the distance to the center of the attracting body.

- 'C' is the velocity of light.

- 'L' is the difference in elevation.

4.1.2 - Combining Equations 3.1.4 and 4.1.1 provides the time dilation predicted by General Relativity in terms of the gravitational potential as defined by this discussion, Θ, between elevations:

Eq. 4.1.2 $Y_{GR} = 1/(1+\Theta)$

Subtracting Equation 3.3.12 from Equation 4.1.2 provides, for the difference, δY, between these predicted time dilations:

Eq. 4.1.3 $\delta Y = \Theta^2/(1+\Theta)$

Or, for small values of Θ:

Eq. 4.1.4 $\delta Y = \Theta^2$

4.1.3 - The gravitational potential, Θ, from an infinite distance to the surface of the Earth is on the order of 10^{-9} and to the surface of the Sun is on the order of 10^{-6}. As a result, a discrepancy on the order of Θ^2 is undetectable in our Solar system at the present or foreseeable state of the art. It may be considered, therefore, that the observations which are cited as a verification of the time dilation predicted by General Relativity also verify the value of Y provided by Equation 3.3.12.

Section 4.2 - The Euclidian Nature of Space in the Gravitational Field

4.2.1 - In order to use the results of astronomical observation to verify the values of X and Z provided in Equations 3.4.2 and 3.4.3. it is necessary to determine what portion of any observed 'refraction' of light by the field results from true refraction caused by a reduction of its velocity and what portion results from the curvature of a possible non-Euclidian space resulting from the field, as asserted by General Relativity. True refraction in Euclidian space requires that the spatial acceleration (the second derivative of position with respect to time) of the light be accompanied by an equivalent inertial force acting on the mass equivalent of its energy and which lies within the Euclidian space and is therefore is observable. An apparent refraction which results from the curvature of a non-Euclidian space will also produce an inertia force. This force, however, will be unobservable since it will occur along an axis which is orthogonal to the observable spatial axes. To observers such as ourselves, gravitational refraction of light which is produced by a curved non-Euclidian space will not be accompanied by an inertia force. (It should be noted that the observed existence of radiation pressure verifies that the energy represented by light possesses inertial mass in accordance with $E=M*C^2$.)

4.2.2 - An illustration of the above in terms which permit visualization may be helpful at this point. Let us consider an ideal, non-rotating perfectly spherical planet having a frictionless surface and located at a sufficiently remote location so as to be free of disturbing influences. The surface of such a planet is gravitationally equipotential and an ideal frictionless spherical ball will either remain stationary at any point, or if it is set in motion, it will roll along a great circle path of the planet indefinitely. Infinitesimally inscribed on the planet are two great circles which cross at right angles and represents the axes of a spherical coordinate system. We can consider that the surface of the planet represents a two dimensional non-Euclidian geometry which is embedded in a three dimensional Euclidian geometry. (A non-Euclidian geometry of N dimensions may be considered to be a subset of a Euclidian geometry of N+1 dimensions.)

4.2.3 - Let us move along one of the two great circles, #1, for a fraction of its circumference, perhaps 30 degrees from the point of intersection with the other great circle, #2, and start the ball rolling at right angles to great circle #1 and note that initially it is moving parallel to great circle #2. It will be observed that, in terms of the two dimensional non-Euclidian geometry represented by the surface of the plane, the path of the ball undergoes a spatial acceleration (second derivative of position with respect to time) towards great circle #2 and eventually crosses it. This observed spatial acceleration is not accompanied by an equivalent inertial force in the two dimensional non-Euclidian space as evidenced by the fact that if one stops the ball at any point along its path, it will remain stationary. An inertial force does occur as the ball moves along the great circle, but this force is at all points normal to the two dimensional non-Euclidian geometry representing the surface of the planet and is therefore undetectable within the 'space' represented by that geometry. A two dimensional observer would conclude that the ball was following the 'null geodesic' which was the 'straight line' of his non-Euclidian universe.

4.2.4 - In terms of our familiar three dimensional space, if we observe that light is 'refracted' by a gravitational field, that refraction will be accompanied by an inertial force to the degree that it results from conventional refraction. The inertial force will be absent to the degree that the observed 'refraction' results from the curvature of a non-Euclidian three dimensional space. (Remember, light possesses inertial mass as evidenced by its ability to exert radiation pressure.) Should all of the observed 'refraction' be consistent with the observed inertial force, it must be concluded that space, in the presence of the gravitational field, is represented by three dimensional Euclidian geometry.

4.2.5 - Consider an ideal thought experiment within a closed system in which two perfect and lossless retroreflectors are located at the ends of ideal, rigid and massless booms of length L which are attached to opposite sides of an ideal non-rotating planet located far from any disturbing influences, as shown in Figure 4.2.1. A beam of photons is assumed to be traveling

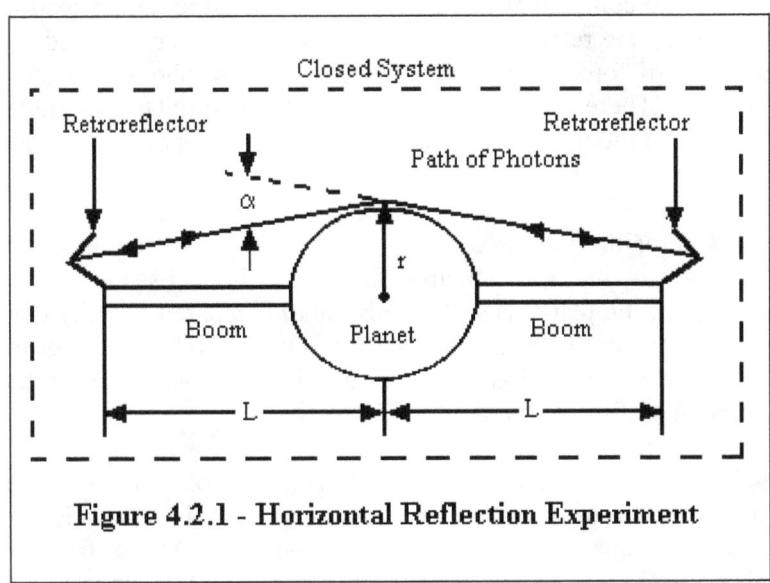

Figure 4.2.1 - Horizontal Reflection Experiment

endlessly back and forth between the retroreflectors along a path which passes close to the surface of the planet at a distance 'r' from its center. (It is assumed that the planet does not possess an atmosphere.) As the beam of photons passes through the planet's gravitational field, it is refracted through an angle 'α' by the field as observed by an external observer. In the diagram, the angle 'α' may be positive, negative, or zero. In accordance with the Law of Conservation of Momentum, the momentum of this closed system must remain constant. If this were not true, it would be possible, in principle to attach a drawbar to the system and use it as a space tug. Such an implementation would be capable of providing energy continuously to the external universe without depletion of its internal resources, a clear absurdity.

4.2.6 - In the above diagram, symmetry requires that the horizontal forces at the retroflectors will balance, and momentum is conserved along that axis. For momentum to be conserved in the vertical direction, it is necessary that the downward component of the reaction forces at the retroreflectior caused by the refraction of the beam of photons be matched by an upward force on the planet exerted by the photons as they pass by. There must, in other words, be a gravitational force exerted on the photons by the planet which is proportional to the angle 'α'.

4.2.7 - For the purposes of this discussion, it will be assumed that the gravitational potential from an infinite distance to the surface of the planet is sufficiently small compared to unity that second order effects may be ignored. It is valid, therefore, to eliminate considerations of upper and lower elevation from the remainder of the discussion in this Section.

4.2.8 - Let us consider that N photons per second of energy E_P strike each retroreflector. (It is assumed that N is sufficiently large that quantum effects may be ignored.) As verified by observations of radiation pressure, each photon imparts an impulse to the surface which it strikes that is equal to its mass equivalent energy times the change in its velocity vector. When the photon is absorbed, its change in velocity is equal to the velocity of light. When it is reflected, its change in velocity is equal to twice the velocity of light times the sine of the angle of incidence. Since, for the case of the retroreflectors, the angle of incidence is effectively equal to 90 degrees, the force, F, on the retroreflector, which is equal to the impulse imported per second from the beam of photons, is given by:

Eq. 4.2.1 $\quad F = 2*N*E_P/C$

Where:

- 'N' is the number of photons striking the retroreflector per second.

- 'E_P' is the energy of the photons striking the retroreflector.

- 'C' is the velocity of light.

The vertical force, F_V', at each retroreflector is therefore:

Eq. 4.2.2 $\quad F_V' = 2*N*E_P*\sin(\alpha/2)/C$

Or, since α is a small angle:

Eq. 4.2.3 $\quad F_V' = N*E_P*\alpha/C$

And for the two retroreflectors:

Eq. 4.2.4 $\quad F_V = 2*N*E_P*\alpha/C$

But, since the beam consists of both arriving and departing photons, the energy per unit length of the beam of photons, E_L, is given by:

Eq. 4.2.5 $\quad E_L = 2*N*E_P/C$

Then:

Eq. 4.2.6 $F_V = E_L * \alpha$

4.2.9 - Now let us consider the bending of the beam of photons by the gravitational field. If the beam of photons is subject to a spatial acceleration of a_N normal to its path, then in an incremental distance, δL, it will change in direction by an incremental angle $\delta\alpha$. To determine the effect, let us consider a particle traveling horizontally at a velocity V for an incremental time δT and subjected to a vertical acceleration a_N, as shown in Figure 4.2.2. As a result of the acceleration, the particle will move in a circular path of radius R, as shown in Figure 4.2.3. In time δT, the particle will travel a distance δL and its path will curve through an angle $\delta\alpha$:

Eq. 4.2.7 $\delta\alpha = \delta L/R$

The acceleration experienced by the particle is, in accordance with the laws of mechanics:

Eq. 4.2.8 $a_N = V^2/R$

And we may write:

Eq. 4.2.9 $\delta\alpha = a_N * \delta L/V^2$

And, since the particle of interest is the photon whose velocity is equal to the velocity of light:

Eq. 4.2.10 $\delta\alpha = a_N * \delta L/C^2$

4.2.10 - The angle through which the beam of photons is deflected as it passes the planet is extremely small and, in determining the normal acceleration experienced by the photon due to the planet's gravitational field, we may ignore the bend,

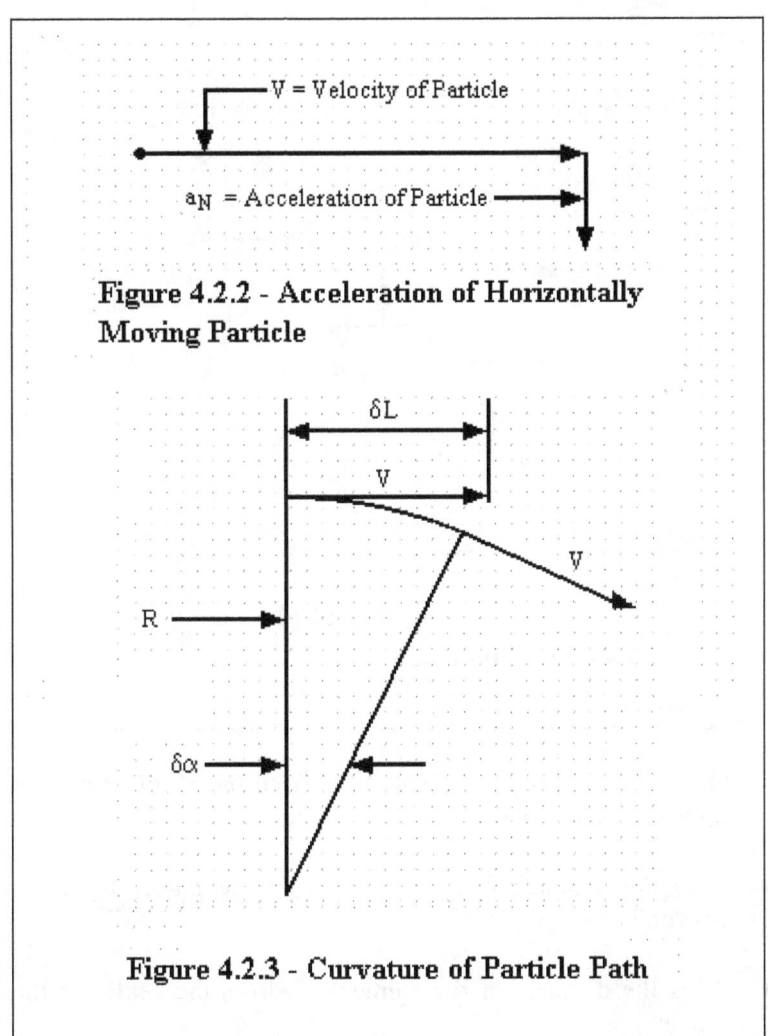

Figure 4.2.2 - Acceleration of Horizontally Moving Particle

Figure 4.2.3 - Curvature of Particle Path

since the effects of the bend are proportional to the cosine of a small angel. Figure 4.2.4 then applies.

Where:

- 'δL' is an incremental length of the photon beam.

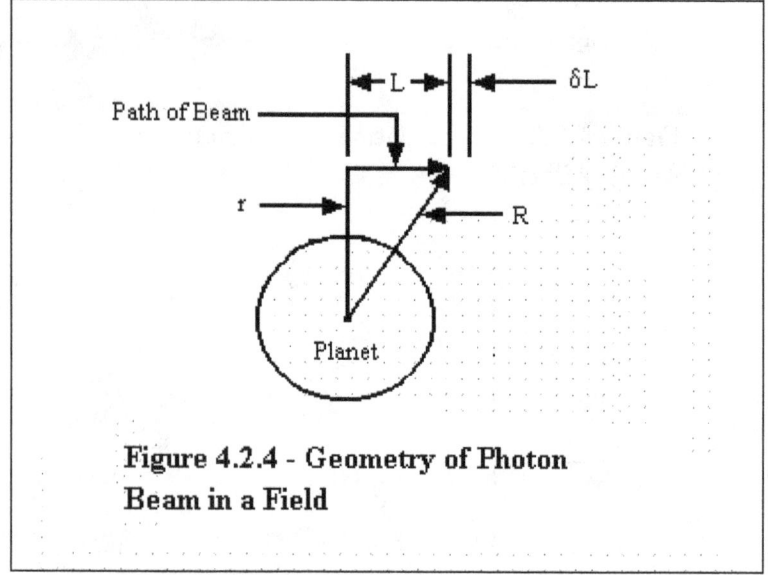

Figure 4.2.4 - Geometry of Photon Beam in a Field

- 'L' is the distance of element δL from the point of closest approach to the planet.

- 'r' is the distance form the center of the planet from the photon beam.

- 'R' is the distance of the element δL from the center of the planet.

If we define a_N' as the acceleration of the element δL at the point of closest approach, then, allowing that the local acceleration is proportional to the local gravitational acceleration as required by the Principle of Relativity, at any point along the beam the normal acceleration, a_N, is given by:

Eq. 4.2.11 $a_N = a_N' * r^3 / R^3$

But:

Eq. 4.2.12 $R = (r^2+L^2)^{0.5}$

Then:

Eq. 4.2.13 $a_N = a_N'*r^3/(r^2+R^2)^{1.5}$

And:

Eq. 4.2.14 $\delta\alpha = a_N'*r^3*\delta L/(C^2*[r^2+R^2]^{1.5})$

Integrating between the limits of +/-L provides:

Eq. 4.2.15 $\alpha = 2*a_N'*r*L/(C^2*[r^2+R^2]^{0.5})$

But, if L is large compared to r:

Eq. 4.2.16 $\alpha = 2*a_N'*r/C^2$

Then combining Equations 4.2.6 and 4.2.16:

Eq. 4.2.17 $F_V = 2*a_N'*r*E_L/C^2$

4.2.11 - Now consider a beam of photons as a gravitating mass having an inertial mass per unit length of E_L/C^2. The incremental force, δF_{NG}, in the vertical direction of Figure 4.2.4 is:

Eq. 4.2.18 $\delta F_{NG} = K_P*a_{GN}*E_L*r^3*\delta L/(C^2*R^3)$

Where:

- 'a_{GN}' is the acceleration of gravity at the point of closest approach of the beam to the planet.

- 'K_P' is the ratio of the gravitational force acting on the photon to the gravitational force acting on the material particle of the same mass equivalent energy. (Note: This does not imply that a photon does or does not possess gravitational mass since K_P may be positive, negative, or zero.)

Combining Equations 4.2.12 and 4.2.18, integrating between the limits of +/-L, and allowing L to be large compared to r provides:

Eq. 4.2.19 $F_{NG} = 2*K_P*a_{GN}*E_L*r/C^2$

4.2.12 - If the Law of Conservation of Momentum is to be valid for the closed system represented by Figure 4.2.1, then:

Eq. 4.2.20 $F_V = F_{NG}$

Combining Equations 4.2.17, 4.2.19, and 4.2.20 provides:

Eq. 4.2.21 $a_N' = K_P*a_{GN}$

It should be noted that, since the quantity to be integrated in Equation 4.2.14 is identical to the quantity to be integrated in the combined Equations 4.2.12 and 4.2.18, the approximation which occurs as a result of ignoring the effect of the changing gravitational potential along the beam cancels. The only approximation remaining in Equation 4.2.21 is the small angle approximation for 'α'.

4.2.13 - Consider the significance of the preceding equation. The term a_N' represents the normal acceleration, where acceleration is defined as the second derivative of position with respect to time, required to cause the beam of photons to bend in terms of three dimensional Euclidian space. The term K_P*a_{GN} represents the normal acceleration, where acceleration is defined as the applied force divided by the mass equivalence of the energy of the beam, required to produce a force which is equal to the net force applied to the retroreflectors. The equality of the preceding equation means that, to the first order at least, any and all gravitationally induced refraction observed for light results from gravitational forces which occur in our familiar three dimensional space. There is no first order effect produced by a component of force acting along an axis normal to our observable three spatial axes. It is concluded, therefore, that *space, in the presence of the gravitational field, must be three dimensional Euclidian.* This conclusion is basic and stands alone, it is independent of the values of X, Y, and Z found previously and of the arguments presented previously. ***The curved space of General Relativity cannot be a valid representation of reality.***

Section 4.3 - The Gravitational Acceleration of the Photon

4.3.1 - In this Section, the Law of Conservation of Momentum as applied to a closed system will be employed, this time to evaluate the gravitational acceleration of the photon. Let us again consider a closed system consisting of the ideal planet and a pair of ideal retroreflectors which endlessly reflect a beam of photons. This time, the retroreflectors will be mounted along a vertical axis with one retroreflector at the planet's surface and one retroreflector mounted directly above it. For this thought experiment, it will be assumed that the support for the upper retroreflector is made of an ideal material whose length, in terms of 'constant' units of measurement, is unaffected by the effects of the gravitational field. (This assumption is necessary to prevent the radiation pressure contained between the retroreflectors from acting on the support so as to produce a spurious force into the thought experiment, as will be described in a later Section.)

4.3.2 - In the thought experiment of Figure 4.3.1, a beam of photons is continuously reflected between the upper and lower retroreflectors without loss. (A sufficient number of photons is assumed so that quantum effects may be ignored.) At each reflection, the photons impart momentum to the retroreflectors. In order for the Law of Conservation of Momentum to be satisfied for the closed system, the algebraic sum of the momentum applied to the upper retrorelector and the lower retroreflector, minus the momentum applied to the photon stream by the gravitational field must be equal to zero, as measured with 'constant' units of measurement. Since the momentum imparted is the integral of the force applied and the time for which it is applied, and since we are interested in the momentum imparted per unit time, we need only consider the force.

4.3.3 - Equation 4.2.1 provides, for the force produced by a beam of photons on a retroreflector:

Eq. 4.3.1 $F = 2*N*E_P/C$

Which we shall rewrite for the upper retroreflector as:

Eq. 4.3.2 $F_U = 2*N*E_P/C_U$

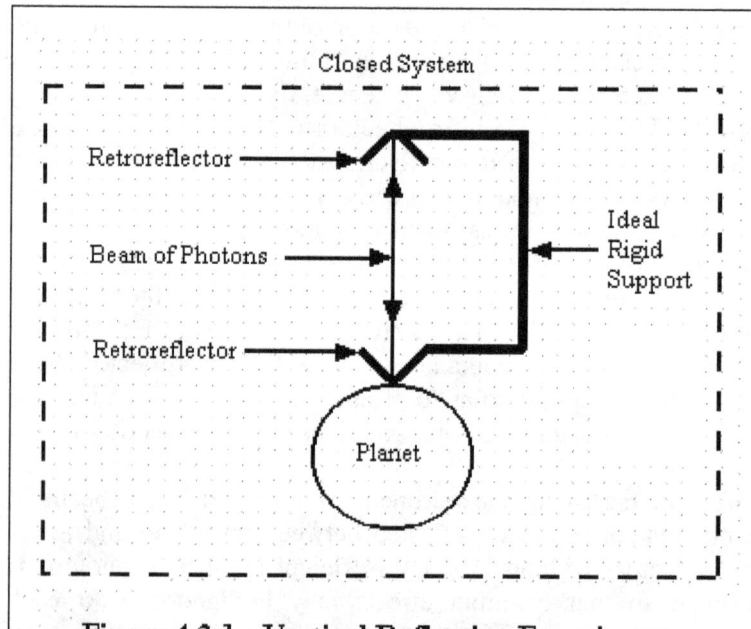

Figure 4.3.1 - Vertical Reflection Experiment

And, for the lower retroreflector, as:

Eq. 4.3.3 $F_L = 2*N*E_P/C_L$

Where:

- 'N' is the number of photons striking the upper retroreflector in one second of time as measured in upper elevation units of measurement and also is the number of photons striking the lower retroreflector in one second of time as measured with upper elevation units of measurement since photons are presumed not to be gained or lost as a result of the change in elevation.

- 'E_P' is the energy of each photon striking the upper retroreflector as measured with upper elevation units of measurement and is also the energy of each photon striking the lower retroreflector as measured with upper elevation units of measurement. It was shown in Section 3.2 that the frequency of the photon does not change between elevations as measured in 'constant' units of time measurement. and Table 3.5.1 shows that Planck's Constant is both constant and invariant between elevations. The energy of the photon, which is Planck's Constant times its frequency, is therefore unchanged between elevations as measured with 'constant' units of measurement for energy.

- 'C_U' is the velocity of light at the upper elevation as measured with upper elevation units of measurement.

- 'C_L' is the velocity of light at the lower elevation as measured with upper elevation units of measurement.

The net force, δF in terms of upper elevation units of measurement, that the beam of photons applies to the retrorelector structure is the difference between the forces at the upper and lower retroreflectors. Subtracting Equation 4.3.2 from Equation 4.3.3 provides:

Eq. 4.3.4 $\delta F = 2*N*E_P*(1/C_L - 1/C_U)$

But:

Eq. 4.3.5 $c_L = C_U = X*C_L/Y$

Then:

Eq. 4.3.6 $\delta F = 2*N*E_P*(X-Y)/(C_U*Y)$

4.3.4 - As shown in Section 4.2, the energy per unit length of the photon beam striking the upper retroreflector, E_{LU}, in terms of upper elevation units of measurement is given by:

Eq. 4.3.7 $E_{LU} = 2*N*E_P/C_U$

And, the energy per unit length of the photon beam striking the lower retroreflector, E_{LL}, in terms of upper elevation units of measurement is given by:

Eq. 4.3.8 $E_{LL} = 2*N*E_P/C_L$

The average energy of the beam, per unit length, in terms of upper elevation units of measurement, is given by:

Eq. 4.3.9 $E_L = N*E_P*(1/C_L + 1/C_U)$

And, from Equation 4.3.5:

Eq. 4.3 10 $E_L = N*E_P*(X+Y)/(C_U*Y)$

The total energy of the beam, E_T, is its average energy times the distance between the mirrors:

Eq. 4.3.11 $E_T = N*E_P*L*(X+Y)/(C_U*Y)$

The apparent 'energy of fall', E_F, of the photons is equal to the difference in force between the upper and lower retroreflectors (Equation 4.3.6) times the distance, L, between them:

Eq. 4.3.12 $E_F = 2*N*E_P*L*(X-Y)/(C_U*Y)$

Dividing Equation 4.3.12 by Equation 4.3.11 provides the photon's equivalent of the gravitational potential, Θ_P, as:

Eq. 4.3.13 $\Theta_P = 2*(X-Y)/(X+Y)$

For the case where the gravitational potential is small compared to unity:

Eq. 4.3.14 $X+Y = 2$

Then:

Eq. 4.3.15 $\Theta_P = X-Y$

Applying the values of X and Y from Table 3.5.1 provides:

Eq. 4.3.16 $\Theta_P = 1/(1-\Theta)-(1-\Theta)$

Which, for values of Θ which are small compared to unity becomes:

Eq. 4.3.17 $\Theta_P = 2*\Theta$

In this derivation, Equation 4.3.17 is rigorously valid only when Θ is infinitesimal. However, it seems reasonable to consider (and is probably required by the Principle of Relativity) that the relationship between Θ_P and Θ will be fixed for all values of Θ. With this assumption, Equation 4.3.17 becomes exact.

4.3.5 - The gravitational potential, Θ, has been defined as the energy released by a change in elevation divided by the energy equivalent of the mass whose elevation has changed. This energy release is the force produced by gravity times the change in elevation. We may write, therefore:

Eq. 4.3.18 $F_{PG}*L = 2*F_P*L$

Or:

Eq. 4.3.19 $F_{PG} = 2*F_P$

Where:

- 'F_{PG}' is the force applied by gravity to the free photon of energy E_P.

- 'F_P' is the force applied by gravity to a material particle of energy E_P.

Since both the material particle and the photon are assumed to have the same energy, they also have the same inertial mass, therefore:

Eq. 4.3.20 $a_{PG} = 2*a_G$

Where:

- 'a_{PG}' is the gravitational acceleration of the free photon.

- 'a_G' is the gravitational acceleration of a free particle.

Section 4.4 - The Observational Verification of X by the Bending of the Path of Starlight

4.4.1 - Eq. 4.3.20 shows that the gravitational acceleration of the free photon in the gravitational field is twice the gravitational acceleration of a material particle. Recognizing that the quantity a_N' of Equation 4.2.21, when applied to the photon, is the quantity a_{PG} of Equation 4.3.20, we find that by combining Equations 4.2.16 and 4.3.20 one obtains:

Eq. 4.4.1 $\quad \alpha = 4*a_G*r/C^2$

Where:

- $'a_G'$ is the acceleration of gravity at the radius of closest approach of a ray of light to the gravitating body.

- $'r'$ is the distance between the ray of light and the gravitating body at the point of closest approach.

- $'C'$ is the velocity of light.

- $'\alpha'$ is the angle, in radians, through which the ray of light is refracted by the gravitational field.

Or, in terms of seconds of arc:

Eq. 4.4.2 $\quad \beta = 8.251*a_G*r*10^5/C^2$

4.4.2 - The acceleration of gravity at a radius, r, from the Sun's center is readily determined from the parameters of the Earth's orbit and the laws of mechanics as:

Eq. 4.4.3 $\quad a_G = 4*\pi^2*r_E^3/(T_Y^2*r^2)$

Where:

- 'r_E' is the radius of the Earth's orbit, assumed as $1.495*10^8$ kilometers.

- 'T_Y' is the number of seconds as a year, assumed as $3.156*10^7$ seconds.

- 'r' is the radius of the Sun, assumed as $6.97*10^5$ kilometers.

Combining Equations 4.4.2 and 4.4.3 provides:

Eq. 4.4.4 $\beta = 1.743$ arc-seconds

4.4.3 - Equation 4.4.4 is in excellent agreement with the reported value of 1.75 arc-seconds for the deflection of a stars apparent position as it occults the Sun. The value of X provided by Equation 3.4.2 is therefore verified by observation since it was shown that space in the presence of the gravitational field is Euclidian thus requiring that the bending of starlight result from conventional refraction alone. As was the case with time dilation (factor Y) described in Section 4.1, observations made within the Solar system are currently incapable of distinguishing higher order effects in Θ.

Section 4.5 - The Gravitational Effects on the Velocity Vector and the Precession of Mercury's Orbit

4.5.1 - In this Section, it well be shown that the gravitational field refracts the velocity vector of a moving object to the same degree that it refracts the path of a ray of light. It will then be shown that this refraction of the velocity vector by the gravitational field of the Sun is the amount required to account for the observed precession of Mercury's orbit. The effect to be considered is independent of the effect of gravitational acceleration on the object.

4.5.2 - Consider the following thought experiment, as shown in Figure 4.5.1. A test mass is suspended from an ideal massless

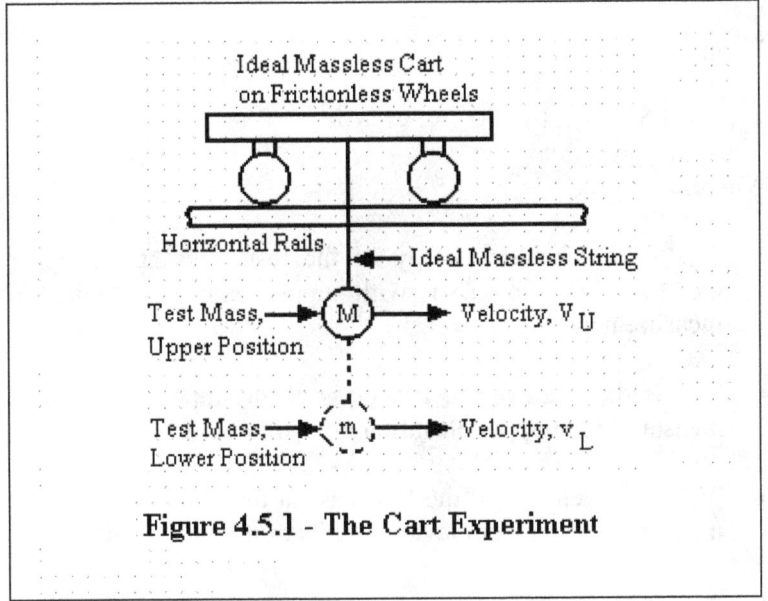

Figure 4.5.1 - The Cart Experiment

cart by an ideal massless string. The cart is supported by ideal frictionless and massless wheels on horizontal rails in a gravitational field. To simplify the discussion, the radius of the field is considered to be infinite. (Care must be taken in assuming an infinite radius for the field since all known fields are finite. In this case, the assumption is valid since its effect merely eliminates the need to consider the requirements of the Law of Conservation of Angular Momentum without affecting the conclusions.)

4.5.3 - With the mass at the upper elevation and moving to the right with a velocity V_U, the kinetic energy E_{KU}, of the mass is given by:

Eq. 4.5.1 $E_{KU} = M_U * V_U^2 / 2$

The energy equivalent of the mass, E_{RU}, is given by:

Eq. 4.5.2 $E_{RU} = M_U * C_U^2$

Then:

Eq. 4.5.3 $2*E_{KU}/E_{RU} = V_U^2/C_U^2$

Where:

- 'E_{KU}' is the kinetic energy of the test mass at the upper elevation as measured with upper elevation units of measurement.

- 'M_U' is the mass of the test mass at the upper elevation as measured with upper elevation units of measurement.

- 'V_U' is the velocity of the test mass at the upper elevation as measured with upper elevation units of measurement.

- 'E_{RU}' is the energy represented by the mass of the test mass at the upper elevation as measured with upper elevation units of measurement.

- 'C_U' is the velocity of light at the upper elevation as measured with upper elevation units of measurement.

(V_U is considered to be sufficiently small so that the effects described by the Special Theory of Relativity may be ignored.)

4.5.4 - Since the ratio of gravitational mass to inertial mass of various materials has been found to be identical to a high degree of accuracy, it is to be expected that kinetic energy which is bound to matter will be affected by the gravitational field to the same degree as is the energy associated with the rest mass. This conclusion follows from the fact that different portions of the rest mass of various materials consists of the mass imposed by kinetic energy. This kinetic energy is contained, for example, in the kinetic energy of electrons in their orbits, the thermal motion of atoms, and perhaps in the resonance states of nuclear particles.

(Special Relativity establishes clearly that kinetic energy possesses inertial mass.)

4.5.5 - If we accept that rest mass equivalent energy and kinetic energy bound to rest mass gravitate to the same degree, then, as the mass is lowered to the lower elevation, the rest mass equivalent energy of the test mass and the velocity of light remain invariant. Then:

Eq. 4.5.4 $e_{RL} = E_{RU}$

Eq. 4.5.5 $c_L = C_U$

Since the kinetic energy of the test mass is assumed to be affected by the gravitational potential to an identical degree as is the energy equivalent of the rest mass of the test mass, and since the massless, frictionless cart and string cannot affect the kinetic energy of the test mass:

Eq. 4.5.6 $e_{KL}/e_{RL} = E_{KU}/E_{RU}$

But:

Eq. 4.5.7 $2*e_{KL}*e_{RL} = v_L^2/c_L^2$

Then:

Eq. 4.5.8 $V_U^2/C_U^2 = v_L^2/c_L^2$

From Table 2.5.1:

Eq. 4.5.9 $v_L = X*V_L/Y$

Eq. 4.5.10 $c_L = X*C_L/Y$

Then:

Eq. 4.5.11 $V_L/C_L = V_U/C_U$

4.5.6 - Observe the significance of the preceding equation, It states that, as a mass changes in elevation, its horizontal velocity, as a fraction of the velocity of light, remains unchanged (assuming zero curvature to the field to negate the effects of the Law of Conservation of Angular Momentum). The velocity vector of a moving object will therefore be refracted by the field to the identical degree as is a ray of light in a vacuum. This refraction of the velocity vector is an effect which is superimposed on the effects of the gravitational attraction on the object itself.

4.5.7 - Equation 4.3.20 shows that the free photon experiences a refraction equivalent to twice the acceleration of gravity. This means that the velocity vector of moving objects must also experience a similar refractional acceleration in addition to the normal acceleration of gravity if Equation 4.5.11 is to be satisfied. This acceleration, in turn, will 'refract' the orbits of planets. Let us consider then, the orbit of Mercury.

4.5.8 - At the semi-lattus rectum of the orbit, the velocity of Mercury around the Sun is 52.14 kilometers per second and its distance form the center of the Sun is 56.0 million kilometers, as provided by standard astronomical references. With the velocity and radius known, the centrifugal acceleration acting on the planet due to its circular path (ignoring the fact that its orbit is not quite circular) is given by:

Eq. 4.5.12 $A = V^2/R$

Or:

Eq. 4.5.13 $A = 4.86*10^{-5}$ kilometers/second/second

Since the gravitational and centrifugal accelerations of an object in orbit are equal in magnitude, the acceleration of gravity due to the Sun at the semi-lattus rectum of Mercury's orbit is also:

Eq. 4.5.14 $A_G = 4.86*10^{-5}$ kilometers/second/second

4.5.9 - As required by Equation 4.3.20, light traveling horizontally at this radius from the Sun will experience an acceleration of twice that amount, or $9.72*10^{-5}$ kilometers/second/second, and, in accordance with the laws of mechanics, the light will bend with a radius of curvature of $9.26*10^{14}$ kilometers. The circumference of a circle having such a radius is $5.82*10^{15}$ kilometers. At a velocity of 52.14 kilometers/second, $3.54*10^4$ centuries are required to traverse a full circle. In one century, 36.6 arc-seconds of such a circle are traversed. One would expect, therefore, that Mercury's orbit would precess by this amount due to the effects of gravitational refraction. However, because the velocity of Mercury in its orbit is not constant around its orbital path, an additional source of precession is present due to the effects of velocity as described by Special Relativity. The amount of this precession is reported in texts written early in the 20th Century to be 4.5 arc-seconds per century. Adding these effects yields a prediction for the unexplained (in terms of Newtonian Theory) precession of Mercury's orbit of 41.1 arc-seconds per century.

4.5.10 - The observed unexplained precession is reported to be 41 arc-seconds per century, and it is concluded that the treatment of the gravitational field presented in this discussion is consistent with the observations of orbital precession. What is also shown in this Section is the fact that kinetic energy which is 'free' gravitates at twice the rate of matter as do photons. Further discussion of this last statement will be provided in Section 4.6.

Section 4.6 - Gravitational Acceleration of the Confined Photon and the Equivalence of Inertial and Gravitational Mass

4.6.1 - In Section 4.3, and ideal thought experiment, as defined by Figure 4.3.1, was devised to allow the gravitational potential of the 'free' photon to be determined as a function of the gravitational potential existing between elevations, as provided by Equation 4.3.17. This thought experiment required that the support structure between the retroreflectors at the upper and lower elevation somehow be immune to the effects of a change in elevation on the units of measurement for length. This approach was useful in determining the gravitational effects on photons which were free to propagate in the gravitational field without being intimately influenced by constraining matter. In this Section, we will consider that the photons are bound to matter, such as would be the case for the radiation in an integrating sphere, or such as would be the case of the electromagnetic fields within atoms and molecules.

4.6.2 - Let us consider again the thought experiment represented by Figure 4.3.1, but this time let us consider that the support member consists of matter which **is** subject to the gravitational effects on the units of measurement for length, as represented by the factor X. If the support structure and the retroreflectors are lowered in the gravitational field so that the upper retroreflector is at the elevation previously occupied by the lower retroreflector, the length of the structure will remain invariant. Its length, as measured with lower elevation units of measurement will be identical to its original length, as measured with upper elevation units of measurement. Its length, however, as measured with upper elevation units of measurement, will be 1/X times as large as its original length. The change in length resulting from the lowering of the support/retroreflector structure through a gravitational potential of Θ is given by:

Eq. 4.6.1 $\delta L = L*(1-1/X)$

Combining Equations 3.4.2 and 4.6.1 provides:

Eq. 4.6.2 $\delta L = \Theta*L$

Where:

- 'L' is the length of the support structure in its original position as measured with upper elevation units of measurement.

The force exerted on each retroreflector, as provided by Equation 4.3.1, is given by:

Eq. 4.6.3 $F = 2*N*E_P/C$

Where:

- 'N' is the number of photons striking each retroreflector per second as measured with upper elevation units of measurement.

- 'E_P' is the energy of each photon striking each retroreflector as measured with upper elevation units of measurement.

- 'C' is the velocity of light as measured in the upper reference frame with upper elevation units of measurement or in the lower elevation with lower elevation units of measurement.

Since, unlike Section 4.3, the arguments which follow do not involve a difference equation, it is not necessary to consider the effects of changes in size of units of measurement resulting from a change in elevation in applying Equation 4.6.3 to small changes in gravitational potential. Equation 4.3.11 provides the energy contained in the beam of photons as:

Eq. 4.6.4 $E_T = N*E_P*L*(X+Y)/(C_U*Y)$

Where:

- 'E_T' is the energy contained in the beam of photons between retroreflectors as measured with upper elevation units of measurement.

- 'N' is the number of photons striking each retroreflector per second in terms of upper elevation units of measurement.

- 'E_P' is the energy of the photon in terms of upper elevation units of measurement.

- 'L' is the distance between retroreflectors in terms of upper elevation units of measurement.

- 'C_U' is the velocity of light between the mirrors in terms of upper elevation units of measurement.

4.6.3 - Since second order effects will not change the conclusions which follow if we allow the value of Θ to be small, we may substitute:

Eq. 4.6.5 $E_T = 2*N*E_P/C$

Shortening the length of the support structure by $1/X$ will compress this beam of photons and increase their energy by X times, requiring the addition of an amount of energy equal to $(X-1)$ times the energy content of the photon beam. To the first order, the addition of energy to the photon beam is:

Eq. 4.6.6 $\delta E_P = \Theta$

The fact that energy must be added to the structure to lower it means that the photon induced gravitational potential on the support structure is negative and is therefore equal to $-\Theta$. Adding this potential to the gravitational potential of the free photon provided by Equation 4.3.17 provides the gravitational potential of the confined photon:

Eq. 4.6.7 $\Theta_{PC} = 2*\Theta-\Theta$

Or:

Eq. 4.6.8 $\Theta_{PC} = \Theta$

4.6.4 - Observe the significance of Equations 4.6.17 and 4.6.8. Together, they state that electromagnetic energy, when free of matter, gravitates at twice the level experienced by the equivalent energy of material particles, but when electromagnetic energy is confined by matter, the change in the size of the unit of measurement for length as a result of a change in elevation causes the radiation pressure to exert a negative gravitational effect on the confining matter. This reduces the net gravitational effect on the matter/electromagnetic energy combination to the same level as is experienced by the equivalent of material particles. The gravitational mass of 'confined photons', such as would occur in an optical integrating sphere or in the electromagnetic fields between atoms and molecules, is thus seen to be identical to the gravitational mass of matter.

4.6.5 - It should be noted that the identity of the gravitational masses of energy represented by confined photons and energy represented by the rest mass of matter only depends on the factor Y. If the value of X differed from that provided by Equation 3.4.2, both the change in length of the support structure and the value Θ_P provided by Equation 4.3.17 would be changed. The effects of these two changes would cancel, and the equality between the gravitational masses of the confined photon and of a material particle would remain.

4.6.6 - Now let us consider the behavior of kinetic energy. In the thought experiments of Figure 4.2.1 and Figure 4.3.1 we will replace the retroreflectors with ideal neutral beam particle accelerators. These accelerators are assumed to be capable of both accelerating electrically neutral material particles to a velocity which is infinitesimally close to the velocity of light and of receiving such particles and slowing them to rest. The process is assumed to be 100% efficient. As described, these accelerators perform the same function for material particles as did the retroreflectors for photons.

4.6.7 - As the velocity of a material particle (object) approaches the velocity of light, the fraction of its inertial mass represented by its rest mass approaches zero. We may consider then, that the inertial mass of the accelerated particles in these revised thought experiments is the inertial mass of their kinetic energy. In Section 4.5 it was shown, both by a thought experiment and by the observed precession of Mercury's orbit, that the velocity vector of a moving object is refracted by a gravitational field to the same degree as is light. The arguments generated for the thought experiments of Figure 4.2.1 and Figure 4.3.1 and of this Section must apply equally to both electromagnetic energy (photons) and kinetic energy. (Neutrinos may be considered [circa 1987 but revised in "The Einstein Hoax" circa 1997] to be the quantum particles of kinetic energy in the same manner as photons are the quantum particles of electromagnetic energy.) Both electromagnetic and kinetic energy are thus seen to gravitate at twice the rate of material particles, but when these energies are confined by matter, they produce a negative gravitational force on the matter which reduces the net gravitational attraction of electromagnetic and of kinetic energy to an equivalence with the gravitational attraction experienced by the energy equivalence of the rest mass of matter. Furthermore, since the relationship between gravitational potential of the free photon and the gravitational potential is, in effect, a 'constant' of the Science of Physics, it will be invariant between reference frames. As a result, the relationship must hold for all values of Θ despite having been derived for small values of Θ.

4.6.8 - If one considers the observed identical relationship between the inertial and gravitational masses of different materials, the reason for the identity becomes obvious, Different forms of matter contain energy represented by rest mass, energy represented by confined electromagnetic fields, and confined kinetic energy in proportions which depend upon the excitation state and type of atoms present and on the combination of these atoms with other atoms. All of these energy forms have been shown to gravitate at the same rate. Since, in accordance with

$E=M*C^2$, they all have the inertial mass, the gravitational and inertial masses of all materials must be equal, as shown by the Eotvos Experiment.

Section 4.7 - Conservation of Energy in the Gravitational Field

4.7.1 - So far it has been shown that the approach taken by this discussion leads to conclusions which are consistent with the observations that have been cited as verification of the validity of General Relativity, namely the effects of the field on time dilation, refraction of starlight, and the precession of planetary orbits, and of the equivalence of inertial and gravitational mass. In addition, it has been shown that non-Euclidian geometry for space violates the requirement that momentum and energy be conserved in a closed system. Hence, the gravitational field, as well as the Universe as a whole, must exist in terms of three dimensional Euclidian space.

4.7.2 - All of the considerations which are presently considered as verifications of General Relativity involve effects which are undetectable without precise instrumentation and sophisticated observation. There remains one effect of the gravitational field which is so powerful that instrumentation is not required. The necessary responses for the observation are built into almost every species by the process by which they evolved on Earth. Every newborn baby instinctively fears the ability of the gravitational field to release energy (fear of falling).

4.7.3 - A treatment of gravitation must be compatible with the Law of Conservation of Energy. Under Newtonian theory, a clear incompatibility occurs. A hydrogen atom which fell from space to the surface of an object which was sufficiently close to the 'Black Hole' state would release nearly an infinite amount of gravitational energy. This energy would be converted to radiation and emerge to 'fry' the surrounding environs. General Relativity does not conclude that the 'frying' would occur. It states that the energy would be released, but in climbing out of the gravitational field, it would be reabsorbed by doing work against the field so that the energy presented to the external environment would be limited to the energy of the original infalling particle. General Relativity answers the requirements of the Law of Conservation of Energy by the assertion that, while the gravitational field does create energy, since that energy cannot be observed outside of the field, the Law of Conservation of Energy has not been violated. (Early writings on the subject by those involved speak of the 'ability of mass (i.e.- energy) to call into being additional mass'.) It would seem far more satisfying to accept the requirement that the Law of Conservation of Energy be valid throughout the entire gravitational field, in absolute terms, as a test of the validity of a gravitational theory.

4.7.4 - Consider an ideal thought experiment in which a test mass of equivalent energy, E_U, as measured with upper elevation units of measurement, is lowered from an upper elevation to a lower elevation through a gravitational potential Θ. At the lower elevation, an energy of fall, $E_U*\Theta$, is added to the equivalent energy of the test mass, E_L, both as measured with upper elevation units of measurement. The energy of fall, $E_U*\Theta$, was released, either from the energy represented by the test mass, from the gravitational field, or from a combination of both. This is shown in Figure 4.7.1 by defining, as a result of the change in elevation of the test mass, $(1-K)*E_U*\Theta$ as the energy release from the test mass and $K*E_U*\Theta$ as the energy released by the field itself.

4.7.5 - The mass of the test mass and its energy equivalent are both constants of the Science of Physics and are therefore invariant between elevations, Thus:

Eq. 4.7.1 $e_L = E_U$

From Table 2.5.1:

Eq. 4.7.2 $e_L = X*Z*E_L$

Combining Equations 3.3.1, 4.7.1, and 4.7.2 provides:

Eq. 4.7.3 $E_L = (1-\Theta)*E_U$

And the change in energy of the test mass, δE, is given by:

Eq. 4.7.4 $\delta E = E_U - E_L$

Eq. 4.7.5 $\delta E = E_U * \Theta$

But the change in energy of the test mass must equal the energy supplied by the test mass to the energy of fall, therefore:

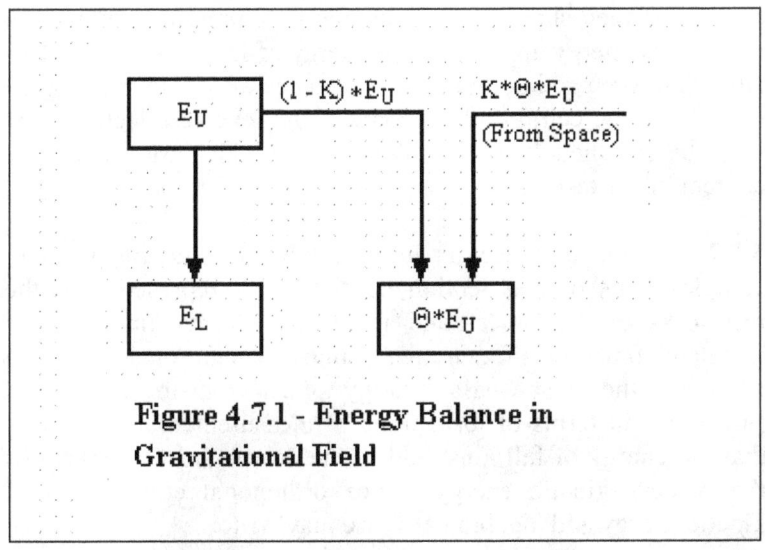

Figure 4.7.1 - Energy Balance in Gravitational Field

Eq. 4.7.6 $E_U * \Theta = (1-K)*E_U * \Theta$

Or:

Eq. 4.7.7 $K = 0$

4.7.6 - Equation 4.7.7 states that the gravitational field makes no contribution to the energy of fall. That energy results entirely from a reduction in the energy content of the test mass, as measured with 'constant' units of measurement. Since the gravitational field exchanges no energy with the falling object, the gravitational field is entirely compatible with the Law of Conservation of Energy throughout its entirety, not just in terms of the external universe. Instead of being a contributor of gravitational energy, the gravitational field is seen to be a modifier of the form of existing energy. It results from the fact that the proximity of energy changes the parameters of space. This change of parameters, in turn, causes a reduction in the size of the units of measurement of energy as the energy represented by the gravitating mass is approached. In accordance with the Principle of Relativity, matter which approaches the gravitating mass must then shed some of the energy represented by its rest mass, This energy appears as the energy of fall in the form of an attracting force acting thought the distance of fall. Increasing the distance to the central mass produces the reverse effect. Energy must be supplied in raising the object to increase the energy content of the test mass.

4.7.7 - When an object is in 'free fall' the situation is more complex. As shown in Section 4.5, the energy represented by the rest mass of the object and its kinetic energy, in terms of 'constant' units of measurement, must remain unchanged. In addition, the rest mass energy must remain unchanged (invariant) in terms of local units of measurement. This means that the energy of fall must add to the downward component of the object's kinetic energy. Since orthogonal components of kinetic energy add algebraically, we may write:

Eq. 4.7.8 $E_{TU} = E_{RU} + E_{KHU} + E_{KVU}$

Where:

- 'E_{TU}' is the total energy of the object at the upper elevation as measured with upper elevation units of measurement.

- 'E_{RU}' is the rest mass equivalent energy of the object at the upper elevation as measured with upper elevation units of measurement.

- 'E_{KHU}' is the horizontal component of the kinetic energy of the object at the upper elevation as measured with upper elevation units of measurement.

- 'E_{KVU}' is the vertical component of the kinetic energy of the object at the upper elevation as measured with upper elevation units of measurement.

At the lower elevation, the horizontal component of the kinetic energy and the energy equivalent of the rest mass are reduced, in terms of 'constant' units of measurement in proportion to 1/X and become equal to:

Eq. 4.7.9 $E_{KHL} = (1-\Theta)*E_{KHU}$

Eq. 4.7.10 $E_{RL} = (1-\Theta)*E_{RU}$

Where:

- 'E_{KHL}' is the horizontal component of kinetic energy at the lower elevation as measured with upper elevation units of measurement.

- 'E_{RL}' is the rest mass equivalent energy of the object at the lower elevation as measured with upper elevation units of measurement.

The change in the rest mass equivalent energy and the horizontal component of kinetic energy represented by Equations 4.7.9 and 4.7.10 between the upper elevation and the lower elevation appears as the energy of fall:

Eq. 4.7.11 $E_F = \Theta*(E_{RU}+E_{KHU})$

Where:

- 'E_F' is the energy of fall as measured at the lower elevation with upper elevation units of measurement.

The energy of fall, E_F, adds to the vertical component of kinetic energy to maintain the total energy unchanged:

Eq. 4.7.12 $E_{KVL} = E_{KVU}+\Theta*(E_{RU}+E_{KHU})$

Or:

Eq. 4.7.13 $E_{KVL} = E_{KVU}+\Theta*(E_{RU}+E_{KHU})$

Where :

- 'E_{KVL}' is the vertical component of kinetic energy at the lower elevation as measured with upper elevation units of measurement.

In terms of lower elevation units of measurement:

Eq. 4.7.14 $e_{RL} = E_{RU}$

Eq. 4.7.15 $e_{KHL} = E_{KHU}$

Eq. 4.7.16 $e_{KVL} = (E_{KVU}+[E_{KHU}+E_{RU}]*\Theta)/(1-\Theta)$

And:

Eq. 4.7.17 $e_{TL} = E_{TU}/(1-\Theta)$

Where:

- 'e_{RL}' is the energy of the test mass at the lower elevation as measured with lower elevation units of measurement.

- 'e_{KHL}' is the horizontal component of the kinetic energy at the lower elevation as measured with lower elevation units of measurement.

- 'e_{KVL}' is the vertical component of the kinetic energy at the lower elevation as measured with lower elevation units of measurement.

- 'e_{TL}' is the total energy at the lower elevation as measured with lower elevation units of measurement.

(It will be noted that no attention was paid to the fact that the vertical component of kinetic energy may result from an upward or downward velocity in the preceding equations. Since, unless an upward velocity exceeds the escape velocity, the object will return to the upper elevation with a downward velocity of identical magnitude. The direction of the initial vertical velocity does not affect the conditions which obtain when the object reaches the lower elevation.)

4.7.8 - Let us now consider the gravitational effects on the free photon. From Table 3.5.1, the conversion factor for Planck's Constant is unity and:

Eq. 4.7.18 $\quad h = H$

The energy of the photon, E_P, is given in terms of its frequency, Ω, by:

Eq. 4.7.19 $\quad E_P = H * \Omega$

For the photon at the upper elevation in terms of local units of measurement:

Eq. 4.7.20 $E_{PU} = H*\Omega_U$

And, for the same photon at the lower elevation in terms of local units of measurement:

Eq. 4.7.21 $e_{PL} = h*\omega_L$

Combining equations 3.2.10 and 3.3.11 provides:

Eq. 4.7.22 $\omega_L = \Omega_U/(1-\Theta)$

Combining equations 4.7.18 through 4.7.22 provides:

Eq. 4.7.23 $e_{PL} = E_{PU}/(1-\Theta)$

And, combining equations 4.7.2 and 4.7.23 provides:

Eq. 4.7.24 $E_{PL} = E_{PU}$

4.7.9 - Equation 4.7.24 shows that the energy of the free (unconstrained by matter) photon does not change, in terms of 'constant' units of measurement, as a result of a change in elevation. The change of energy which is observed, in terms of local units of measurement, results from the change in the size of the units for both energy and time caused by the change in gravitational potential. That this is the case should not be surprising. It has been shown that as a material particle 'falls' in a gravitational field, part of its rest mass equivalent energy is released as the energy of fall in the form of an increase of the kinetic energy of the particle. (Its downward velocity increases.) Unlike the material particle, the photon (and the neutrino) must travel at the velocity of light. If any energy of fall were subtracted from the photon, that loss of energy would reduce its frequency in terms of 'constant' units of measurement. That energy could not be returned to the photon as an increase in its velocity since the velocity of the photon must equal the local velocity of light. Instead, it would of necessity be returned as an increase in frequency, again as measured in terms of 'constant' units of measurement, so that no net change would occur. (The same argument holds for the neutrino.)

4.7.10 - Section 4.3 shows that, where photons are constrained by matter, the gravitational field extracts energy from the photons through the Doppler effect when the elevation is lowered. Such photons lose energy in proportional to the gravitational potential between elevations in the same manner as does the rest mass equivalent energy of matter. This energy appears as the energy of fall:

Eq. 4.7.25 $e_{PCL} = E_{PCU}$

Eq. 4.7.26 $E_{PCL} = E_{PCU}*(1-\Theta)$

Eq. 4.7.27 $E_{PCF} = \Theta*E_{PCU}$

Eq. 4.7.28 $e_{PCF} = \Theta*E_{PCU}/(1-\Theta)$

Where:

- 'E_{PCU}' is the energy the constrained photon at the upper elevation as measured with upper elevation units of measurement.

- 'E_{PCL}' is the energy of the same photon at the lower elevation as measured with lower elevation units of measurement.

- 'E_{PCF}' is the energy of fall of the same photon between elevations as measured with upper elevation units of measurement.

- 'e_{PCL}' is the energy of the same photon at the lower elevation as measured with lower elevation units of measurement.

- 'e_{PCF}' is the energy of fall of the same photon between elevations as measured with lower elevation units of measurement.

4.7.11 - In Section 4.5 it was shown that the velocity vector of a moving object must be refracted by the gravitational field as is the path of a ray of light and that this refraction is equivalent to a gravitational acceleration twice that experienced by a material particle. Unlike the acceleration of a material particle, however, this acceleration does not result in an interchange of energy. As will be shown in Section 4.8, the force represented by this acceleration results from the effects of the gravitational field and the requirement that momentum must be conserved. That the gravitational effects on kinetic energy which is unconstrained by matter should not involve an energy interchange is not unreasonable from a mechanistic point of view. As is the case with the unconstrained photon, any energy of fall which was released would cause a reduction of the kinetic energy in terms of 'constant' units of measurement. This energy, in terms of 'constant' units of measurement, then adds to the remaining kinetic energy to increase it to its previous level.

4.7.12 - In Section 4.3, it was shown that the free photon experiences a gravitational force (acceleration) twice that experienced by a material particle of equivalent energy. As the photon travels vertically in the gravitational field, this force would seem to act through the distance between the upper and lower elevations. Since the product of the applied force and the distance through which the force acts is an energy, one might expect that the 'free falling' photon would release an energy of fall equivalent to twice its energy times the gravitational potential. Equation 4.7.24 shows that this energy release does not occur. If one examines the reduction in elevation from the frame of reference of the photon, one finds that Special Relativity tells us that the photon has traveled though zero distance since the 'interval', δs, for the photon's travel between points is:

Eq. 4.7.29 $\delta s = [C^2 * \delta T^2 - \delta x^2]^{0.5}$

Where:

- 'δs' is the 'interval' (Special Relativity) between the upper and lower elevation.

- 'δx' is the distance between the upper and lower elevation in the 'stationary' frame of reference represented by the gravitational field.

- 'δT' is the time experienced by the photon in traveling between elevations as measured in the 'stationary' frame of reference represented by the gravitational field.

- 'C' is the velocity of light.

And, for the photon:

Eq. 4.7.30 $\delta s = 0$

Since, in its own frame of reference the photon experiences the force of gravity acting though a distance of zero, no energy exchange occurs. In the frame of reference of the gravitational field, accepting that the force of gravity also travels at the velocity of light, there is no time for an energy interchange to occur. The apparent dichotomy does not exist.

Section 4.8 - Conservation of Momentum in the Gravitational Field

4.8.1 - The Law of Conservation of Momentum is one of the basic laws of the Science of Physics. It is necessary, therefore, to examine the effect of the gravitational field on momentum. If one considers only the transfer of momentum during the interaction between two or more moving objects, the Law of Conservation of Momentum is rather trite. The unit of measurement for momentum is the product of force and time.

During the interaction of two or more moving objects the vector sum of the forces applied to the object as a result of the interaction must be instantaneously equal to zero in accordance with Newton's Second Law of Motion. Since the time of application of these forces is identical for the interacting objects, the vector sum of the impulses applied to them must be zero for the entire interaction. Momentum is, of necessity, conserved since the change in momentum is the integrated impulse. One might consider, therefore, that for interaction between objects (particles), the Law of Conservation of Momentum is a convenient restatement of Newton's Second Law of Motion rather than a profound and independent law in its own right. (An identical conclusion follows for the Law of Conservation of Angular Momentum.)

4.8.2 - The Law of Conservation of Momentum takes on independent meaning when the transportation of kinetic energy by ballistic objects is involved. Consider the case where a projectile of mass M is accelerated uniformly from rest at a rate, A, for a period of time, T. (The arguments which follow are valid for non-uniform rates of acceleration but the treatment is more tedious without a corresponding improvement in validity or lucidity. For simplicity, it is assumed that the velocities involved are sufficiently low that the effects described by Special Relativity may be ignored.) The distance, L, traveled by the projectile during the acceleration time is given by:

Eq. 4.8.1 $L = A*T^2/2$

But:

Eq. 4.8.2 $A = F/M$

Where:

- 'F' is the accelerating force.

Then:

Eq. 4.8.3 $L = F*T^2/(2*M)$

But:

Eq. 4.8.4 $M = E_R/C^2$

Where:

- $'E'_R'$ is the energy equivalent of the mass, M.

- $'C'$ is the locally measured velocity of light.

Then:

Eq. 4.8.5 $L = F*T^2*C^2/(2*E_R)$

Multiplying both sides of the preceding equation by the accelerating force, F, provides:

Eq. 4.8.6 $F*L = F^2*T^2*C^2/(2*E_R)$

But:

Eq. 4.8.7 $F*L = E_K$

Where:

- $'E_K'$ is the kinetic energy supplied to the projectile during acceleration.

Then:

Eq. 4.8.8 $2*E_R*E_K = F^2*T^2*C^2$

The momentum supplied to the projectile is the product of the accelerating force and the time for which it is applied.

Then:

Eq. 4.8.9 $U = F*T$

Where:

- 'U' is the momentum of the projectile.

Substituting Equation 4.8.9 into Equation 4.8.8 provides:

Eq. 4.8.10 $2*E_R*E_K = U^2*C^2$

4.8.3 - By analogous derivation, a similar expression may be provided for angular momentum:

Eq. 4.8.11 $2*E_R*E_K = J^2*C^2/R^2$

Where:

- 'J' is the angular momentum of a rotating mass.

- 'R' is the radius of gyration of that rotating mass.

However, the angular momentum of a rotating mass may be considered to be the linear momentum of that mass moving in a circular path of radius equal to the radius of gyration and at an angular velocity equal to the angular velocity of the rotating mass. Therefore:

Eq. 4.8.12 $U = J/R$

Combining Equations 4.8.11 and 4.8.12 provides Equation 4.8.10.

4.8.4 - In Equation 4.8.10, it will be noted that the left side of the expression consists of the product of two energies, the energy represented by the rest mass of the assumed ballistic object and the kinetic energy that the object possesses as a result of its velocity. The right side of the expression is the square of the product of the momentum of the object and the velocity of light. While both quantities on the left side consist of energy and individually and in totality obey a conservation law, their product does not necessarily do so since a transfer of energy between the rest mass and the kinetic energy would change the product of the energies without changing their sum. And as been shown in Section 4.7, the movement of a material object to a region of space (i.e.- a change in elevation) where the velocity of light, in terms of 'constant' units of measurement, is different, causes a change in the rest mass energy (in terms of 'constant' units of measurement). For a freely moving object, this change in rest mass energy produces a transfer of energy to (or from) the vertical component of the kinetic energy. The sum of the energies must remain the same, but the product changes.

4.8.5 - If we consider the case represented by the thought experiment of Figure 4.5.1, it was shown that the lowering of a mass which possessed kinetic energy moved it to a region of space where the velocity of light was lower. No vertical component of kinetic energy was acquired by the test mass because the energy of fall was absorbed by the lowering mechanism. In this case, both the rest mass equivalent energy and the kinetic energy were reduced by the factor $1/X$. The product of the rest mass energy and the kinetic energy were also reduced the square of that factor, and the conserved quantity was the product of the momentum and the velocity of light. In systems where the velocity of light is unchanged in terms of constant units of measurement (our normal experience) one might state that momentum is conserved. If one considers an object in free fall, the horizontal component of the kinetic energy and the rest mass equivalent energy are also reduced by the factor $1/X$, in terms of 'constant' units of measurement. The difference in energy is added to the vertical component of the kinetic energy rather than being absorbed by the lowering mechanism of Figure 4.5.1. The sum of the rest mass equivalent energy and the kinetic energy is unchanged but the product of those energies does change in terms of 'constant' units of measurement. In this process, it would appear that the product of momentum times the velocity of light is not conserved, but this is not the case. The force of gravity imparts momentum to the falling object to make up the difference.

4.8.6 - If one considers conservation of momentum as applied to the so-called massless particles (photons, neutrinos, etc.), the Law of Conservation of Momentum is clearly the Law of Conservation of the Momentum-Velocity product. For these particles, which must travel at the local velocity of light, the equivalent of the rest mass equivalent energy product of Equation 4.8.10 would be the square of the energy of the particle. The factor of '2' in Equation 4.8.10 would no longer appear since it resulted from the acceleration of the material object to achieve a displacement and does not apply to a particle such as the photon which springs into being at its final velocity (the velocity of light). For the photon, therefore, Equation 4.8.10 becomes:

Eq. 4.8.13 $E_P = U_P * C_P$

Section 4.9 - Gravitation and the Principle of Equivalence

4.9.1 - The Principle of Equivalence was a conceptual tool used in generating a second order refinement (General Relativity) of first order (Newtonian) gravitational theory. In this Section the Principle of Equivalence will be examined. It will be shown that the Principle of Equivalence, as it is currently employed, conflicts with the Principle of Relativity, and causes General Relativity, which is based upon both of these principles, to have a second order error with respect to time dilation, a first order error with respect to energy, and to require the existence of an otherwise undetectable 'fourth spatial dimension' (shown in Section 4.2 to be incompatible with the requirement that momentum be conserved within a closed system). It will then be shown that the reason that the Principle of Equivalence yields incorrect results is not inherent. It yields incorrect results because it is applied incorrectly in the generation of the General Theory of Relativity.

4.9.2 - To be useful, the Principle of Equivalence must be more profound than a simple conclusion that the inertial and gravitational masses of given material objects (particles) are equal to each other. Such an interpretation requires only that things (inertial mass, gravitational mass) proportional to the same thing (energy content) are proportional to each other. This is true but trite. It is necessary that the Principle of Equivalence be interpreted to mean that a uniform gravitational field is equivalent to an 'acceleration' field produced by an accelerated reference frame. This means, in turn, that if an observer in a completely closed chamber from which external observation was impossible were to observe that all objects, when released from constraint, were to accelerate towards one wall of that chamber, and those same objects, when constrained, experienced a force impelling them to that same wall, he would have no way of distinguishing between the following alternatives:

- The chamber is in gravityless space and is experiencing a continuous inertial acceleration. The observed force is produced by this acceleration, and the objects which appear to fall are freed of that acceleration while in 'free fall'.

- The chamber is resting on the surface of a gravitating body and the observed acceleration or force was produced by a gravitational field, assuming, of course, that the radius of the field were sufficiently large that it was impossible to detect the effects of a gravitational gradient.

4.9.3 - If we are to accept the premise that there is no favored reference frame, the Laws of the Science of Physics must be invariant between the two alternatives. If we consider a static position in a gravitational field to be one reference frame, then it should be possible to describe a reference frame which is accelerated with respect to the first one such that the apparent force of gravity does not exist. Conventially, this concept is illustrated by Dr. Einstein's Elevator/Chamber Model of Figure 4.9.1 and Figure 4.9.2.

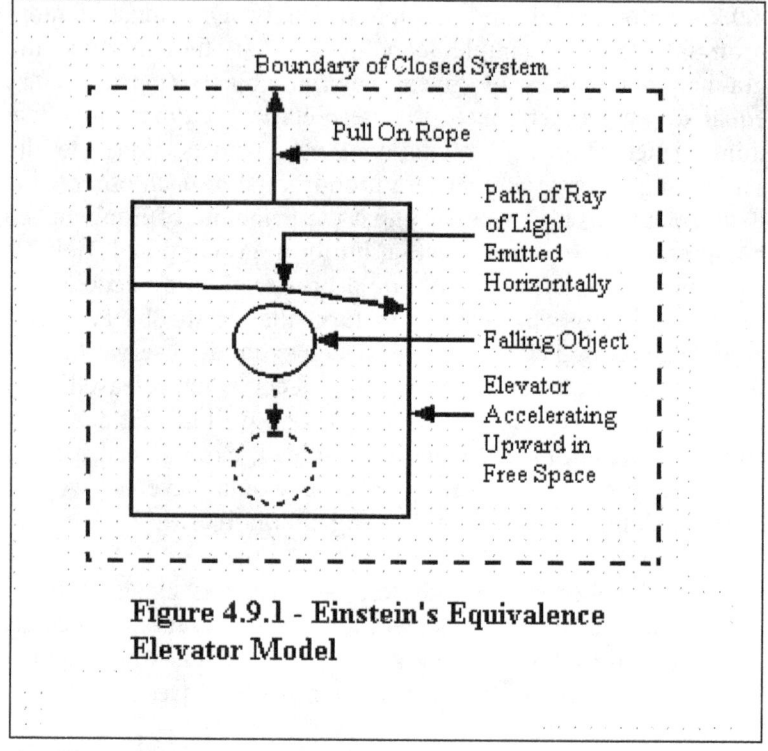

Figure 4.9.1 - Einstein's Equivalence Elevator Model

4.9.4 - In each of the above models, the experimenter releases a material object and observes that it undergoes, relative to the chamber, an acceleration towards the floor. He also observes that a ray of light, projected horizontally, curves towards the floor. If he assumes that the effects are due to a spatial acceleration Figure 4.9.1, then it is obvious that the path of the ray of light must be a straight line since, while the ray of light is moving across the chamber, it is unaffected by the acceleration of the chamber. Similarly, the physical object which was observed to fall actually maintains a constant velocity during the time it takes the floor to accelerate upwards to meet it. In terms of constant units of measurement (i.e.- those of the unaccelerated external reference frame) the rest mass of the object is unchanged by the fall.

4.9.5 - Since the mass energy of the object and the velocity of light are both 'constants' of the Science of Physics, they must be invariant between reference frames as well as being constant, as required by the previous paragraph. In order for a quantity to

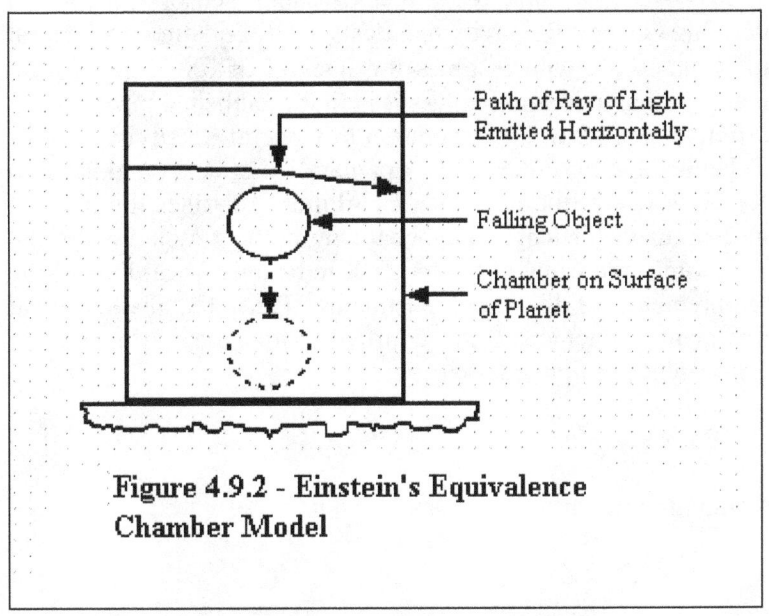

Figure 4.9.2 - Einstein's Equivalence Chamber Model

be both constant and invariant between reference frames, the conversion factor for the units of measurement of that quantity must be unity between those reference frames. Thus:

Eq. 4.9.1 $X*Z = 1$

Eq. 4.9.2 $X/Y = 1$

Combining Equations 3.2.12 and 4.9.1 provides:

Eq. 4.9.3 $Y = 1/(1+\Theta)$

Combining Equations 4.9.1 and 4.9.3 provides:

Eq. 4.9.4 $X = 1/(1+\Theta)$

Eq. 4.9.5 $Z = (1+\Theta)$

4.9.6 - At this point, the effect of using Tensor Calculus in deriving General Relativity manifests itself. Equating the integral of K times δX with K times the integral of δX is not a legal mathematical operation unless it is first established that K is not a function of X. In the generation of General Relativity, Tensor Calculus is used to derive the gravitational effects (analogous to K). As a result, the mathematics arbitrarily defines the units of measurement for length as unchanging in the gravitational field. This forces the value of the factor X to be unity regardless of the requirements of the input postulates. Thus, the mathematical treatment concludes, in conflict with the Principle of Equivalence as applied, that:

Eq. 4.9.6 $X = 1$

Combining Equations 4.9.1 and 4.9.6 provides:

Eq. 4.9.7 $Z = 1$

Those readers familiar with General Relativity will recognize that the conversion factors represented by Equations 4.9.3, 4.9.6 and 4.9.7 are identical with the latter day interpretations of General Relativity as applied in the context of this discussion. (It will be noted that Equations 4.9.2, 4.9.3 and 4.9.6 are in conflict as a result of the illegal mathematical operation.)

4.9.7 - Another difficulty presents itself. The ray of light is observed to follow a curved path by an observer in the chamber, and yet, to follow such a path in the model illustrated by Figure 4.9.2, the light would have to be refracted by the gravitational field and refraction cannot occur under the combination of conditions represented by Equations 4.9.3 and 4.9.4. The solution to this dilemma, on the part of General Relativity, was to conclude that space is distorted by the gravitational field. This distortion produces a curvature of the path of a ray of light by causing the lower portion of the ray to travel further than the upper portion. Assuming that Equation 4.9.4 is the correct one is equivalent to defining the null geodesic as the straight-line of a non-Euclidian geometry. The null geodesic, which is the path of a ray of light in a vacuum, represents the least time path between two points in the field. For this case it can be shown that the curvature of space calls extra space into being in proportion to $(1+\Theta)^2$. Assuming that Equation 4.9.6 is the correct one is equivalent to defining the line represented by an ideal massless string stretched between the two points in the field as the straight line of the non-Euclidian geometry. This line has one-half the curvature of the null-geodesic and represents the shortest measurable distance between the points. For this assumption, half of the curvature of the ray of light results from conventional refraction and the other half the generation of additional distance in proportion to $(1+\Theta)$ by the curvature of space. Both of these assumptions yield identical observations for the bending of starlight and the precession of orbits. A choice could only be made between them by a direct observation of the transformation for force, Z. As shall be shown, it is not necessary to make such a choice.

4.9.8 - If the conclusions of General Relativity are valid, they must obey the Principle of Relativity since that principle is one of the postulates of General Relativity. In the arrangement of Figure 3.3.1, it follows, therefore, that the same relationship must apply to the transformations for time as a function of the gravitational potential between the upper and middle elevations, between the middle and the lower elevations, and between the upper and lower elevations. Equation 4.9.1 shows that, under General Relativity, the transformation for energy is unity, therefore, under General Relativity:

Eq. 4.9.8 $\Theta_T = 2*\Theta'$

From Equation 4.9.3:

Eq. 4.9.9 $Y' = 1/(1+\Theta')$

Eq. 4.9.10 $Y_T = 1/(1+\Theta)$

The conversion factor for time between the upper and the middle elevation times the conversion for time between the middle and the lower elevations must equal the conversion for time between the upper and the lower elevations, therefore:

Eq. 4.9.11 $Y_T = Y'^2$

Combining Equations 4.9.8 through 4.9.11 requires that:

Eq. 4.9.12 $(1+\Theta')^2 = (1 + 2*\Theta')$

Which is obviously untrue.

4.9.9 - The preceding inequality has considerable significance. It reveals again that the gravitational transformation for time provided by General Relativity is not multiplicatively commutative and therefore the conclusions of General Relativity are in conflict with the requirements of the Principle of Relativity upon which General Relativity is based. It follows that General Relativity is in internal conflict with its own postulate structure and therefore cannot be correct. A bit of reflection reveals that this incompatibility does not arise from the specific form of Equation 3.2.13. *Any relativistic theory for which the gravitational transformation for energy is unity will be found to violate both the Principle of Relativity and the Principle of Equivalence. The gravitational transformation for energy must be the reciprocal of the gravitational transformation for time. General Relativity does not meet this test.*

4.9.10 - If one subtracts the correct expression for Y, as provided by Equation 3.3.12 from Equation 4.9.3, one finds that the error,

Y_E, in the prediction by General Relativity for the time dilation is given by:

Eq. 4.9.13 $Y_E = \Theta^2/(1-\Theta)$

Or, for the weak fields (Θ small compared to unity) currently observable:

Eq. 4.9.14 $Y_E = \Theta^2$

(The value of Θ at the Earth's surface from an infinite distance is 10^{-9}.)

4.9.11 - In terms of energy, the error produced in the predictions of General Relativity by the misuse of the Principle of Equivalence as applied is large and quite obvious, as is shown by subtracting Equation 4.9.1 from Equation 3.3.11. The error for the energy dilation, $(X^*Z)_E$, becomes:

Eq. 4.9.15 $(X^*Z)_E = \Theta/(1+\Theta)$

Or, for small values of Θ:

Eq. 4.9.16 $(X^*Z)_E = \Theta$

This error shows up under General Relativity as an increase in the total energy (rest mass energy plus the energy of fall), whereas it has been shown in Section 4.7 that no such increase in total energy occurs. This increase in total energy was recognized early on. It was considered in early writings that the "gravitational field called into being additional mass (energy)!" The contradiction of such a conclusion with the Law of Conservation of Energy was explained by concluding that since the extra energy could not escape from the gravitational field to where it could be measured by an external observer, the Law of Conservation of Energy was not violated. Such a conclusion is, of course, false. The fact that the additional energy cannot be observed by an external observer does not prevent him from obtaining the necessary information from an observer within the field since there is no impediment to their communication. There is no way of avoiding the conclusion that both Newtonian Theory and General Relativity violate the Law of Conservation of Energy.

4.9.12 - The conclusion that the gravitational field is able to 'create' energy is implicit in the conceptual model upon which the Principle of Equivalence is based. It is a fundamental principle that, when one employs a conceptual model in a thought experiment, either the entire model be contained within a closed system or all pertinent factors which cross the boundaries of the system must be included in the evaluation. If one examines the thought experiment of Figure 4.9.1, he will find that the force required to cause the elevator to accelerate upwards is applied across the boundary of the model through the rope attached to the elevator from an external source. In Figure 4.9.2, the force of gravity also acts across the boundary of the model. To be valid both models must contain the source of the force so that the complete energy and momentum interchanges represented by the models is described. In the case of the model of Figure 4.9.2, this requirement is easily met by enlarging the boundaries of the closed system containing the chamber to include the planet producing the gravitational field. In the case of the model of Figure 4.9.1, the situation is more complex. The force applied to the elevator crosses through the boundary of the model through the pull on the rope and continuously imparts momentum to the system. Furthermore, since the force on the rope is imparting a collinear spatial acceleration to the elevator, the rope crossing the boundary is adding energy to the supposedly closed system because the second integral of that acceleration with respect to time is the distance through which that force acts. Since momentum is not involved with the arguments which follow, the Law of Conservation of Momentum for a closed system may be ignored. The fact that energy is added to the system through the rope, however, cannot be ignored because energy is one of the principle parameters of gravitation. The model of Figure 4.9.1 can be applied in the form of the Principle of Equivalence providing one compensates for the energy added through the rope through a process of 'renormalization'. When such compensation is performed, the Principle of Equivalence and the Principle of Relativity yield, in combination, results which are identical to the results obtained earlier in this discussion.

4.9.13 - In the model of Figure 4.9.1, let us consider that the elevator contains two test masses of identical rest mass equivalent energy, as measured with local units of measurement, as shown in Figure 4.9.3. Initially, test masses #1 and #2 are located at the ceiling. The floor consists of an ideal energy absorbing material such that, when an object impacts the floor, all of the kinetic energy of that object (in terms of the units of the velocity reference frame represented by the floor) is absorbed by the floor instantaneously, leaving none of the kinetic energy remaining in the object in the form of heat. As in the thought experiment of Figure 4.9.1, the elevator is in gravityless space and is being accelerated upwards to produce an apparent gravitational field within the elevator.

4.9.14 - The Special Theory of Relativity tells us that any velocity reference frame may be considered to be 'stationary' and observations made in reference frames which are moving with respect to the so-called 'stationary' reference frame may be converted to observations which would be made in the 'stationary' reference frame by the use of the appropriate Lorentz Transformations. Let us then consider that, at time zero, the velocity of the elevator is zero and test mass #2 is released from its attachment to the ceiling. Following its release, the floor accelerates upwards to contact test mass #2. As a result of the upwards acceleration of the elevator during the period of time required for the floor to come in contact with test mass #2, test mass #1 has acquired kinetic energy from its upwards acceleration which is not acquired by test mass #2. The kinetic energy would have been acquired by test mass #2 if it had not been released is, instead, imparted to the energy absorbing floor when it contacts test mass #2. Since the energy absorbing floor is

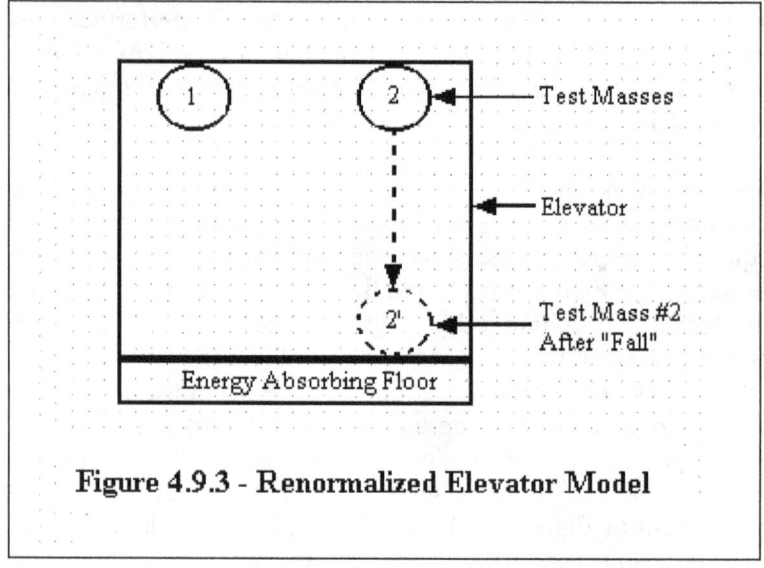

Figure 4.9.3 - Renormalized Elevator Model

ideal, all of this energy will be absorbed by it and none of it will be imparted to test mass #2.

4.9.15 - In the frame of reference of the ceiling and of the floor at the time test mass #2 contacts the floor, the energy imparted by the rope does not exist as kinetic energy in test mass #2. The only energy contained in test mass #2 in terms of the elevator frame of reference is the rest mass equivalent energy it originally contained, since, in the elevator frame of reference, the velocity of test mass #2 is zero when it contacts the floor. In the frame of reference of test mass #2, however, at an instant prior to its contact with the floor, test mass #1 has acquired kinetic energy equivalent to the product of the apparent gravitational potential between the ceiling and the floor and its equivalent rest mass energy. This increase in total energy of test mass #1 manifests itself as the increase in mass resulting from the velocity acquired by the elevator during the time that test mass #2 was in 'free fall', in accordance with the requirements of Special Relativity. Since rest mass #2 was not in contact with the elevator while it was in 'free fall, we may conclude that when test mass #2 arrives at the floor and deposits its energy of fall to the floor, its rest mass equivalent energy is identical with the rest mass equivalent energy it had (in terms of 'constant' units of measurement) when it left the ceiling. Test mass #1, which remained at the ceiling, has acquired a rest mass equivalent energy which is larger than the energy it had when test mass #2 was released. When test mass #2 has arrived at the floor and surrendered its 'energy of fall' to the energy absorbing floor, its rest mass equivalent energy is identical to the rest mass equivalent energy of test mass #1 (which remained at the ceiling) when both energies are measured with local units of measurement. (These energies must be invariant.) In terms of 'constant' units of measurement for energy, the rest mass equivalent energy of test mass #2 must be less than the rest mass equivalent energy of test mass #1 by the energy added to test mass #1 during the time that test mass #2 was in 'free fall'.

4.9.16 - Renormalizing the energy conditions so that the rest mass equivalent energy of test mass #1 (ceiling) in terms of the velocity reference frame which exists for the elevator at the instant that test mass #2 strikes the floor is equal to unity means that test mass #2, after it imparts its 'energy of fall' to the floor, will have a rest mass equivalent energy which is smaller by Θ than the rest mass equivalent energy of the test mass #1 (ceiling) since, during its 'free fall' it was deprived of its share of the energy being supplied by the rope, as represented on a normalized basis as Θ In terms of 'constant' units of measurement, therefore, we may write:

Eq. 4.9.17 $E_{2F} = (1-\Theta)*E_{1C}$

And, since the rest mass equivalent mass of test masses #1 and #2 were identical at the ceiling, and the rest mass equivalent energy of test mass #2 must be invariant between reference frames:

Eq. 4.9.18 $e_{2F} = E_{1C}$

Where:

- 'E_{1C}' is the rest mass equivalent energy of test mass #1 at the ceiling in terms of the units of measurement for energy which exist at the ceiling at the instant that test mass #2 contacts the floor.

- 'E_{2F}' is the rest mass equivalent energy of test mass #2 at the floor in terms of the units of measurement for energy which exist at the ceiling at the instant that test mass #2 contacts the floor.

- 'e_{2F}' is the rest mass equivalent energy of test mass #2 at the floor in terms of the units of measurement for energy which exist at the floor at the instant that test mass #2 contacts the floor.

Or:

Eq. 4.9.19 $e_{2F} = E_{2F}/(1-\Theta)$

And, from the definition of the conversion factor for energy, X*Z:

Eq. 4.9.20 $X*Z = 1/(1-\Theta)$

4.9.18 - If one now considers a photon which travels vertically from the ceiling to the floor, it is apparent that in Figure 4.9.1, the photon does not change energy enroute since it has no contact with the chamber. In traveling from the ceiling to the floor, however, it moves to a region where the units of measurement for energy are $(1-\Theta)$ times as large as they are at the region where the photon originated. In terms of local units of measurement, therefore, the energy of the photon is $1/(1-\Theta)$ times as large at the floor as it is at the ceiling, therefore:

Eq. 4.9.21 $e_{PL} = E_{PU}/(1-\Theta)$

Where:

- 'E_{PU}' is the energy of a photon at the ceiling as measured with the units of measurement existing at the ceiling.

- 'e_{PL}' is the energy of the same photon at the floor as measured with the units of measurement for energy existing at the floor. The time of travel of the photon between the ceiling and the floor is presumed to be so short that the change in velocity of the elevator during transit may be ignored.

Since the energy of the photon is Planck's Constant times its frequency, the frequency of the photon as observed at the lower elevation with local units of measurement will therefore be $1/(1-\Theta)$ times as large as the frequency of the same photon as observed at the upper elevation with local units of measurement in order to meet the requirements of Equation 4.9.20. (A sufficiently large number of photons are assumed as to reduce quantum uncertainties to a negligible level.) Since frequency is cycles per unit of time, and since cycles are discrete entities which are not subject to the effects of a change in size of units of measurement, it follows that the units of measurement for time must obey the following transformation law:

Eq. 4.9.22 $Y = (1-\Theta)$

It will be noted that Equations 4.9.20 and 4.9.22 are identical to Equations 3.3.11 and 3.3.12 previously derived.

4.9.19 - Equation 4.9.22 states that the units of measurement for time differ between the ceiling and the floor. Special Relativity tells us that the Lorentz Transformation for length in a direction transverse to the velocity vector between reference frames having a relative velocity is unity. It follows, therefore, that the units of measurement for length in a horizontal direction in the model of Figure 4.9.3 are identical at the ceiling and the floor. Since, in a horizontal direction in this model, there is no simultaneity uncertainty, the transformation for velocity in the horizontal direction is equal to the transformation for length (unity) divided by the transformation for time (Equation 4.9.21) and is given by:

Eq. 4.9.23 $X/Y = 1/(1-\Theta)$

If the velocity of light in the horizontal direction is to be invariant between the ceiling and the floor, as required by the Principle of Equivalence, it must be slower at the floor than at the ceiling, in terms of 'constant' units of measurement, in proportion to $(1-\Theta)$. A difference in the velocity of light as a function of elevation causes light to be refracted towards the region of lower velocity, as discussed in Section 4.2. If we consider a beam of light projected horizontally across the chamber, the amount of refraction observed on the beam by the velocity gradient provided in Equation 4.9.23 is equal to the bending of the path that would result from an acceleration equal to the acceleration applied to the chamber. Superimposed upon this refraction induced acceleration would be the acceleration of the chamber itself. An observer within the chamber who assumed that the chamber was at rest on the surface of a planet, would conclude, therefore, that photons accelerate in a gravitational field at twice the rate of material objects. For such an acceleration to occur it is necessary that the gravitational transformation for velocity, X/Y be equal to:

Eq. 4.9.24 $X/Y = 1/(1-\Theta)^2$

Combining Equations 4.9.22 and 4.9.24 provides:

Eq. 4.9.25 $X = 1/(1-\Theta)$

And, combining Equations 4.9.20 and 4.9.25 provides:

Eq. 4.9.26 $Z = 1$

It will be noted that Equations 4.9.25, 4.9.22, and 4.9.26 generated by applying the Principle of Equivalence correctly in combination with the Principle of Relativity are respectively identical with Equations 3.4.2, 3.3.12, and 3.4.3 generated by an entirely different line of reasoning.

4.9.20 - While the correct application of the Principle of Equivalence in combination with the Principle of Relativity allows valid results to be obtained for the gravitational field, it is incorrect to assume that these results indicate the premise of the Principle of Equivalence represents actual reality. If what we experience as a gravitational field is actually an acceleration field, then it is necessary for that acceleration field to deal with gravitational fields as they actually exist. Every gravitational field that is observed is finite in extent and is radial. If one considers the gravitational acceleration experienced by observers in, for example, London, England and Melbourne, Australia, he notes that they are experiencing a gravitational acceleration of one 'g' in approximately opposite directions. To maintain the Principle of Equivalence, it is necessary to describe an acceleration field which would cause spatial acceleration in a direction as to result in their separation without producing that separation. For the gravitational field of the Earth to be an inertial acceleration field would seem to require that the Earth be rotating about an axis perpendicular to the three observable spatial axes at an imaginary angular velocity which was equal in magnitude to the orbital velocity which would occur at that elevation. (The imaginary angular rotation rate is required to make the centrifugal force attract rather than repel.) Such a model would involve high rates of imaginary shear and high imaginary coriolis accelerations between elevations. While the fact that these effects are 'imaginary' might prevent them from being observed, and therefore permit the use of such a model as a mathematical abstraction, it would seem difficult to accept the concept as a valid representation of reality.

4.9.21 - In generating the General Theory of Relativity, Dr. Einstein's use of the Principle of Relativity in combination with the Principle of Equivalence represented excellent insight. Unfortunately, his failure to renormalize the treatment to compensate for the energy added across the boundary of the thought model upon which the Principle of Equivalence was based led to contradictions which could not be reconciled within the confines of a three dimensional Euclidian Universe and led to the inclusion of an additional spatial dimension to our familiar three spatial dimensions. As we have seen, this was not only unnecessary, but it violates the Laws of Conservation of Momentum and of Energy as applied to a closed system and must therefore be invalid. Even more unfortunate is the fact that it has led generations of cosmologists and theoretical physicists astray in their attempts to reconcile General Relativity with their own work.

Section 4.10 - A Speculation on the Gravitation of and Nature of Rest Mass

4.10.1 - In previous Sections it was shown that the momentum effects associated with the slowing of the velocity of light in the gravitational field produce a gravitating force on free energy (photons, neutrinos, the kinetic energy of particles moving at a velocity incrementally close to the velocity of light, etc.) which was twice the gravitational force experienced by the same amount of energy bound to matter. (An example of the latter would be the observed weight of the energy contained in an optical integrating sphere flooded with light.) It was shown that, when such energy was bound to matter, the constraining effect of the matter produced a negative gravitational force on the matter equal to one-half of the positive force on the energy so as to cause the confined energy to gravitate with the same force as does a rest mass of the same equivalent energy. In this Section, a speculation will be provided in which it is suggested that the gravitational mass of material particles may be an inertial effect resulting from the differential momentum of energy traveling at the velocity of light in a region of space where the velocity of light has a gradient in terms of constant units of measurement. This has already been shown to be the case with electromagnetic and kinetic energy.

4.10.2 - For all practical purposes, matter consists of protons, neutrons, and electrons in various combinations. Let us examine, for example, the proton. It is a particle having an observed radius of 10^{-15} meters, a mass of $1.67*10^{-27}$ kilograms, and is currently believed to contain three subatomic particles called quarks. Multiplying the mass of the proton by the square of the velocity of light provides its rest mass equivalent energy as $1.503*10^{-10}$ joules. If we apportion this energy equally between the three quarks which make up the photon, each quark has an energy of $5.01*10^{-11}$ joules:

Eq. 4.10.1 $E_K = 5.01*10^{-11}$

The wavelength associated with this energy may be obtained from the conventional expression for the energy of a photon in terms of its wavelength and the velocity of light:

Eq. 4.10.2 $L_K = H*C/E_K$

Where:

- 'E_K' is the energy of each quark in the proton in terms of local units of measurement, in joules.

- 'L_K' is the wavelength of the quark in terms of local units of measurement, in meters.

- 'H' is Planck's Constant in terms of local units of measurement, equal to 6.626*10-34 joule-seconds.

- 'C' is the velocity of light in terms of local units of measurement, equal to $3*10^8$ meters per second.

Solving for the wavelength provides:

Eq. 4.10.3 $L_K = 3.97*10^{-15}$ meters

Suppose that the energy represented by a quark was traveling in a circular path at the velocity of light and that the circumference of that path was equal to the wavelength represented by that energy. Then the radius of the circular path would be its length divided by $2*\pi$, or:

Eq. 4.10.4 $r_K = 6.315*10^{-11}$ meters

Dividing Equation 4.10.4 by the radius of the proton $1*10^{-15}$ meters provides:

Eq. 4.10.5 $r_K/r_P = 0.6315$

Where:

- 'r_K' is the radius of the quark path, meters.

- 'r_P' is the radius of the proton, meters.

This is a familiar number. It is quite close to (1-1/e') and seems consistent with the radius that one might expect for a circular radiation performing the function of the quark within the particle (proton or neutron) formed by three quarks in combination. (The term e' in this case represents the natural number 2.73 which is the base number for Natural Logarithms.)

4.10.3 - The same argument might be applied to the electron with a slight modification. The electron has been found to have no internal structure. Suppose that an electron also consisted of energy traveling at the velocity of light in a circular path and that the nature of the electron was to cause the velocity of light to be reduced in proportion to the distance from its center, (While its is customary to regard the velocity of light as unchangeable, the fact that you are reading these words means that the relatively weak electromagnetic fields between the atoms of the lenses of your eyes is slowing the velocity of light by at least 20%, Perhaps it is not too difficult to entertain that the much stronger fields that must exist within an electron have a much more pronounced effect on the velocity of light.) Under such a supposition the electron would rotate without shear between different radii and would exhibit no discrete structure and no well defined boundary. Since the mass of the electron is 1/1836 times the mass of the proton, and the proton consists of three quarks while the electron is a single particle, the wavelength represented by the energy of the nominal velocity of light would be 612 times the wavelength of the quark as provided by Equation 4.10.3. The nominal radius of the electron, r_E, would then be :

Eq. 4.10.6 $r_E = 3.87*10^{-13}$ meters

4.10.4 - If matter were constructed in such a fashion, then all matter would consist of energy traveling at the velocity of light and gravitation would result from the effect on the momentum of that energy produced by a gradient in the velocity of light between physical locations. That gradient, in turn, would be produced by the proximity of energy. Does this, or something similar, represent reality? As was pointed out, it is a speculation. (Circa 1987)

Part 5 - The Complete Gravitational Field

List of Topics

Sections	Section Titles	Page
5.0	Introduction	351
5.1	The Gravitational Object	354
5.2	The Optical Properties of the Gravitational Field	369
5.3	The Temperature Boundaries of a Gravitating Object	384
5.4	The Minimum Time of Fall to a 'Collapsed' Object	388
5.5	Gravitation and Cosmology	392

Section 5.0 - Introduction

5.0.1- With the gravitational conversion factors determined and observationally verified, it is possible to examine the gravitational field itself. The first step will be to examine a gravitational object by applying Newton's Law of Gravitation, in terms of local units of measurement, to a model of such an object which has been artificially idealized to simplify the discussion while maintaining the necessary relevance to reality. Application of the gravitational transformations to this model shows that, providing the object cannot rid itself completely of the energy of fall produced by its contraction by radiation to space, it contracts, in terms of its own reference frame, until the gravitational potential from infinity (external units of measurement) has reached 0.5. It then expands towards an infinite radius as the gravitational potential (external units of measurement) increases from 0.5 to 1.0. The apparent expansion occurs because, beyond a gravitational potential of 0.5, the units of measurement for length contract faster than does the radius of the object, as measured with external (constant) units of measurement. At some externally observed radius which is marginally greater than the radius at which the gravitational potential (external units) reaches unity, the locally observed radius has increased to the point where the gravitationally induced pressure (local units of measurement) has declined to the level where it is balanced by radiation pressure. Further contraction of the object, in terms of the external universe, ceases. Observed externally, contraction ceases at a radius marginally greater than the Horizon Radius of General Relativity. [**Note:** At the time this text was copyrighted, the author was under the misapprehension that the Schwarzchild Radius was the radius of the 'event horizon' of the postulated 'Black Hole'. The Schwarzchild Radius actually represents the radius at which a ray of light can circle endlessly and is twice the 'black hole' radius under General Relativity and is three times that radius under the treatment provided in this discussion. To avoid confusion, this text differs from its originally copyrighted version by correcting this misdesignation and replacing the term "Schwarzchild Radius" with the term "Horizon Radius". This correction does not alter the meaning of the discussion. It should also be noted that the designation is correct in the subsequent text, "The Einstein Hoax", copyrighted in 1997.] If the object, as observed externally, is warmer than the external space, it will then contract slowly as its loss of energy reduces its Horizon Radius. The singularity commonly designated as a 'Black Hole' does not occur in Nature. As observed internally, the object will have an enormous radius. To both the internal and the external observer, the collapsed object will have an extremely low matter density, consisting mostly of 'empty' space.

5.0.2- As observed externally, the collapsing gravitational object appears to contract to a radius of 6.75 times the Horizon Radius at a gravitational potential of 1/3, at which point contraction apparently ceases. This apparent cessation of contraction results from the refractive optical properties of the gravitational field, and occurs when the actual radius of the object (external units of measurement) is equal to or less than three times the Horizon Radius. To the internal observer viewing the external universe, total internal reflection commences at the horizon when the radius has reduced to three times the Horizon Radius (external units of measurement) and proceeds towards the zenith as the Horizon Radius is approached. This total internal reflection might be expected to produce a 'greenhouse effect' which in combination with the reduction in size of the units of measurement for energy within the object, causes the internally observed equilibrium temperature of the object with the internal space to be enormously higher than the observed temperature of space and strongly impedes the loss of energy to space during the final stages of gravitational contraction. In reality, the anticipated 'greenhouse effect' does not occur because, as shall be shown, the externally observed refraction effects of the field act to increase the capture area of the object.

5.0.3- While it is conventional to assume that the final stages of gravitational contraction occur with extreme rapidity, this is not the case in terms of the external universe. It is shown that the time, in the external universe, for a photon to reach the surface of a collapsed object approaches infinity as the gravitational potential approaches unity, This occurs because, when the gravitational potential becomes large enough, the gravitational field reduces the velocity of light faster than the radius is reduced, and the photon never quite arrives at the Horizon Radius. Ignoring, for the moment, the time dilation effects described by Special Relativity, an observer riding along with the photon would, however, consider that the time to arrive at the surface of the object was essentially equivalent to the externally observed distance of travel divided by the velocity of light since his units of measurement of time were increasing in size as the field became more intense. It should be noted that, when correctly applied, General Relativity yields an identical conclusion. Under General Relativity, the velocity of light does not change, but instead, additional space is created faster than the photon approaches the object, and again, the photon requires an infinite time to reach the Horizon Radius. If 'Black Holes' could form, they still would not exist since the apparent age of the universe is not infinite.

5.0.4- If one examines the gravitationally collapsing object from the viewpoint of an internal observer at the point in time where the contraction is virtually complete, a large number of parallels to what we observe in our own universe are apparent. The observer would conclude that at some point in the past there was a 'big bang' at which time most, if not all of the matter in the universe was created. Since the 'big bang', that matter would seem to have been flying apart with the radius of the object becoming enormous and with the density and temperature of the object becaming very low. A good approximation of what an observer would see can be obtained by looking up on a clear moonless night.

Section 5.1-The Gravitational Object

5.1.1- In order for a treatment of gravitation to be complete, it must cover the gravitational field over its complete range, from Θ equals zero to Θ equals unity, not merely over small changes in elevation. To provide the description, it will be assumed that the values of X, Y, and Z which were determined for extremely small values of Θ are valid for all values of Θ. It must be remembered that such an assumption represents an interpretation far beyond the limits of observation and therefore may not represent reality.

5.1.2- Observable gravitational fields are generated by a central body which is nominally spherical in shape and which contain rest mass distributed non-uniformly throughout its volume. The matter which comprises this rest mass is compressible. Furthermore, the gravitational potential increases toward the center. These effects greatly complicate analysis, particularly as the gravitational potential begins to approach unity. The complications may be by-passed by choosing a rather artificial model of the central mass, such as shown in Figure 5.1.1. The conclusions reached with this model will be valid for distances from the center of the model which are as large or larger than its radius.

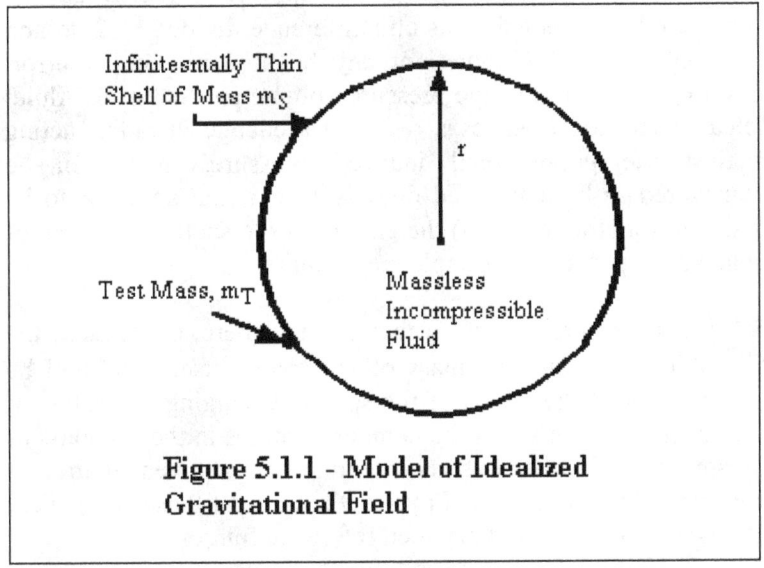

Figure 5.1.1 - Model of Idealized Gravitational Field

5.1.3- Consider a hollow spherical shell of locally measured mass m_S, of locally measured radius r, and of infinitesimal thickness. Resting upon this shell is an infinitesimal test object of locally measured mass m_T. Since such a shell would collapse of its own gravitational attraction, it is considered to be filled with a hypothetical fluid which is incompressible, has a specific heat of zero, and is devoid of gravitational mass, The advantages of such a model are that all of the mass and the entire volume of the sphere is at the same gravitational potential as the surface, enabling the model to be described in terms of a single value of Θ, and insuring that gravitational energy release by contraction (or absorbed by expansion) is not stored in (or released by) the central fluid or the shell as heat. Since with this model, gravitational contraction (or expansion) can only occur as a result of the withdrawal (or addition) of fluid, gravitational energy appears as the pressure-volume product of the fluid withdrawn (or added) to change the radius of the sphere. In this model, the classical integration of Newtonian gravitation over the entire volume is valid. All of the mass may be considered to be concentrated at the center of the sphere, its radius may be considered to be equal to its circumference divided by $2*\pi$, and Newton's Law of Gravitation may be applied without error. Energy, in the form of the pressure-volume product of the fluid, released (or absorbed) as a result of a change in radius acting against the gravitational induced pressure, may then be considered to be conducted directly to external space or to be added to (subtracted from) the energy of the shell in the form of heat, as the particular example may require.

5.1.4- As measured at the surface of the sphere, the mass of the test object, m_T, and the mass of the sphere itself, m_S, will be independent of the radius of the sphere. Changing the radius of the sphere will not affect the number of atoms in the test mass or in spherical shell and, since the mass of a given atom is a constant of the Science of Physics, the mass of these component atoms must be invariant between reference frames.

5.1.5- In Section 4.5, it was seen that the observed orbit of Mercury is consistent with the predictions of Newtonian Gravitational Theory when one corrects the planet's path for the effects of orbital velocity (Special Relativity) and the effects of the gravitationally induced 'refraction' of the planet's velocity vector. Newton's Law of Gravitation remains observationally valid and may be employed as a Law of the Science of Physics under the conditions which apply to the model of Figure 5.1.1. (Newton's Law of Gravitation will therefore be invariant between reference frames. The gravitational constant, however, may itself be a function of Θ, and, if so, a modification of the conclusions which follow would be required.)

5.1.6- Under Newtonian gravitation, the radius of the attracting mass has no effect upon the attractive force provided that the distance to the center of the mass is at least as great as the radius of the mass. As a result, as the hypothetical sphere of Figure 5.1.1 contracts from an infinite radius, it may be considered to generate a gravitational field at each radius which does not change as the radius of the sphere is reduced to a lower value. Consequently, the solution which is obtained for the gravitational field at a given radius of the sphere is valid for that distance from the center of the sphere for all values of the sphere's radius which are equal to or less than that distance.

5.1.7- The gravitational force acting on the test object, m_T, is provided by Newton's Law of Gravitation as:

Eq. 5.1.1 $\quad f = -g * m_S * m_T / r^2$

Where:

- 'f' is the gravitational force as measured at the surface of the sphere in terms of local units of measurement.

- 'r' is the radius of the sphere as measured at its surface in terms of local units of measurement.

- 'g' is the gravitational constant as measured in terms of local units of measurement.

- 'm_S' is the mass of the sphere as measured in terms of local units of measurement.

- 'm_T' is the mass of the test object as measured in terms of local units of measurement.

It is convenient to modify the preceding equation so as to express the force in terms of the energy equivalents of the masses through the use of the ergo-gravitational constant, d. In terms of local units of measurement:

Eq. 5.1.2 $d = g/c^4$

And:

Eq. 5.1.3 $e = m*c^2$

Where:

- 'c' is the invariant (locally measured) velocity of light.

- 'e' is the invariant (locally measured) energy equivalent of the locally measured mass.

Combining Equation 5.1.1, 5.1.2, and 5.1.3 provides:

Eq. 5.1.4 $f = d*e_{SL}*e_{TL}/r^2$

Where:

- 'e_{SL}' is the locally measured energy equivalent of the mass of the spherical object at the lower elevation.

- 'e_{TL}' is the locally measured energy equivalent of the mass of the test object at the lower elevation.

And, since the gravitational conversion factor for the ergo-gravitational constant has been shown (Table 3.5.1) to be unity:

Eq. 5.1.5 $f = D*e_{SL}*e_{TL}/r^2$

(It will be noted that, for the shell model provided, Equation 5.1.5 will also be valid of the energies represented by the test object and/or the spherical object are in the form of radiation rather than physical matter. Radiation will be confined by the shell and will therefore gravitate at the normal gravitational acceleration, in a manner analogous to the effect on radiation confined in an integrating sphere, rather than at twice the normal gravitational acceleration, as has been shown to be the case for unconfined radiation.)

5.1.8- Decreasing the radius of the sphere by an incremental, δr, causes work to be done on the test object by the gravitational field, releasing an amount of work locally measured as δe_F:

Eq. 5.1.6 $\delta e_F = f * \delta r$

Where:

- 'e_F' is the locally measured energy of fall of the test object.

Then:

Eq. 5.1.7 $\delta e_F = -D * e_{SL} * e_{TL} * \delta r / r^2$

From Table 3.5.1, in terms of external units of measurement:

Eq. 5.1.8 $\delta e_F = \delta E_F / (1 - \Theta)$

Eq. 5.1.9 $r = R / (1 - \Theta)$

And:

Eq. 5.1.10 $\delta r = \delta R / (1 - \Theta)$

Then combining Equations 5.1.7 through 5.1.10 provides:

Eq. 5.1.11 $\delta E_F / (e_{TL} * [1-\Theta]^2) = -D * e_{SL} * \delta R / R^2$

If none of the energy of fall is radiated to space (i.e.- the pressure-volume product of the central fluid which was removed to allow the sphere to contract is discharged as heat to the spherical shell and test mass), the locally measured energy of the test object and the spherical object will be equal to the energy of these objects prior to the contraction (as measured with external units of measurement) multiplied by the gravitational conversion factor for energy provided by Table 3.5.1. Then:

Eq. 5.1.12 $e_{TL} = E_{TL}/(1-\Theta)$

And:

Eq. 5.1.13 $e_{SL} = E_{SL}/(1-\Theta)$

Where:

- 'E_{TL}' is the locally measured energy equivalent of the test object as measured with external units of measurement.

- 'E_{SL}' is the locally measured energy equivalent of the spherical object as measured with external units of measurement.

And since all of the energy of contraction is returned to the test object and spherical object:

Eq. 5.1.14 $E_{TL} = E_T$

Eq. 5.1.15 $E_{SL} = E_S$

Combining Equations 5.1.12 and 5.1.14 and Equations 5.1.13 and 5.1.15 provides:

Eq. 5.1.16 $e_{TL} = E_T/(1-\Theta)$

Eq. 5.1.17 $e_{SL} = E_S/(1-\Theta)$

Where:

- 'E_T' is the energy of the test object prior to the contraction as measured with external units of measurement.

- 'E_S' is the energy of the sphere prior to contraction as measured with external units of measurement.

Combining Equation 5.1.11, 5.1.16, and 5.1.17 provides:

Eq.5.1.18 $\delta E_F/E_T = -D*E_S*\delta R/R^2$

But, from the definition of gravitational potential:

Eq. 5.1.19 $\delta E_F/E_T = \delta\Theta$

Combining Equations 5.1.18 and 5.1.19 provides:

Eq. 5.1.20 $\delta\Theta = -D*E_S*\delta R/R^2$

Integrating Equation 5.1.20 between the limits of zero and Θ for Θ and between infinity and R for R provides:

Eq. 5.1.21 $\Theta = D*E_S/R$

Or:

Eq. 5.1.22 $R = D*E_S/\Theta$

Setting Θ equal to its maximum value of unity provides:

Eq. 5.1.23 $R_S = D*E_H$

Or:

Eq. 5.1.24 $R_H = R_S/\Theta$

Where:

- 'R_H' is the radius of the object in terms of external units of measurement under conditions where no energy of fall is radiated to space.

- 'R_S' is the radius of the gravitating object under conditions of complete gravitational collapse (Θ = unity) where no energy of fall is radiated to space, as measured with external units of measurement. R_S is identical to the Horizon Radius associated with General Relativity.

Combining the radius provided by Equation 5.1.22 with the conversion factor for length provided by Table 3.5.1 provides the radius of the gravitating object as a function of Θ in terms of the units of measurement existing at the surface of the object:

Eq. 5.1.25 $r_H = D*E_S/(\Theta*[1-\Theta])$

Or, in terms of the Horizon Radius:

Eq. 5.1.26 $r_H = R_S/(\Theta*[1-\Theta])$

5.1.9- At the other extreme is the condition where all of the energy of fall is radiated to space. Under this condition, the energy of the spherical object does not remain unchanged in terms of the external units of measurement (infinite distance), but rather, remains unchanged in terms of local units of measurement and is reduced in proportion to $(1-\Theta)$ in terms of external units of measurement.then:

Eq. 5.1.27 $E_{SL} = E_S*(1-\Theta)$

Adjusting for the reduction of energy defined by Equation 5.1.27 and analogously with Equation 5.1.22 one obtains:

Eq. 5.1.28 $R_C = D*E_S*(1-\Theta)/\Theta$

Or:

Eq. 5.1.29 $R_C = R_S*(1-\Theta)/\Theta$

And, in terms of the units of measurement at the surface of the sphere:

Eq. 5.1.30 $r_C = D*E_S/\Theta$

Or:

Eq. 5.1.31 $r_C = R_S/\Theta$

Where:

- 'R_C' is the radius of the object under conditions where all of the energy of fall is radiated to space, in terms of external units of measurement.

- 'r_C' is the same radius in terms of local units of measurement.

5.1.10- It will be noted that, in the case where the gravitating mass radiates all of its energy of contraction to external space, the radius of the gravitationally collapsed mass ($\Theta = 1$) is zero as measured with external units of measurement and is equal to the Horizon Radius as measured with local units of measurement. In the case where the gravitating mass cannot radiate any of its energy of contraction to space, the radius of the gravitationally collapsed mass ($\Theta = 1$) is equal to the Horizon Radius as measured with the external units of measurement. In terms of the local units of measurement, the radius reaches a minimum of four times the Horizon Radius at a value of Θ equal to 0.5. For values of Θ above 0.5, the locally measured radius increases as Θ increases, becoming infinite when Θ equals unity! The implications of this result will be discussed in more detail in Section 5.5 (Cosmology).

5.1.11- Let us now consider the pressure inside the sphere for these two cases. If we define the test mass as a portion of the shell subtending as solid angle of 'N' steradians, then, in terms of local units of measurement, in the area of the shell represented by the test mass for $N*r^2$ while the area of the entire shell is $4*\pi*r^2$. Then:

Eq. 5.1.32 $e_T = N*e_S/(4*\pi)$

Combining Equations 5.1.5 and 5.1.32 provides the gravitational force on the test mass:

Eq. 5.1.33 $f = -D*N*e_S^2/(4*\pi)$

Since pressure is force per unit area and is positive when the force provided by Equation 5.1.33 is negative, the pressure, p, becomes:

Eq. 5.1.34 $p = D*e_S^2/(4*\pi*r^4)$

Where the gravitational energy is completely radiated to space, the energy represented by the spherical mass is invariant, then:

Eq. 5.1.35 $p_C = D*E_S^2/(4*\pi*r^4)$

Combining Equations 5.1.30 and 5.1.35 provides:

Eq. 5.1.36 $p_C = \Theta^4/(4*D^3*E_S^2*\pi)$

And, in terms of external units of measurement for pressure using the conversion factors for force and length provided by Table 3.5.1:

Eq. 5.1.37 $P_C = \Theta^4/(4*D^3*E_S^2*\pi*[1-\Theta]^2)$

Where:

- 'p_C' is the pressure in the gravitating mass when all energy of contraction has been removed as measured with local units of measurement.

- 'P_C' is the pressure in the gravitating mass when all energy of contraction has been removed as measured with external units of measurement.

In the case where no energy is radiated to space:

Eq. 5.1.38 $e_S = E_S/(1-\Theta)$

And Equation 5.1.34 becomes:

Eq. 5.1.39 $p_H = D*E_S^2/(4*\pi*[1-\Theta]^2*r^4)$

And, combining Equations 5.1.25 and 5.1.39 provides:

Eq. 5.1.40 $p_H = \Theta^4*(1-\Theta)^2/(4*D^3*E_S^2*\pi)$

And, in terms of external units of measurement for pressure:

Eq. 5.1.41 $P_H = \Theta^4/(4*D^3*E_S^2*\pi)$

Where:

- 'p_H' is the pressure in the sphere with no removal of energy of contraction as measured with local units of measurement.

- 'P_H' is the pressure in the sphere with no removal of the energy of contraction, as measured with external units of measurement.

Equation 5.1.40 shows that, under conditions of no energy removal, the locally measured pressure within the massive object increases to a maximum as Θ increases from zero to 2/3 and falls to zero as Θ increases from 2/3 to unity. As is the case for the locally measured radius, this result has profound cosmological implications.

5.1.12- Finally, let us consider the energy density of the contracted mass for the two cases, in terms of both local and external units of measurement. For the case where all of the energy of fall is radiated to space, combining Equations 5.1.28 and 5.1.30 with the expression for the volume of a sphere provides:

Eq. 5.1.42 $E_{D1} = 0.75*\Theta^3/(D^3*E_S^2*[1-\Theta]^3*\pi)$

And:

Eq. 5.1.43 $e_{D1} = 0.75*\Theta^3/(D^3*e_S^2*\pi)$

And for the case where no energy is radiate to space, combining Equations 5.1.13 and 5.1.25 with the expression for the volume of a sphere provides:

Eq. 5.1.44 $E_{D2} = 0.75*\Theta^3/(D^3*E_S^2*\pi)$

And:

Eq. 5.1.45 $e_{D2} = 0.75*\Theta^3*(1-\Theta)/(D^3*e_S^2*\pi)$

Where:

- 'E_{D1}' is the energy density of the object with all of the gravitational energy radiated, as measured with external units of measurement.

- 'e_{D1}' is the energy density of the object with all gravitational energy radiated, as measured with local units of measurement.

- 'E_{D2}' is the energy density of the object with no energy radiated, as measured with external units of measurement.

- 'e_{D2}' is the energy density of the object with no energy radiated, as measured with local units of measurement.

5.1.13- In Section 5.2, the effects of the gravitational field on the refraction of light will be discussed. It will be shown that, as the gravitational potential approaches unity, it becomes progressively more difficult for energy to escape to space so that a collapsing gravitational object eventually approaches the conditions represented by the second case (no heat loss). This effect should have significant cosmological implications.

5.1.14- Consider now the effects of the gravitational field as observed externally. If we define the energy which comprises the gravitational object in terms of the units of measurement existing at an infinite radius as E, then the radius, the internal pressure, and the energy density, in terms of those same units of measurement are identical to the results provided in terms of Equation 5.1.23, 5.1.24, 5.1.41, and 5.1.44:

Eq. 5.1.46 $R = D*E/\Theta$

Eq. 5.1.47 $R = R_S/\Theta$

Eq. 5.1.48 $P = \Theta^4/(4*D^3*E^2*\pi)$

Eq. 5.1.49 $E_D = 0.75*\Theta^3/(D^3*E^2*\pi)$

Where:

- 'E' is the energy represented by the gravitating object in terms of external units of measurement.

- 'R' is the radius of the object in terms of external units of measurement.

- 'R_S' is the Horizon Radius associated with the energy content of the object.

- 'P' is the pressure within the object in terms of external units of measurement.

- 'E_D' is the energy density within the object in terms of external units of measurement.

Note: - Equation 5.1.45 shows that, with the ability of a collapsing gravitational object to radiate its energy severely limited, as is the case in the final stages of gravitational contraction, the internally observed energy density approaches zero as a limit. While this conclusion may seem strange, it also follows from General Relativity. Under General Relativity, as the object collapses, the gravitational field 'calls into being' additional space to accommodate the postulated gravitationally induced curvature. As the contraction approaches the condition where the escape velocity is equal to the velocity of light, the rate at which additional space is 'called into being' approaches infinity faster than the rate at which the externally observed radius approaches zero.

5.1.15- The size of the units of measurement listed in Table 3.5.1 are provided in Table 5.1.1 for a value of unity for the gravitational potential, Θ.

Table 5.1.1 - Sizes of Units of Measurement at Θ of Unity

Quantity	Size of Unit of Measurement
Length	Zero
Time	Infinite
Force	Unchanged
Charge	Zero
Energy	Zero
Momentum	Infinite
Angular Momentum	Unchanged
Planck's Constant	Unchanged
Velocity	Zero
Acceleration	Zero
Mass	Infinite
Gravitational Constant	Zero
Dielectric Constant of Space	Unchanged
Permeability of Space	Infinite
Existence	Unchanged
Ergo-Gravitational Constant	Unchanged

Section 5.2- The Optical Properties of the Gravitational Field

5.2.1- The ability of the gravitational field to refract both light and the velocity vectors of moving objects requires that the refractive properties of the field be examined in detail. These properties are readily determined from the conversion factors provided in Table 3.5.1 in combination with the conventional laws of physical optics. It will be shown that, in terms of the conditions that obtain at an infinite distance:

- A photon can circle endlessly at three times the Horizon Radius (R=3*D*E) about an object of equivalent energy 'E' due to gravitational refraction. (The path is divergently unstable.)

- An object as small or smaller than three times the Horizon Radius will appear to have a radius of 6.75 times the Horizon Radius.

- Photons (or particles) emitted from the surface having a radius smaller than three times the Horizon Radius must be emitted closer to the zenith in order to escape to space as the radius approaches the Horizon Radius. The "greenhouse effect" which this effect might be expected to produce is compensated by the large "gravitational refraction" induced increase in oprical size of the collapsed object described later in this Section.

(The conclusions of this Section may require revision if the gravitational constant is not invariant between reference frames.)

5.2.2- Consider the path of a photon traveling horizontally in a gravitational field. In accordance with the laws of physical optics, the path of a ray of light will be refracted by a medium in which the velocity of propagation of light possesses a gradient having a component normal to its path. The conversion factor for velocity provided in Table 3.5.1 and the requirement that the velocity of light be invariant between reference frames requires that the space which surrounds a gravitating object behave as if it were in such a medium. Figure 5.2.1 illustrates the path of a ray of light of width δR initially traveling horizontally in a gravitational field. In that diagram, the photons comprising a ray of light which is traveling tangentially to the gravitational field at a distance R from its center have a velocity V_U at elevation 1 and a velocity V_L at elevation 2. The distance between these elevations is the incremental distance δR.

Note:- All of the units of measurement employed in this Section are the units of measurement which exist at an infinite distance from the gravitating object.

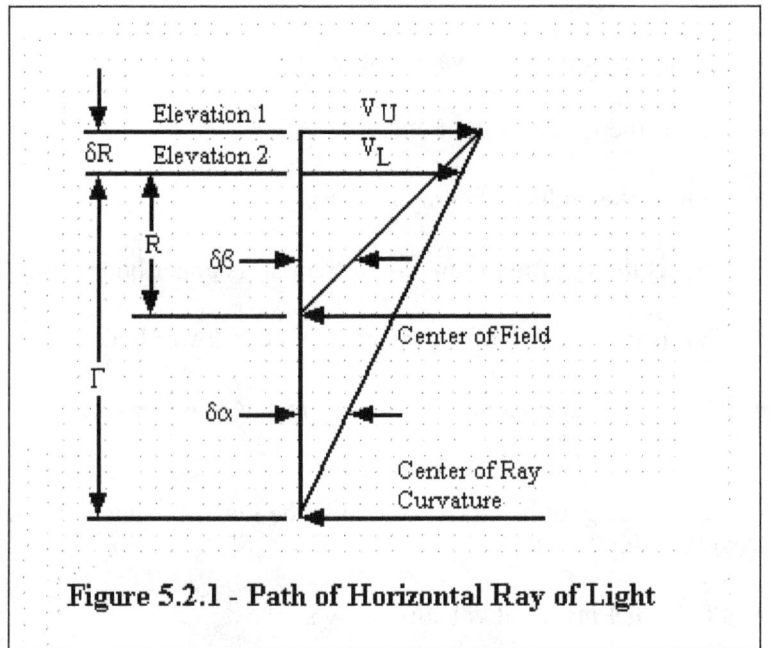

Figure 5.2.1 - Path of Horizontal Ray of Light

From the geometry of Figure 5.2.1:

Eq. 5.2.1 $(\Gamma+\delta R)/\Gamma = V_U/V_L$

But, from Table 3.5.1:

Eq. 5.2.2 $V_U = (1-\Theta_U)^2 * C'$

Eq. 5.2.3 $V_L = (1-\Theta_L)^2 * C'$

Then:

Eq. 5.2.4 $V_U/V_L = (1-\Theta_U)^2/(1-\Theta_L)^2$

Where:

- 'Γ' is the radius of curvature of the ray of light.

- 'R' is the radius of the field.

- 'δR' is the width of the ray of light.

- 'V_U' is the velocity of the ray of light at its upper boundary.

- 'V_L' is the velocity of the ray of light at its lower boundary.

- 'Θ_U' is the gravitational potential at the upper boundary of the ray of light.

- 'Θ_L' is the gravitational potential at the lower boundary of the ray of light.

- 'C' ' is the invariant velocity of light.

Defining:

Eq. 5.2.5 $\delta\Theta = \Theta_U - \Theta_L$

Eq. 5.2.6 $\Theta_L = \Theta$

Then:

Eq. 5.2.7 $V_U/V_L = (1-\Theta-\delta\Theta)^2/(1-\Theta)^2$

Since $\delta\Theta$ may be considered to be an infinitesimal, Equation 5.2.7 may be expanded and simplified with higher order terms of $\delta\Theta$ ignored. Then, combining Equations 5.2.1 and 5.2.7 provides:

Eq. 5.2.8 $\delta R/\Gamma = -2*\delta\Theta)/(1-\Theta)$

From Equation 5.1.47:

Eq. 5.2.9 $\quad R = R_S/\Theta$

Then:

Eq. 5.2.10 $\quad \delta R = -R_S * \delta\Theta/\Theta^2$

Combining Equations 5.2.8 and 5.2.10 provides:

Eq. 5.2.11 $\quad \Gamma = R_S*(1-\Theta)/(2*\Theta^2)$

Since Γ represents the radius of curvature of a horizontal ray of light and R represents the radius of the field at the elevation of that ray, both in terms of the gravitational potential Θ, setting R equal to Γ and combining Equations 5.2.9 and 5.2.11 provides the value of Θ at which a photon would travel in a continuous circle about the central object:

Eq. 5.2.12 $\quad \Theta_{CP} = 1/3$

Substituting this value of Θ into Equation 5.2.9 provides:

Eq. 5.2.13 $\quad R_{CP} = 3*R_S$

Where:

- 'Θ_{CP}' is the gravitational potential at which photons can circle continuously about the gravitational object.

- 'R_{CP}' is the radius at which photons can circle continuously about the gravitating object.

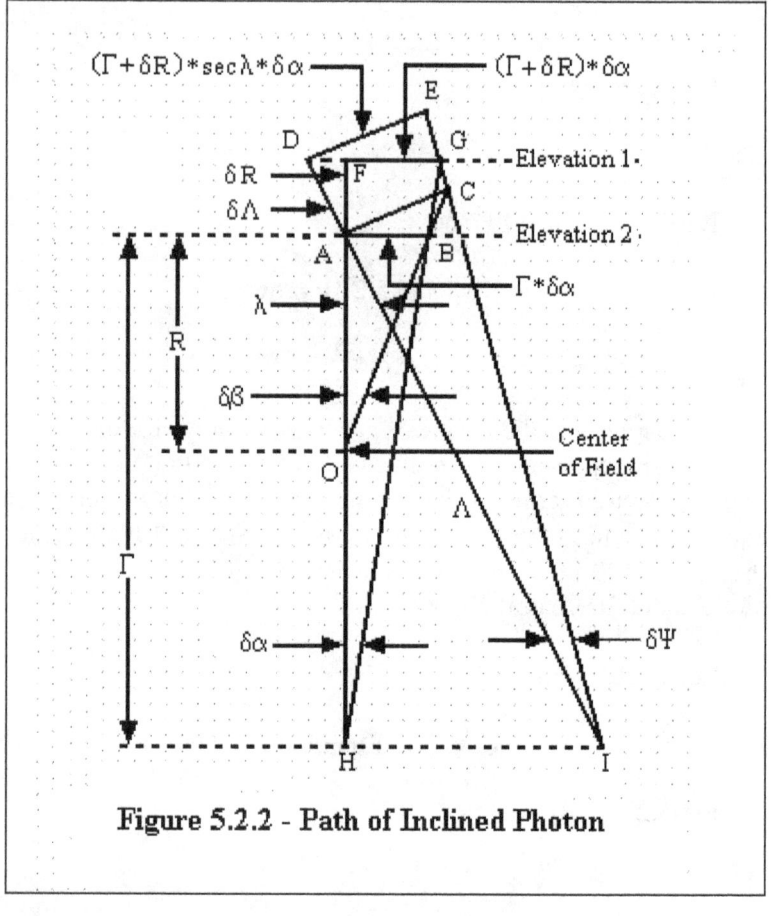

Figure 5.2.2 - Path of Inclined Photon

Equation 5.2.13 states that a photon will circle endlessly at a radius of three times the Horizon Radius. The path, however, is divergently unstable, and the slightest departure from a locally horizontal path will cause the photon either to escape to space or to spirally converge on the gravitating object.

5.2.3- The fact that photons can circle endlessly about a gravitating object due to gravitational refraction at a radius equal to three times the Horizon Radius does not represent a barrier to their escape from an object. In order to determine the effects involved, it is necessary to combine the laws of physical optics with the refractive index (i.e.- the gravitational transformation for velocity) of the field. Figure 5.2.2 expands the diagram of Figure 5.2.1 to include the behavior of a photon which is traveling at an angle of λ to the horizontal. It will be used to first show that the incremental refraction of a ray of light, as a function of the angular motion of the wave front about the center of the field, is independent of the elevation angle of that ray.

5.2.4- Consider two rays of light passing through a point A in Figure 5.2.2. Ray 1 travels horizontally and changes direction by an incremental angle $\delta\alpha$ as the central angle changes by an angle $\delta\beta$. Ray 2 travels at an angle λ to the horizontal and changes direction by an angle $\delta\Psi$ as the central angle changes by angle $\delta\beta$. Considering ray 1 first, as the central angle changes by $\delta\beta$, a photon at elevation 2 travels a distance of $\Gamma*\delta\alpha$ and a photon at elevation 1 travels a distance of $(\Gamma+\delta R)*\delta\alpha$. For ray 2, as the central angle changes by $\delta\beta$, a photon at elevation 2 travels a distance $\Gamma*\sec(\lambda)*\delta\Psi$ and a photon at elevation 1 travels a distance of $(\Gamma+\delta R)*\sec(\lambda)*\delta\Psi$.

Therefore:

Eq. 5.2.14 $AC = \Gamma*\sec(\lambda)*\delta\Psi$

Eq. 5.2.15 $DE = (\Gamma+\delta R)*\sec(\lambda)*\delta\Psi$

Eq. 5.2.16 $FG = (\Gamma+\delta R)*\delta\alpha$

Eq. 5.2.17 $\delta\Lambda = \sec(\lambda)*\delta R$

Since triangles IDE and IAC are similar:

Eq. 5.2.18 $\Gamma*\sec(\lambda)*\delta\Psi/\Lambda = (\Gamma+\delta R)*\sec(\lambda)*\delta\Psi/(\Lambda+\delta\Lambda)$

From Figure 5.2.2:

Eq. 5.2.19 $\Lambda = \Gamma * \sec(\lambda)$

But:

Eq. 5.2.20 $\delta\Psi = \Gamma * \sec(\lambda) * \delta\Psi / \Lambda$

Then:

Eq. 5.2.21 $\delta\Psi = \delta\alpha$

Equation 5.2.21 shows that the incremental angle through which a photon is refracted as a result of an incremental change in its angular position in the gravitational field is independent of the elevation angle of its path.

5.2.5- It is next necessary to establish that the elevation angle of the photon's path as a function of its instantaneous radius in the field and of the angle through which its path crossed a predetermined elevation, R. It may be recognized that the distance AB of Figure 5.2.2 may be provided in two ways:

Eq. 5.2.22 $AB = R * \delta\beta$

And:

Eq. 5.2.23 $AB = \Gamma * \delta\alpha$

Then:

Eq. 5.2.24 $R * \delta\beta = \Gamma * \delta\alpha$

Combining Equations 5.2.9, 5.2.11, 5.2.21 and 5.2.24 provides:

Eq. 5.2.25 $\delta\Psi = 2 * \Theta * \delta\beta / (1 - \Theta)$

5.2.6- As the photons of ray 2 move from point A to point C, the central angle changes by $\delta\beta$, causing the 'horizontal' to curve away from the path of the photons by the angle $\delta\beta$. At the same time, the path of the photons curves towards the horizontal by the angle $\delta\Psi$. The angle between the path of the photons and the horizontal, λ, changes by $\delta\lambda$, where:

Eq. 5.2.26 $\delta\lambda = \delta\alpha - \delta\Psi$

Combining Equations 5.2.25 and 5.2.26 provides:

Eq. 5.2.27 $\delta\lambda = (1-3*\theta)*\delta\beta/(1-\Theta)$

At this point, a simplification of Figure 5.2.2 may be made. As the central angle changes by $\delta\beta$, the photons traveling at angle λ move from point A to point C. The radius of the photon's location increases by δR. If the angle $\delta\beta$ is infinitesimal, triangle ABC may be assumed to be a right triangle, then, from Figure 5.2.3:

Eq. 5.2.28 $\delta R = R*\tan(\lambda)*\delta\beta$

Combining Equations 5.2.9, 5.2.27, and 5.2.28 provides:

Eq. 5.2.29 $\tan(\lambda)*\delta\lambda = (R-3*R_S)*\delta R/([R-R_S]*R)$

Integrating between the limits of λ_1 and λ_2 for λ and between the limits of R_1 and R_2 and simplifying provides:

Eq. 5.2.30 $\cos(\lambda_1)/\cos(\lambda_2) = R_2^3*(R_1-R_S)^2/(R_1^3*[R_2-R_S]^2)$

5.2.7- Equation 5.2.30 permits the description of the optical properties of the gravitational field. Consider first the diameter of the gravitating object as observed from an external point far removed from the object. The observed boundary of the spherical object is defined by rays of light which were emitted tangentially from its surface. In terms of Equation 5.2.30, λ is zero. The geometry for the observation at the observer's distance,

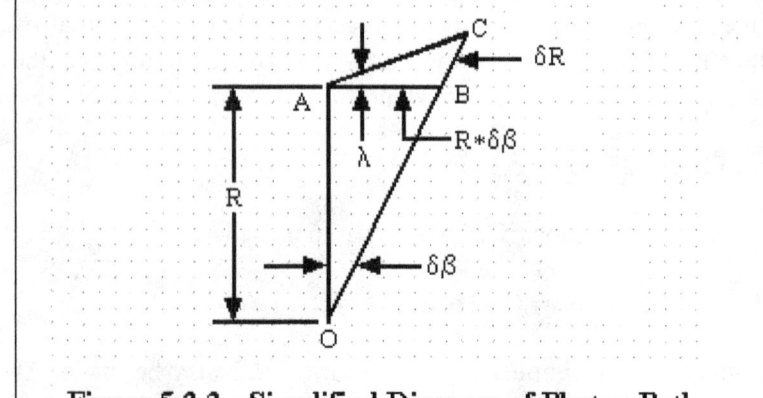

Figure 5.2.3 - Simplified Diagram of Photon Path

R_O, is provided in Figure 5.2.4. The observed radius of the object, R_V, is provided by:

Eq. 5.2.31 $R_V = R_O * \cos(\lambda)$

Where:

- 'λ' is the elevation angle, with respect to the normal to the line of sight to the center of the object, of the light defining the edge of the object as seen at the observer's location.

5.2.8- In the case where the path of the ray of light from the edge of the object is refracted by the object's gravitational field, the apparent diameter of the object is increased by the bending of the light rays as shown in Figure 5.2.5. The angle λ of Figure 5.2.5 is the angle λ_2 of Equation 5.2.30. Recognizing that, for the case in question, angle λ_1 equals zero, R_1 equals R, and R_2 equals R_O, combining Equations 5.2.30 and 5.2.31, we obtain:

Eq. 5.2.32 $R_V = R^3 * (R_O - R_S)^2 / (R_O^2 * [R - R_S]^2)$

And, as the distance of the observer, R_O, approaches infinity, R_V approaches:

Eq. 5.2.33 $R_V = R^3/(R-R_S)^2$

5.2.9- The value of R_V in the preceding equation is a minimum when R is equal to 3 times R_S. As shown by Equation 5.2.13, this value of R represents the radius at which a photon will circle endlessly due to refraction, and represents the smallest radius at which a photon emitted tangentially can escape from the gravitational field. Equation 5.2.33 is applicable, therefore, for values of R equal to or greater than $3*R_S$. Substituting this value of R into Equation 5.2.33 provides the radius, as seen by an external observer, of a gravitational object having a gravitational potential at its surface equal to or greater than 1/3 (Equation 5.2.9):

Eq. 5.2.34 $R_{VM} = 6.75*R_S$

Where:

- 'R_{VM}' is the minimum visual radius of a gravitational object having a radius equal to or less than $3*R_S$.

5.2.10- Equation 5.2.34 provides the apparent diameter of a highly collapsed object (Θ equal to or greater than 1/3) as observed with light that escaped tangentially from a spherical surface having a radius of $3*R_S$. In order for photons to be emitted from such a surface, however, they must be supplied by the object within the surface. To evaluate that supply mechanism, it is necessary to again employ Equation 5.2.30. this time setting λ_1 equal to zero, λ_2 equal to λ, R_1 equal to $3*R_S$, and R_2 equal to R. Then:

Eq. 5.2.35 $\lambda = \arccos(6.75*R_S*[R-R_S]^2/R^3)$

5.2.11- A photon emitted from the surface of a gravitating object having a radius of less than $3*R_S$ at an angle of α (to the xenith)

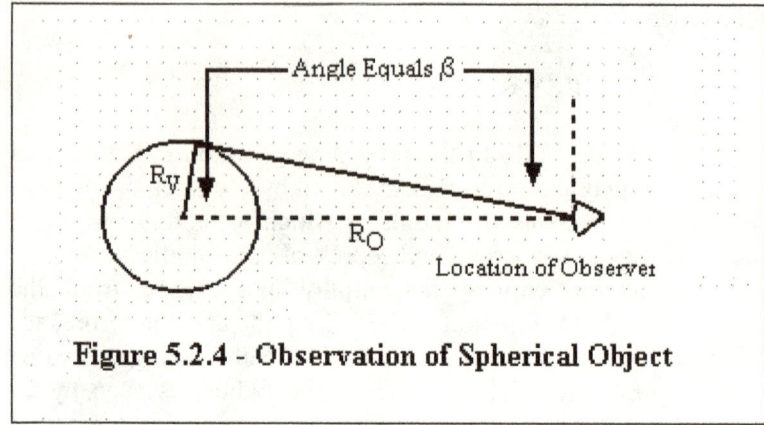

Figure 5.2.4 - Observation of Spherical Object

will arrive at α radius of $3*R_S$ traveling horizontally, as required by Equation 5.2.34. Photons emitted from the surface at an angle less than α (to the zenith) will cross the radius of $3*R_S$ at an angle greater than zero and will escape to external space. The spherical surface at a radius of $3*R_S$ does not represent a barrier to the escape of photons. It represents, rather, the radius at which total internal reflection begins. As the object contracts below this radius, a region of total internal reflection begins at the horizon and proceeds towards the zenith as the object contracts to its limiting value of R_S, as shown in Figure 5.2.6. (The effect is similar to that observed when looking upwards from the bottom of a swimming pool.) The sky is seen only a region near the vertical. An observer at A looking upward at elevation angles less than α (to the zenith) will see light coming in from "outer space". At angles equal to or greater than α, the light will have been emitted from some other portion of the gravitating object.

5.2.12- The optical effects of the gravitational field cause it to behave as a trap for all forms of energy. The capture area for particles traveling from space to the gravitating object is equal to the area of a sphere having the radius provided by Equation 5.2.33 for values of R greater than $3*R_S$ and having a radius of $6.75*R_S$ for values of R less than $3*R_S$, as shown by Figure

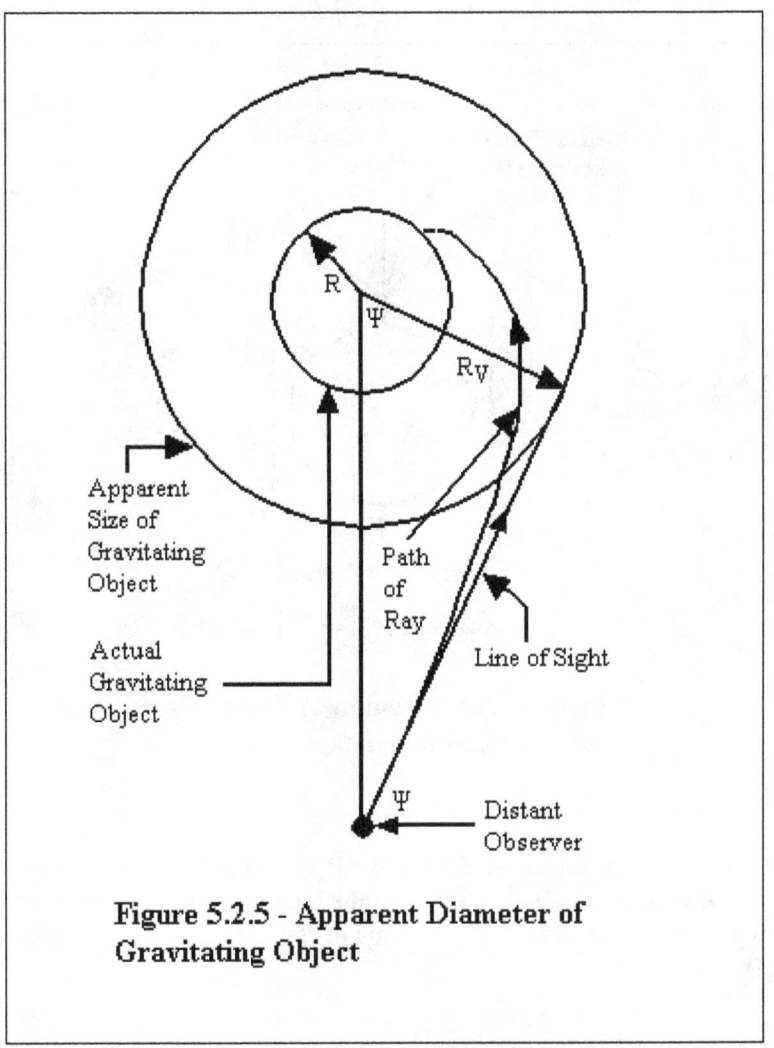

Figure 5.2.5 - Apparent Diameter of Gravitating Object

5.2.7. The emitting area for particles leaving the object, on the other hand, is equal to the area of a sphere of radius R. Dividing Equation 5.2.33 by R, squaring, and combining with Equation 5.2.9 provides the capture ratio for conditions where R is greater than or equal to $3*R_S$. In terms of the gravitational potential, Θ:

Eq. 5.2.36 $P_{CR1} = 1/(1-\Theta)^4$

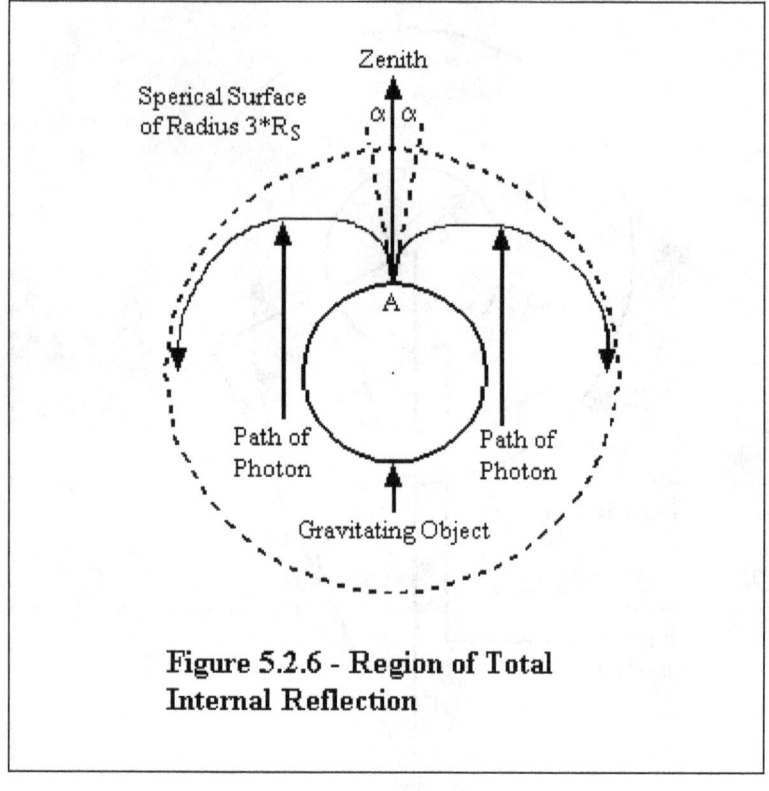

Figure 5.2.6 - Region of Total Internal Reflection

And dividing Equation 5.2.33 by R, squaring, and combining with Equation 5.2.9 also provides the capture ratio for conditions where R is less than $3*R_S$ in terms of the gravitational potential, Θ:

Eq. 5.2.37 $P_{CR2} = (6.75*\Theta)^2$

Where:

- 'P_{CR1}' is the area based component of the particle capture ratio for a gravitating object having a radius greater than or equal to 3*RS.

- 'P_{CR2}' is the area base component of the particle capture ratio for a gravitating object having a radius less than $3*R_S$.

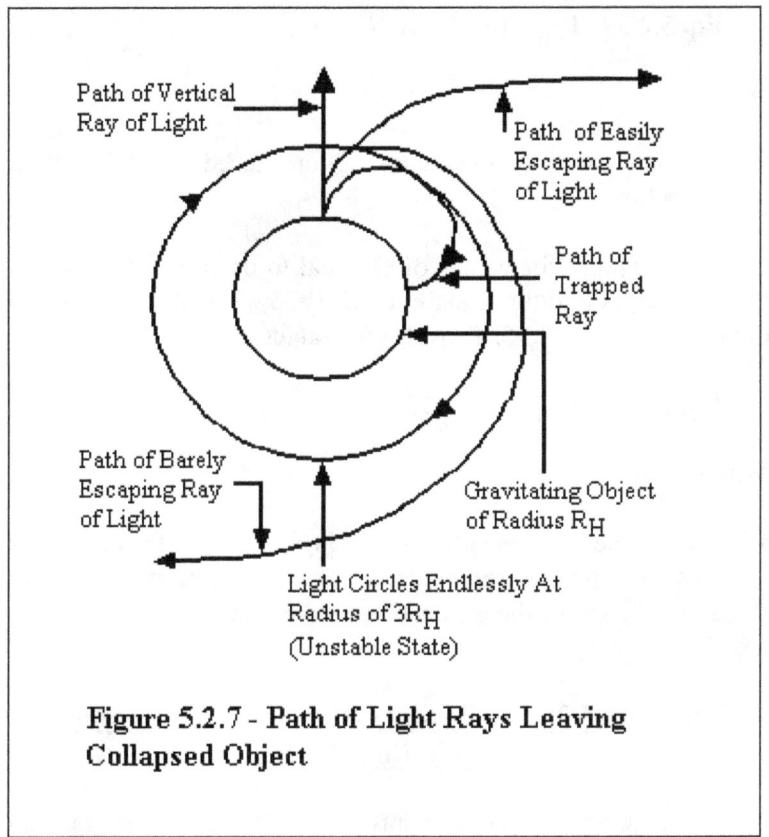

Figure 5.2.7 - Path of Light Rays Leaving Collapsed Object

5.2.13- The total internal reflection of particles emitted from a gravitating object in directions away from the zenith under conditions where the gravitational potential at the surface is greater than 1/3 also serves to increase the capture ratio by reducing the effectiveness of the emitting surface. In terms of the angle α, as shown in Figure 5.2.6, it can be shown for values of Ψ between 0 and 90 degrees is given by:

Eq. 5.2.38 $P_{EA} = \sin^2(\alpha)$

Combining Equations 5.2.9, 5.2.35 and 5.2.38 provides:

Eq. 5.2.39 $P_{EA} = (6.75*\Theta)^2 *(1-\Theta)^4$

Where:

- 'P_{EA}' is the emission angle restriction caused by total internal reflection.

The total capture for values of Θ equal to or greater than 1/3 is obtained by dividing Equation 5.2.39 by Equation 5.2.37 to obtain Equation 5.2.36. Thus, for all values of Θ:

Eq. 5.2.40 $P_{CR} = 1/(1-\Theta)^4$

Where:

- 'P_{CR}' is the capture ratio (incoming/outgoing) for energy or particles approaching or departing from a gravitating object as a function of the gravitational potential from infinity, Θ.

Section 5.3- The Temperature Boundaries of a Gravitating Object

5.3.1- The temperature of the gravitating object must lie between two limits. At the cold limit, it must be at least as hot as its equilibrium temperature with the surrounding space since it liberates heat due to the energy of fall as it contracts. At the hot limit, it can be no hotter than the temperature at which the thermally induced kinetic energy of its individual composite particles is sufficient to create new particles. Above this temperature, any further energy of contraction tends to produce additional matter to share the energy rather than to increase the temperature. Temperature rise ceases, just as the temperature rise of water on a stove ceases when its temperature reaches the boiling point. Additional energy then goes to produce a change of state of the water rather than a rise in temperature. In this case, the change of state is the creation of additional matter.

5.3.2- For the gravitating object to be in radiant equilibrium with space, it is necessary for the radiation emitted from the object through the gravitationally induced photon trap (Section 5.2) surrounding the object to equal the radiation collected from space surrounding that object. Accepting that radiant energy flux varies with the fourth power of the absolute temperature, Equation 5.2.40 requires that the ratio of the temperature of the object to the temperature of space, in terms of external units of measurement, be equal to the fourth root of that equation, thus:

Eq. 5.3.1 $\Phi_{LMIN} = \Phi_E/(1-\Theta)$

And, since the temperature is proportional to energy per degree of freedom, in accordance with the conversion factor for energy from Table 3.5.1:

Eq. 5.3.2 $\Phi_{MIN} = \Phi_E/(1-\Theta)^2$

Where:

- 'Φ_{LMIN}' is the minimum temperature of the contracted object as measured with external units of measurement.

- 'Φ_E' is the temperature of external space as measured with external units of measurement.

- 'Φ_{MIN}' is the minimum temperature of the contracted object as measured with local units of measurement.

5.3.3- At the other extreme is the condition where the mean thermal velocity is sufficient for the Lorentz Transformation associated with that velocity to be equal to (or greater than) 0.5. At such a thermal velocity, the kinetic energy of a material particle is equal to its rest mass equivalent energy. A collision between like particles has sufficient energy to create an additional particle pair. Once this temperature is reached, the further addition of energy, in terms of local units of measurement, will go to the creation of additional particles rather than an increase in temperature since an increase in particle quantity represents a higher entropy state than does an increase in temperature. To provide a simple means of estimating the upper temperature boundary of the of the contracting gravitating object, it will be assumed that the gravitational pressure is sufficient to force all free leptons (electrons, positrons) to combine with protons and anti-protons to form neutrons. the fact that thermal kinetic energy tends to be equally divided between particles of different mass insures that particles of higher rest mass energy than unexcited hadrons (protons, neutrons) will be at too low a temperature for replication.

5.3.4- Setting the value of the Lorentz Transformation equal to 0.5 allows the mean velocity of the hadron at its replication temperature to be estimated at 0.866 times the velocity of light. Allowing that the locally measured velocity of light is 10^9 feet per second, the mean velocity of air molecules at room temperature is 1350 feet per second, and the average molecular weight of an air molecule is 29 (75% nitrogen, 25% oxygen), allows one to estimate the replication temperature of the unexcited hadron as:

Eq. 5.3.3 $\Phi_{MAX} = 4.25*10^{12}$ degrees kelvin

Where:

- 'Φ_{MAX}' is the maximum locally measured temperature of the collapsing object.

The maximum temperature of the object, as measured locally in terms of external units of measurement, would then be :

Eq. 5.3.4 $\Phi_{LMAX} = 4.25*(1-\Theta)*10^{12}$ degrees kelvin

As a result of the 'gravitational greenhouse' effect defined by Equation 5.2.20, the temperature of the object as observed in external space with external units of measurement would then be:

Eq. 5.3.5 $\quad \Phi_{EMAX} = 4.25*(1-\Theta)^2*10^{12}$ degreees kelvin

Where:

- 'Φ_{LMAX}' is the maximum temperature of the object, as measured locally, in terms of external units of measurement.

- 'Φ_{EMAX}' is the maximum temperature of the object, as measured externally, in terms of external units of measurement.

Setting Φ_E equal to Φ_{EMAX} allows the value of $(1-\Theta)$ to be determined at which the maximum temperature of the object would be in equilibrium with the temperature of space:

Eq. 5.3.6 $\quad (1-\Theta)_{ES} = 4.85*\Phi*10^{-7}$

Where:

- '$(1-\Theta)_{ES}$' is the value of the gravitational transformation at which the temperature of space is in equilibrium with the temperature of the object.

5.3.5- At values of $(1-\Theta)$ less than $(1-\Theta)_{ES}$, the collapsed object is colder than the space which surrounds it and it becomes an absorber rather than a supplier of energy. The object is cooled by the conversion of thermal energy to matter, just as water is cooled by boiling. In our familiar space, where Θ_E is on the order of 3 degrees Kelvin, this point is reached when $(1-\Theta)$ is on the order of 10^{-6}. In terms of the contraction of an object into the collapsed state, this point is reached quite rapidly, in astronomical terms, once the conditions for gravitational collapse are present. The remainder of the gravitational contraction may be considered to occur under the conditions of no heat loss, as defined in Section 5.1.

Section 5.4 - The Minimum Time of Fall to a 'Collapsed' Object

5.4.1- Current literature deals extensively with the effects produced by gravitation when an object is so massive that its gravitationally induced pressure exceeds the compressive strength of nuclear particles and the object no longer has any nuclear energy to convert to heat to enable it to resist contraction. It is assumed under these conditions that, since no other stronger forces are known, that the escape velocity from its surface will equal the velocity of light. Beyond this point, no energy or matter can escape from the object (other than effects such as those predicted by Dr. Hawking) and it will appear to an external observer to be a 'black hole' in space whose only external properties are the gravitational field it leaves behind.

5.4.2- As Sections 5.2 and 5.3 show, as the gravitational contraction approaches its end limit ($\Theta = 1$), it becomes virtually impossible for the object to radiate its remaining energy of contraction to space. As a result of the inability to effectively radiate energy in the final stages of the contraction, the locally measured internal pressure is that provided by Equation 5.1.40, which reaches a maximum and then falls to zero as Θ approaches unity. As a result, gravitational collapse always ceases short of the 'black hole' state, the surface of which is commonly called the 'event horizon'.

5.4.3- This section will consider the minimum time which is required for an object to fall from external space to an infinitesimal distance from the hypothetical 'event horizon'. (The 'event horizon' is an end limit condition which can be approached but not attained.) For the purpose of this discussion, the falling object to be considered will be a photon since no object can travel faster. A derivation will be provided for the time, in terms of the units of measurement existing at an infinite distance, for a photon to travel from an upper elevation at radius R_U to a lower elevation at a radius of R_L, both as measured in terms of the units of measurement existing at an infinite distance. The subjective time of travel for the photon will also be provided by integrating the elapsed time of travel through each incremental elevation using the units of measurement of that elevation, and ignoring the time dilation effects predicted by Special Relativity. (These effects would insure a null result for an otherwise useful determination.)

5.4.4- Since the velocity of light is a velocity and that velocity is a constant of the Science of Physics and has a value of C', it follows, from Table 3.5.1, that:

Eq. 5.4.1 $c_L = C_L/(1-\Theta)^2$

And:

Eq. 5.4.2 $c_L = C'$

And, since the photon is traveling vertically:

Eq. 5.4.3 $C_L = \delta R/\delta T$

Then, combining Equations 5.4.1 through 5.4.3 provides:

Eq. 5.4.4 $\delta T = \delta R/(C'*[1-\Theta]^2)$

From Equation 5.1.47:

Eq. 5.4.5 $R = R_S/\Theta$

And:

Eq. 5.4.6 $\delta R = -R_S*\delta\Theta/\Theta^2$

Combining Equations 5.4.3 through 5.4.6 provides, in terms of external units of measurement:

Eq. 5.4.7 $C'*\delta T/R_S = -\delta\Theta/(\Theta*[1-\Theta]^2)$

And, in terms of instantaneous local units of measurement, from Table 3.5.1:

Eq. 5.4.8 $C'*\delta t/R_S = -\delta\Theta/(\Theta^2*[1-\Theta])$

Integrating Equation 5.4.7 between the limits of Θ_U and Θ_L and for Θ, zero and T for T, and defining:

Eq. 5.4.9 $\eta_{LL} = \Theta_L*(1-\Theta_L)$

Eq. 5.4.10 $\eta_{UU} = \Theta_U*(1-\Theta_U)$

Eq. 5.4.11 $\eta_{LU} = \Theta_L*(1-\Theta_U)$

Eq. 5.4.12 $\eta_{UL} = \Theta_U*(1-\Theta_L)$

provides the minimum time for a particle or an object to travel between the upper and lower gravitational potentials in terms of external units of measurement as:

Eq. 5.4.13 $T = R_S*[1-2*\Theta_U]*\eta_{LL}/[C'*\eta_{LU}*\eta_{UL}]$
             $-R_S*[1-2*\Theta_L]*\eta_{UU}/[C'*\eta_{LU}*\eta_{UL}]$
             $+2*R_S*\log[\eta_{LU}/\eta_{UL}]/C'$

Integrating Equation 5.4.8 between the same limits for Θ and between the limits of zero and t for t provides the minimum subjective time (as observed at the particle in terms of its instantaneous local units of measurement) for a particle to travel between gravitational potentials:

Eq. 5.4.14 $t = (R_S/C')*([\Theta_L-\Theta_U]/[\Theta_L*\Theta_U]+\log[\eta_{LU}/\eta_{UL}]$

Equation 5.4.14 yields rather unsurprising results. The first term in the expression is the time for light to travel the distance represented by the difference in elevations in the absence of the field. The second term introduces a small modification due to the relativistic effects of the field. (Equation 5.4.14 assumes that compensation has been made for the time dilation effects predicted by Special Relativity as a result of the velocity of fall.)

5.4.5- Equation 5.4.13 is more interesting. If one examines the first term, one finds that, from a gravitational potential close to zero (R/R_S very large) to a gravitation potential on the order of 0.5, ($R/R_S = 2$), the minimum time of travel, in terms of external units of measurement is essentially equal to the distance divided by the velocity of light. From a gravitational potential of 0.5 to a gravitational potential which approaches unity (R/R_S infinitesimally approaching unity), the time of travel in terms of external units of measurement approaches infinity. The third term in the equation is significant in the region near a gravitational potential of 0.5 and is unimportant at gravitational potentials which are close to zero or close to unity. As Θ approaches unity, the minimum time of travel, T', between a radius of twice the Horizon Radius and the actual radius approaches:

Eq. 5.4.15 $T' = R_S/(C'*[1-\Theta])$

5.4.6- Equation 5.4.15 states that the time for a photon to travel from the proximity of a collapsed gravitational object to the surface of that object approaches infinity as the radius of that object approaches the Horizon Radius (the hypothetical 'Black Hole' state). It follows, therefore, that the time of formation of a 'Black Hole', if such an object were physically realizable, is infinite in terms of the external universe, since such an object cannot form in less time than it would take for a photon to reach its surface. ***No "Black Hole" has formed since the Universe began!***

Note:- The requirement for infinite time for the formation of a 'Black Hole' also follows from the predictions of General Relativity. Under this theory, extra space is 'called into being' by the gravitational field as a result of increasing gravitational potential at a rate which, for high gravitational potentials, is faster than the rate that the radius reduces with increasing gravitational potential. As a result, under General Relativity, a photon falling to the surface of a hypothetical 'Black Hole' would have to travel an infinite distance at the (finite) velocity of light, and would require an infinite time to do so.

Section 5.5 - Gravitation and Cosmology

5.5.1- In this Section, the formation of a large gravitating object will be considered from its initial state as a gaseous cloud to its final state as an object whose gravitational potential, in terms of external units of measurement, is infinitesimally close to unity. The consideration will be both in terms of an external observer and in terms of a hypothetical observer within the contracting object.

5.5.2- It seems generally accepted that a gravitating object, be it a planet, a star, a quasar, or a galaxy, starts out as a diffuse cloud of gaseous and microparticle material in which the balance between the thermally induced pressure which would cause the cloud to expand and the gravitational pressure which would cause it to contract is disturbed. As a result, a portion of that cloud begins to contract under its own gravitation. As it contracts, it liberates energy of fall from a conversion of its rest mass equivalent energy (in terms of constant units of measurement). The energy of fall becomes kinetic energy stored in the particles which compose the cloud and the temperature of the cloud increases. The increase in temperature, in turn, increases the thermal pressure within the cloud and resists the pressure produced by the gravitational attraction. The increase in temperature also causes the cloud to radiate energy to space, and, to make up for this loss, the cloud must contract further.

Eventually the process continues to the point where the cloud becomes an opaque object from which energy can only radiate from the surface, and a proto-star is formed. The proto-star continues to lose heat by radiation and must continue to contract so as to maintain the necessary internal temperature to balance the pressure produced by gravity. If the object is massive enough, the internal temperature reaches a point where hydrogen begins to fuse, releasing some of the rest mass equivalent energy of the object without the need for further gravitational contraction, and a star is born.

5.5.3- If the star is small, it remains relatively stable for a long period of time as the hydrogen at its core is converted to helium and release the energy required to replace the loss by radiation. When sufficient amount of helium is formed at the core, the energy release process can no longer proceed at the existing core temperature, and the core must contract to release enough gravitational energy to raise the core temperature to the level where helium will undergo fusion. The rate of energy release at this stage is much higher, and the star, in order to radiate this energy, expands enormously to become a 'red giant'. After a relatively short period, the rapid rate of energy release subsides, and the star contracts to become a 'white dwarf', gradually cooling over an enormous period of time.

5.5.4- When the star is larger, the evolution of the star is more dramatic. The process of nuclear energy conversion continues until the core consists of a sizable percentage of iron. At this point no further nuclear energy is available for release since iron has the lowest mass per nucleon of all of the elements and can release no further energy by the nuclear process. The core of the star then collapses cataclysmically under gravitational pressure and releases enough gravitational energy to cause it to explode as a nova or supernova, leaving a neutron star at the center of an expanding cloud of gaseous material.

5.5.5- When the object is so massive as to be incapable of blowing itself apart, either because the explosion cannot proceed against the attraction of gravity, or because, if it does proceed, the remnants fall back on themselves, the object has no choice but to continue to contract. This further contraction releases gravitational energy to supply the heat energy needed to resist contraction. As the object contracts, it becomes more and more difficult for it to rid itself of heat, both because its reduced surface area and because of the 'gravitational greenhouse' effect described in Section 5.2. As the object contracts, it eventually reaches the point where the gravitational potential at its surface exceeds 0.5 and additional contraction results in the creation of additional matter rather than an increase of temperature. Since the pressure in a gas is proportional to both its temperature and the number of particles per unit volume, this additional matter is as effective in resisting the effects of contraction as an increase in temperature would have been, and further contraction results in the creation of additional matter. (Note that the creation of this additional matter does not violate the conservation laws. The energy represented by this additional matter, in terms of external units of measurement, is unchanged. If parity is truly a conserved entity, the total amount of parity present in the original matter will subdivide to be shared among the additional particles, and the polarity of the original matter will be duplicated to maintain the object as matter instead of anti-matter.) Once this point is reached, the locally measured temperature of the object no longer increases, and to an external observer, it appears to decrease as the object contracts further!

5.5.6- The extreme case would occur at the center of a newly formed galaxy. Most of the stars that initially form a galaxy will have highly elliptical orbits about the center of mass of the galaxy, and the ellipticity of their orbits will tend to a random distribution. Eventually, the stars will exchange momentum through gravitational attraction and sort themselves into stars with essentially circular orbits which do not collide and into stars with elliptical orbits which have a high probability of collision. The collision probability will be highest near the center of gravity of the galaxy and will form an immensely massive central object. This object, which may have a mass of millions of

Suns, will attempt to collapse gravitationally, releasing radiation energy to space at an enormous rate, To an observer outside of the galaxy, the central object would be expected to have the characteristics which current literature ascribes to a quasar.

5.5.7- To an external observer, the object appears to contract rapidly to a radius of 6.75 times the Horizon Radius (at a surface gravitational potential of 1/3) and to be extremely hot as a result. From that point on, the object appears to remain almost constant in size, shrinking only as a result of a reduction in its Horizon Radius due to a reduction of its energy content by radiation. From this point on, however, it appears to cool to the temperature of the surrounding space as its surface gravitational potential increases towards unity.

5.5.8- To an internal observer, if it were possible for there to be one, the process would appear quite different. At the surface, once the gravitational potential exceeded 1/3, a region of complete internal reflection would begin to obscure the sky at the horizon. The obscuration would increase as the gravitational potential increased and leave only a minute area of sky visible at the zenith as the gravitational potential approached unity. (At a gravitational potential of unity, the obscuration would be complete.) The temperature would rise to about $4*10^{12}$ degrees K as the gravitational potential approaches 0.5 and remain at that temperature until the gravitational potential was on the order of 0.999999 (assuming that the external universe had a temperature of 3 degrees K, at which point it would be in thermal equilibrium with the external universe. For higher gravitational potentials, the object would be cooler than the surrounding space, and heat energy would flow into the object.

5.5.9- As the gravitational potential increased to 1/2, the object would appear to the internal observer to be contracting. Once this gravitational potential was exceeded, however, the reduction in size of the internal units of measurement for length associated with the increase in gravitational potential would more than compensate for the reduction in externally observed radius, and the object would appear to cease its contraction and start expanding. It would reach an infinite radius and a zero energy density at a gravitational potential of unity, as shown in Section 5.1. As the internally observed radius approached infinity, the Law of Conservation of Angular Momentum would insure that the angular rate of rotation of the object approached zero.

5.5.10- To an observer who commenced his observations late in the process with no knowledge of what went on previously, the interior of the collapsing object would appear to be a cold, virtually empty space in which the observable matter was retreating in all directions at a velocity proportional to its distance. Extrapolating backwards in time, he would conclude that at some time in the past, all of the matter in the object existed at virtually a single point in space at an energy density which approached infinity. He would conclude that this quasi-singularity exploded with sufficient violence to produce the enormous and almost empty volume around him. If he were a sophisticated observer, he might note that when he extrapolated backwards in time, he reached a point where the radius was smaller than the Horizon Radius indicated by the observed mass of the object and wonder how this might be. He would notice that the distribution of matter observed in the object is clustered rather than uniform since the apparent retreat of the various portions of the object from each other results from shrinkage of the matter rather than from an expansion of the object. (A similar effect occurs when a lake dries up. The mud at the bottom contracts as it dries and leaves cracks.) He would observe that occasionally particles would fall in from the external space and would have kinetic energies which were enormously and inexplicably high compared to the rest mass equivalent energy of the particle itself. He would observe that the object had an angular velocity which approximated zero and perhaps conclude that Mach's Principle was at work, Finally, he would conclude as a result of observing the apparent expansion of the object, that a non-Newtonian repulsive force acted at long distances which overcame some of the predicted attractive force of Newtonian Gravitational Theory.

5.5.11- (Note:- This text was written in 1987, its conslusions have been modified by subsequent work as described in "The Einstein Hoax"). If we consider our universe, we find that, according to the best available estimates, it has a mass equivalent to between 10^{79} and 10^{80} nucleons. A simple calculation shows that the Horizon Radius for such a mass is between $1.3*10^9$ and $1.3*10^{10}$ light years. Cosmologists suggest that our early universe became transparent when its radius exceeded 10^5 light years and it has been expanding ever since. Apparently our universe began as an object which was much smaller than it would be as a 'black hole' and expanded to a size which is larger than its 'black hole' size. One might question how such a result might possibly be. We find that the distribution of matter is fractionated so that galaxies are clumped into super-galaxies and into super-super-galaxies with vast empty spaces in between. We find that quasars seem to exist or to have existed in the centers of galaxies. We find that particles with inexplicably high velocities exist and have called them cosmic rays. (Perhaps the observed relativistic mass distribution of cosmic rays might be a measure of the gravitational potential of our universe with respect to the space external to it.) The observed expansion of our universe suggests that, when considered conventionally, at long distances a non-Newtonian repulsive force acts on matter, It would seem if one wished to observe the appearance of the interior of an object in the final stages of gravitational collapse, he would only have to step outside on a clear moonless night and look up!

5.5.12- At this point (1987) a speculation seems in order. Consider that our universe may be the interior of a burned out quasar in the center of one of many galaxies in an incredibly ancient universe. An enormous amount of time would have passed in that universe because its units of time are infinitesimal compared to our own, and, as a result, the temperature of that universe would be vanishingly close to zero. In terms of that universe, our own universe might have a diameter on the order of the 0.1 light years that astronomers seem to ascribe to quasars rather than the perhaps $3*10^{10}$ light years that we observe. In that external universe, light would travel a distance equal to the diameter of our universe in 0.1 years. Due to the gravitational dilation of the units of measurement for time that we experience, the units of measurement for time in our universe would then be $3*10^{11}$ times as large as in the external universe. As a result, the elapsed time observed in our universe for light across it at the velocity of light in the external universe is on the order of 10 microseconds! It has been suggested that quantum effects propagate instantaneously. If such effects are not subject to the reduction in velocity of propagation that the gravitational field of our universe imposes on light, these effect would also be capable of crossing our entire universe in 10 microseconds. We should then consider the velocity of these effects to be infinite, much as our ancestors considered the velocity of light to be infinite.

Note:- If one properly applies General Relativity to cosmology, one reaches similar conclusions. General Relativity requires that the gravitating object call into being additional space as it contracts, Once the gravitational potential exceeds 1/2 (1 in terms of the parameters of General Relativity), space is created faster than contraction occurs and the object also appears to expand to an infinite radius as the contraction goes to completion.

Appendix 2 - Correction of Residual Deficiencies in The Special Theory of Relativity - (1993)

Note:- Appendix 2 was originally written in 1993 and is included to provide a backup for "The Einstein Hoax".

Summary:-

This paper derives the Lorentz Transformations for Force parallel to and transverse to the relative velocity vector between velocity reference frames. In the process it demonstrates that both the accepted Lorentz Transformation for Force in the transverse direction and the interpretation of the Right Angle Lever Thought Experiment, as described in many texts, are incorrect. A final result is the determination of the manner in which the kinetic energy associated with a moving object (particle) is stored. It is also shown that the reason that it is considered that Special Relativity is not valid for accelerated reference frames is that it defines mass in different terms than those commonly used in mechanics. The difficulty with acceleration in Special Relativity vanishes when the conventionally employed definition of mass is used and it is recognized that inertial mass is the incremental impulse required to produce an incremental change in velocity rather than the integral of that incremental impulse between the velocity limits of zero and 'V' as is the current practice.

Index

Section	Section Titles	Page
1	Introduction	400
2	Groundwork of Discussion	401

3	A Comparison of the Velocity Difference Between Velocity Reference Frames B and C as Observed in Reference Frame B and as Observed in Reference Frame A.	402
4	Determination of the Lorentz Transformation for Incremental Mass and for Force Between Reference Frames Having Relative Velocity	407
5	The Balance of Moments Applied to a Right Angled Lever in Velocity Reference Frame B Moving with Velocity V with Respect to Velocity Reference Frame A as Observed in Reverence Frames A and B	413
6	The Conventional Lorentz Transformation for Transverse Force as Related to the Right Angle Lever Thought Experiment	416
7	The Lorentz Transformation for Parallel and Transverse Force as Related to a Compressed Spring Thought Experiment.	419

Section 1 - Introduction

1.1- The purpose of this paper is to develop a system of force, length, and time transformations between reference frames having a relative velocity equivalent to the mass, length, and time system of Lorentz Transformations currently in use. The reason for revising the system of Lorentz Transformations is that, unlike force, mass is not directly observable. The mass of an object (particle) is a property which can only be determined by a measurement which involves force, length, and/or time.

Section 2 - Groundwork of Discussion

2.1 - At first glance, the conversion to a force, length, and time system of transformations would appear to have already been accomplished. Minkowski has already provided the Lorentz Transformations for Force in directions parallel to and transverse to the relative velocity vector between reference frames. Unfortunately, an error has been made in the development of the Lorentz Transformation for Transverse Force. This error was revealed by the classical Right Angle Lever Thought Experiment (described later) but, instead of this thought experiment resulting in the correction of the error, the error was rationalized by an explanation which ignored basic mechanical principles.

2.2- The procedure employed by this paper in developing the correct Lorentz Transformation for Force in directions parallel to and transverse to the relative velocity vector between reference frames is, for the parallel and transverse directions:

- To determine the incremental energy required in the 'stationary' and the 'moving' reference frames to produce an incremental change in velocity.

- To determine the force required in the 'stationary' and the 'moving' reference frame required to impart that energy.

- To determine the Lorentz Transformation for Force by comparing the force in the 'stationary' and the 'moving' reference frame.

2.3- The first step in the procedure consists of determining the Lorentz Transformations for Parallel and Transverse Velocities between the 'stationary' and the 'moving' reference frames. This is accomplished by employing the equation for the addition of velocities provided by Special Relativity to provide the relationship between a small velocity as observed in the 'moving' reference frame and that same small velocity as observed in the 'stationary' reference frame. It is then shown that the Lorentz

Transformation for Incremental Velocity can be approximated by the by the Lorentz Transformation for Length divided by the Lorentz Transformation for Time with negligible error.

2.4- The Lorentz Transformation for Parallel and for Transverse Forces are then determined by determining the incremental impulse required to produce an incremental velocity change in the 'stationary' and 'moving' reference frames to determine the Incremental Mass (mass in the classical mechanical sense) of the object (particle) for both of these reference frames and the Lorentz Transformations for that 'Incremental Mass' between those reference frames. The Lorentz Transformations for Force are determined by dividing the product of the Lorentz Transformations for 'Incremental Mass' and for incremental velocity by the Lorentz Transformation for Time.

2.5- The results obtained agree with the Lorentz Transformation for Parallel force as derived by Minkowski. The results are reciprocal to the conventionally accepted Lorentz Transformations for Transverse Force, however, they are in agreement with a rigorous treatment of the classic Right Angle Lever Thought Experiment. In order to deal with the discrepancy, the textbook treatment of the Right Angle Lever Thought Experiment is discussed. It is shown where the conventional textbook treatment of that thought experiment is deficient.

Section 3 - A Comparison of the Velocity Difference Between Velocity Reference Frames B and C as Observed in Reference Frame B and as Observed in Reference Frame A.

3.1- Consider three collinear velocity reference frames designated as A, B, and C. As observed in reference frame B, reference frame C has a velocity of $+\delta v$ and reference frame A has a velocity of $-v_A$. As observed in reference frame A, reference frame B has a velocity of $+V_B$ and reference frame C has a velocity of $+V_C$. What is to be determined is the velocity,

δV, of reference frame C with respect to reference frame B as observed in reference frame A:

Eq. 3.1.1 $\delta V = V_C - V_B$

3.2- Special Relativity provides the velocity of reference frame C with respect to reference frame A in terms of the velocity of reference A and C with respect to reference frame B. In terms of the previous paragraph, this expression is:

Eq. 3.2.1 $V_C = (-v_A + \delta v)/(1 - v_A * \delta v/C^2)$

But, since the relative velocity between reference frames A and B must be equal and opposite in their respective reference frames if a preferred reference frame is not to be directly observable:

Eq. 3.2.2 $v_A = -V_B$

Then:

Eq. 3.2.3 $V_C = (V_B + \delta v)/(1 + V_B * \delta v/C^2)$

Combining Equations 3.1.1 and 3.2.3 provides:

Eq. 3.2.4 $\delta V = \delta v * (1 - V_B^2/C^2)/(1 + V_B * \delta v/C^2)$

And, if δv is sufficiently small compared to the velocity of light, one may write without significant error:

Eq. 3.2.5 $\delta v = \delta V/(1 - V_B^2/C^2)$

Where:

- 'δv' is the velocity of reference frame C with respect to reference frame B as observed in reference frame B.

- 'δV' is the velocity of reference frame C with respect to reference frame B as observed in reference frame A.

- The upper case letter 'V' refers to observations of velocity made with the units of measurement of reference frame A. The lower case letter 'v' refers to observations of velocity made with the units of measurement of reference frame B.

3.3 - The conventional Lorentz Transformations for Length and Time are given by:

Eq. 3.3.1 $l_P = L_P/(1-V^2/C^2)^{0.5}$

Eq. 3.3.2 $l_T = L_T$

Eq. 3.3.3 $t = T*(1-V^2/C^2)^{0.5}$

Where:

- The symbol 'l' refers to a length in reference frame B as observed in reference frame B.

- The symbol 'L' refers to the same length in reference frame B as observed in reference frame A.

- The symbol 't' refers to a time duration in reference frame B as observed in reference frame B.

- The symbol 'T' refers to that same duration of time in reference frame B as observed in reference frame A.

- The subscript 'P' refers to directions parallel to the velocity vector between reference frames A and B.

- The subscript 'T' refers to directions transverse to the velocity vector between reference frames A and B.

3.4 - It will be noted that Equation 3.2.5 is the same as that which would be obtained by dividing the Lorentz Transformation for Parallel Length by the Lorentz Transformation for Time to obtain the Lorentz Transformation for Velocity (length/time). It follows, therefore that Equation 3.2.5 may be restated as:

Eq. 3.4.1 $v_P = V_P/(1-V_B^2/C^2)$

Where:

- 'v_P' refers to a small velocity within reference frame B as observed in reference frame B in a direction parallel to the relative velocity between reference frames A and B.

- 'V_P' refers to that same velocity as observed in reference frame A. The Lorentz Transformation between reference frames A and B for incremental velocities in reference frame B which are transverse to the velocity vector between reference frames A and B can be obtained by dividing the Lorentz Transformation for Transverse Length, Equation 3.3.2, by the Lorentz Transformation for time, Equation 3.3.3 to obtain:

Eq. 3.4.2 $v_T = V_T/(1-V_B^2/C^2)^{0.5}$

The error in the determination of the Lorentz Transformation for Transverse Velocity given by Equation 3.4.2 is negligible compared to the already negligible error in the determination of Equation 3.4.1 since the simultaneity correction represented by the denominator of Equation 3.2.4 does not apply.

3.5- Since acceleration is defined as the rate of change of velocity with respect to time, the parallel and transverse transformations for acceleration may be obtained by combining Equations 3.3.2, 3.4.1, and 3.4.2. Providing the velocities represented by Equations 3.4.1 and 3.4.2 represent an incremental change in velocity within reference frame B which is changing uniformly with respect to time, accelerations observed in each reference frame are given by:

Eq. 3.5.1 $A_P = \delta V_P/\delta T$

Eq. 3.5.2 $a_P = \delta v_P/\delta t$

Eq. 3.5.3 $A_T = \delta V_T/\delta T$

Eq. 3.5.4 $a_T = \delta v_T/\delta t$

Where:

- 'A' is an acceleration in reference frame B as observed in reference frame A.

- 'a' is the same acceleration in reference frame B as observed in reference frame B.

The Lorentz Transformations for Acceleration then become:

Eq. 3.5.5 $a_P = A_P/(1-V_B^2/C^2)^{1.5}$

Eq. 3.5.6 $a_T = A_T/(1-V_B^2/C^2)$

It should be noted that, unlike the equations for the velocity transformations, the acceleration transformations of Equations 3.5.5 and 3.5.6 are not limited to incremental values. Large accelerations for infinitesimal periods of time produce the infinitesimal changes in velocity required to satisfy the approximation represented by Equation 3.3.4. It should also be noted that Equations 3.5.5 and 3.5.6 are compatible with the classical meaning of acceleration as the second derivative of position with respect to time.

Section 4 - Determination of the Lorentz Transformation for Incremental Mass and for Force Between Reference Frames Having Relative Velocity

4.1 - In this Section, the Lorentz Transformations for Parallel and for Transverse Force will be determined to allow the description of phenomena defined by Special Relativity to be treated in terms of a force (F), length (L), and time (T) system of units rather than the conventional mass (M), length (L), and time (T) system. The reason for this change is the fact that, unlike force, length, and time, mass cannot be observed directly but must be inferred from measurements which involve force, length, and time. The mass of an object (particle) can only be measured in terms of:

- Inertial properties which involve the incremental impulse (force-time product) required to produce an incremental velocity (length/time ratio) change. (An incremental velocity change occurs when photons undergo grazing angle reflection.)

- The energy (force-length product) released when an object (particle) is annihilated. The released energy retains the inertial properties of the original object (particle).

- Gravitational properties which involve the force-length^2 product which exists between it and another mass having a known magnitude. (The need for the second known mass makes this a secondary method of observation which relies on one of the two above methods for calibration.)

4.2 - This discussion will be based upon the classical definition of inertial mass. This mass, M_I, designated as the incremental mass of an object (particle), is defined as the infinitesimal impulse required to produce an infinitesimal velocity change. The Lorentz Transformation for Incremental Mass in the parallel and transverse directions and the Lorentz Transformations for Parallel and Transverse force is derived below.

4.3 - Consider again two reference frames, A and B, having a relative velocity V and with a material object (particle) in reference frame B. As observed in reference frame B, the object has a mass of 'm'. If that object is brought to rest in reference frame A it will impart a momentum to reference frame A which is equal to M*v where 'M' is the mass of the object as observed in reference frame A as provided by the conventional Lorentz Transformation for mass:

Eq. 4.3.1 $\quad M = m/(1-V^2/C^2)^{0.5}$

4.4 - Let us now consider that the velocity of the object in reference frame B is changed incrementally in a direction parallel to the velocity vector between reference frames A and B. What is to be determined first is the incremental mass of the object as observed in reference frame A, as revealed by the incremental impulse required to produce the incremental velocity change. The first step is to differentiate Equation 4.3.1 with respect to V to provide:

Eq. 4.4.1 $\quad \delta M_P = m*V*\delta V_P/(C^2*[1-V^2/C^2]/C^2]^{1.5})$

The relationship between energy and mass is:

Eq. 4.4.2 $\quad E = M*C^2$

Differentiating Equation 4.4.2 provides:

Eq. 4.4.3 $\quad \delta E_P = C^2*\delta M_P$

Then, combining Equations 4.4.1 and 4.4.3:

Eq. 4.4.4 $\quad \delta E_P = m*V*\delta V_P/(1-V^2/C^2)^{1.5}$

But δE_P is the energy, as observed in reference frame A, which was added to (or removed from) the object in order to produce the incremental change in velocity. This energy is given by:

Eq. 4.4.5 $\delta E_P = F_P * \delta L_P$

Where:

- 'δL_P' is the distance the object travels in reference frame A during the incremental velocity change which occurs in the incremental time δT.

- 'F_P' is the force applied to the object to change its velocity, all as observed in reference frame A. Equation 4.4.5 may then be written as:

Eq. 4.4.6 $\delta E_P = F_P * V * \delta T$

Combining Equations 4.4.4 and 4.4.6 provides:

Eq. 4.4.7 $F_P = m * \delta V_P / (\delta T * [1 - V^2/C^2]^{1.5})$

The incremental mass, M_{IP}, of the object in reference frame B as observed in reference frame A is given by:

Eq. 4.4.8 $M_{IP} = F_P * \delta T / \delta V_P$

Therefore:

Eq. 4.4.9 $M_{IP} = m / (1 - V^2/C^2)^{1.5}$

4.5 - Since the Lorentz Transformation for the derivative of a quantity is identical to the Lorentz Transformation for the quantity itself, combining Equation 4.4.7 with Equations 3.3.2 and 3.4.1 provides:

Eq. 4.5.1 $F_P = m * \delta v_P / \delta t$

But, since the terms in lower case letters refer to observations of quantities in reference frame B made within reference frame B, the conventional laws of mechanics apply:

Eq. 4.5.2 $f_P = m*\delta v_P/dt$

Then the Lorentz Transformation for Force in a direction parallel to the relative velocity vector becomes:

Eq. 4.5.3 $f_P = F_P$

It will be noted that Equation 4.5.3 is identical to the transformation provided by Minkowski.

4.6 - The determination of the Lorentz Transformation for a force in a direction perpendicular to the relative velocity vector between reference frames A and B for the incremental mass is accomplished in a similar manner, Consider an object in reference frame B which is subjected to an incremental impulse in a direction perpendicular to the relative velocity vector between reference frames A and B. As a result of this impulse, the object acquires an incremental velocity transverse to the velocity vector between A and B of dv_T, as observed in reference frame B, and of dV_T as observed in reference frame A. Equation 3.2.5 provides:

Eq. 4.6.1 $\delta v_T = \delta V_T/(1-V^2/C^2)^{0.5}$

4.7 - Prior to the incremental acceleration, the mass of the object in reference frame B as observed in reference frames A and B is given by Equation 4.3.1. Following the incremental transverse acceleration, the relative velocity, V_1, between the object and reference frame A has increased and is equal to:

Eq. 4.7.1 $V_1 = (V^2 + \delta V_T^2)^{0.5}$

The mass of the object, as observed in reference frame A is now equal to:

Eq. 4.7.2 $M_1 = m/(1-v^2/C^2-\delta V_t^2/C^2)^{0.5}$

The incremental change in mass of the object as observed in reference frame A is given by:

Eq. 4.7.3 $\delta M_T = M_1 - M$

Or, since δV_T represents an infinitesimal velocity change, we may write without significant error:

Eq. 4.7.4 $\delta M_T = m*C*\delta V_T^2/(2*[C^2-V^2])^{1.5}$

Differentiating Equation 4.4.2 for the transverse direction provides:

Eq. 4.7.5 $\delta E_T = C^2 * \delta M_T$

Then:

Eq. 4.7.6 $\delta E_T = m*\delta V_T^2/(2*[1-V^2/C^2]^{1.5})$

Assuming for simplicity that the force, F, applied to the object to produce the incremental velocity change, as observed in reference frame A, was unchanged during the velocity change, the energy supplied, as observed in reference frame A, is given by:

Eq. 4.7.7 $\delta E_T = F*\delta V_T*\delta T/2$

Where δT is the time which the force was applied. Then:

Eq. 4.7.8 $F_T*\delta T = m*\delta V_T/(1-V^2/C^2)^{1.5}$

Or, since in terms of reference frame A:

Eq. 4.7.9 $F_T*\delta T = M_{IT}*\delta V_T$

Then:

Eq. 4.7.10 $M_{IT} = m/(1-V^2/C^2)^{1.5}$

Combining Equation 4.4.9 with Equation 4.7.10 provides:

Eq. 4.7.11 $M_{IT} = M_{IP}$

Equation 4.7.11 shows that, as is the case with relativistic mass, the instantaneous mass of an object is not a vector quantity. Therefore, we may write, for the instantaneous mass in any direction, M_I:

Eq. 4.7.12 $M_I = m/(1-V^2/C^2)^{1.5}$

4.8 - Equation 4.7.10 allows the transformation for force in the transverse direction to be determined since, in reference frame B:

Eq. 4.8.1 $f_T * \delta t = m_I * \delta v_T$

And in reference frame A:

Eq. 4.8.2 $F_T * \delta T = M_I * \delta V_T$

Combining the Lorentz Transformation for Transverse Length (3.3.2) and for Time (3.3.3) with Equations 4.7.10, 4.8.1 and 4.8.2 provides:

Eq. 4.8.3 $f_T = F_T * (1-V^2/C^2)^{0.5}$

4.9 - It will be noted that Equation 4.8.3 is the reciprocal of the conventionally accepted Lorentz Transformation for Transverse Force. The next Section will discuss the significance of the reciprocal relationship between Equation 4.8.3 and the conventionally accepted Lorentz Transformation for Transverse Force in terms of the Moving Right Angle Lever Thought Experiment. Applied in a straight forward manner, that thought experiment will now yield results with Equations 4.5.3 and 4.8.3. The arguments which have been employed to reconcile the thought experiment with the conventional (and erroneous) Lorentz Transformation for Transverse Force will be discussed in a later Section.

4.10 - It will also be noted that, by incorporating the concept of instantaneous mass, M_I, (mass in the classical sense), Special Relativity becomes valid for accelerated reference frames as well as for reference frames having a constant velocity. This result obtains because the Lorentz Transformation for Instantaneous Mass is identical to the Lorentz Transformation for Momentum (M*V) which is a component of the four vector mass transformation currently in use. The conventional Lorentz Transformation for Mass may be derived from the Lorentz Transformation for Incremental Mass by integrating the incremental impulse required to produce the relative velocity of the incremental mass with respect to the observer's reference frame and dividing that integral by the velocity of the object (particle).

Section 5 - The Balance of Moments Applied to a Right Angled Lever in Velocity Reference Frame B Moving with Velocity V with Respect to Velocity Reference Frame A as Observed in Reverence Frames A and B

5.1 - Consider a right angled lever, as shown in Figure 5.1.1, located in reference frame B which is moving with velocity V with respect to reference frame A. One arm of the lever is parallel to the relative velocity between A and B and the other arm is transverse to that velocity. For simplicity, both lever arms are considered to be of equal length as observed in reference frame B. The definitions provided in Sections 3 and 4 are retained.

5.2 - A force, F_P, parallel to the relative velocity is applied to the transverse arm and generates a force, F_{HP}, equal in magnitude and opposite in direction to F_P at the pivot pin resulting in a couple which causes a torque to be applied to the lever. Another force, F_T, transverse to the relative velocity is applied to the parallel arm of the lever and generates a force, F_{HT}, equal in magnitude and opposite in direction to F_T at the pivot pin resulting in a couple which causes a torque to be applied to the lever opposing the torque generated by F_P. The lever is observed not to rotate in either reference frame.

5.3 - As observed in reference frame A, energy is added to the transverse arm of the lever by force F_P at a rate given by:

Eq. 5.3.1 $\delta E_P = F_P * V * \delta T$

and energy is added to the transverse arm of the lever by the force at the hinge pin at a rate given by:

Eq. 5.3.2 $\delta E_{HP} = F_{HP} * V * dT$

The total rate that energy is added to the transverse arm of the lever is given by:

Eq. 5.3.3 $\delta E = \delta E_P + \delta E_{HP}$

But, since F_P and F_{HP} are equal in magnitude and opposite in direction:

Eq. 5.3.4 $\delta E / \delta T = 0$

5.4 - From the preceding, in terms of reference frame A, energy enters the lever at the end of the transverse arm, flows laterally through the transverse arm, and exits the lever through the hinge pin. None of this energy remains in the lever. Since the transverse velocity of the lever is zero in both reference frames (A and B), no energy enters or leaves the parallel arm in either reference frame.

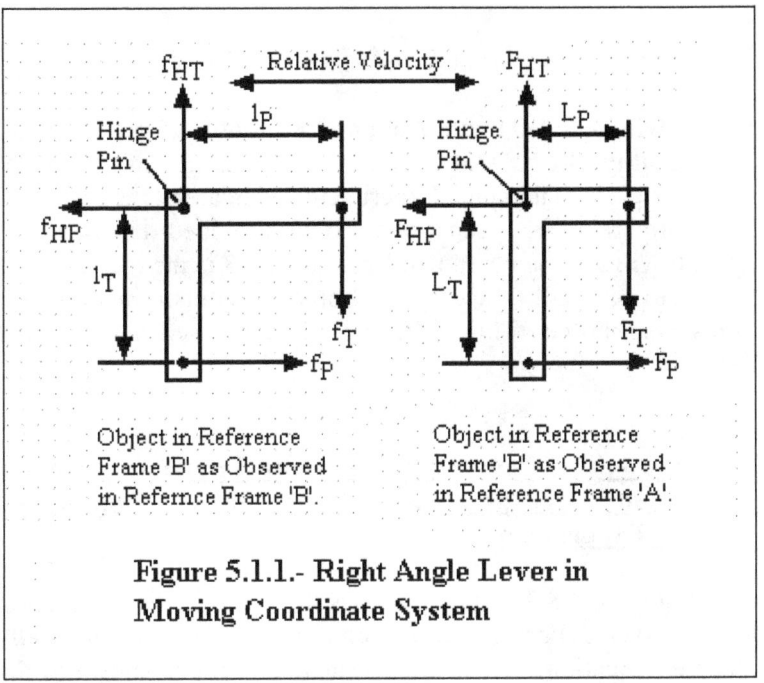

Figure 5.1.1.- Right Angle Lever in Moving Coordinate System

5.5 - The angular momentum of the lever is the product of its moment of inertia and its angular velocity, Since the lever is observed not to rotate in either reference frame A or B, its angular velocity remains at zero in both reference frames as does the rate of change of that angular velocity. The zero rate of change of angular velocity means that the rate of change of angular momentum is also zero in both reference frames. Since the rate of change of angular momentum is zero in both reference frames, the net torque applied to the lever in both reference frames must also be zero. We may write, therefore:

Eq. 5.5.1 $F_T * L_P = F_P * L_T$

Eq. 5.5.2 $f_T * l_P = f_P * l_T$

Combining Equations 3.3.1, 3.3.1, 4.5.3, 5.5.1, and 5.5.2 provides:

Eq. 5.5.3 $\quad f_T = F_P*(1-V^2/C^2)^{0.5}$

5.6 - It will be noted that Equations 4.8.3 and 5.5.3 are identical. This identity shows that a rigorous application of the Right Angle Lever Thought Experiment validates the Lorentz Transformation for Transverse Force provided by Equation 4.8.3. It will also be noted that Equation 4.8.3 is the reciprocal of the conventionally accepted Lorentz Transformation for Transverse Force. This difference is discussed in the next Section.

Section 6 - The Conventional Lorentz Transformation for Transverse Force as Related to the Right Angle Lever Thought Experiment

6.1 - Equation 4.8.3, which provides the Lorentz Transformation for Transverse Force, is the reciprocal of the conventionally accepted equation for that transformation. As was shown in the previous Section, Equation 4.8.3 is consistent with the results of a rigorous analysis of the Right Angle Lever Thought Experiment. It would appear, therefore, that an error was made in the derivation of the conventional Lorentz Transformation for Transverse Force, and rather than using the paradox introduced into the Right Angle Lever Thought Experiment by this error as an indication of its existence and responding by correcting the error, a rather convoluted and faulty reasoning process was employed to explain away the Right Angle Lever Paradox.

6.2 - Considerable effort was expended in the resolution of this conflict in the years following the publication of the Special Theory of Relativity. This result of this effort produced some rather surprising arguments. These arguments, which occur in more than one recognized text, are typified by the description of the Right Angle Lever Thought Experiment provided in one of the accepted texts on the subject which shall be unnamed (see note at the end of this Section). Summarized and converted to the terminology of this discussion, these statements are:

Eq. 6.2.1 $f_T = F_T/(1-V^2/C^2)^{0.5}$

The transformation of Equation 6.2.1 is the inverse of Equation 4.8.29.

As observed in reference frame A, the external forces applied to the right angle lever produce a torque, T_1 as given by:

Eq. 6.2.2 $T_1 = F_P * L_T * (1-V^2/C^2)$

The force, F_P, is doing work on the lever at the rate:

Eq. 6.2.3 $\delta E = F_P * V * \delta T$

The angular momentum, J, of the lever is increasing at the rate:

Eq. 6.2.4 $\delta J = -F_P * L_T * (1-V^2/C^2) * \delta T$

The increase in angular momentum produces a counter torque, T_2, of:

Eq. 6.2.5 $T_2 = -F_P * L_T * (1-V^2/C^2)^{0.5}$

The net torque, T_N, applied to the lever is given by:

Eq. 6.2.6 $T_N = T_1 + T_2$

Or:

Eq. 6.2.7 $T_N = 0$

And, as a result, the lever does not rotate.

6.3 - In analyzing the Right Angle Lever Thought Experiment, as described in conventional texts, one must conclude that:

- No energy is added to the lever as observed in reference frame A as a result of the application of the application of the force F_P. The energy added by F_P flows along the transverse arm of the lever and exits the lever as a result of the equal and opposite force, F_{HP}, existing at the hinge pin (Eq. 5.2). The effects of the force, F_{HP}, have been ignored in textbooks on Special Relativity.

- Since the lever is observed not to rotate in either reference frame its angular momentum is constant at a value of zero since its angular momentum is the product of its moment of inertia and its angular velocity. The rate of change of its angular momentum is therefore also equal to zero.

  - Any net torque applied to the Right Angle Lever by the forces F_P and F_T cannot be compensated by a change in angular momentum of the right angle lever. The torque resulting from a change in angular momentum is equal to zero.

6.4 - The Right angle Lever Thought Experiment requires that the torques applied to the lever by F_P and F_T must be equal and opposite unless:

- As observed in reference frame A, the flow of energy along the transverse arm of the lever produces a torque couple acting on it which is independent of the couple represented by F_P and F_{HP}.

  - A parallel force exerted on the transverse arm by the flow of energy would not satisfy the requirement. It would result in a readily observable inequality between F_P and F_{HP}.

  - For such a couple to exist, there must be an entity for the couple to react against. No such entity has been proposed.

6.5 - Since the torques applied to the lever by F_P and F_T must be equal and opposite, it follows that Equation 4.8.3 is the correct expression for the Lorentz Transformation for Transverse force and the presently accepted transformation (Equation 6.2.1) is incorrect. (The author does not know the nature of the error in the derivation of Equation 6.2.1 since he has not examined the derivation, nor is he interested. The fact that the error(s) produce an invalid result is sufficient to show that at least one error exists. It is unfortunate that theoreticians have not spent as much time searching for the error(s) as they did in explaining away the Right Angle Lever Paradox.)

Note:- The text referred to is not named because the error involved is of a type which would not be excused if made by a student of freshman physics. For the error to be included in texts which are used for teaching and which are written by individuals possessing PhD's in Physics is appalling. It is not, however, the author's wish to embarrass individuals by naming them.

Section 7 - The Lorentz Transformation for Parallel and Transverse Force as Related to a Compressed Spring Thought Experiment.

7.1 - The revision of the Lorentz Transformation for Transverse Force provided by Equation 4.8.29 satisfies two requirements:

- It is consistent with a rigorous analysis of the Right Angle Lever Thought Experiment.

- It is consistent with the effects which must be observed when an object (particle) having a velocity with respect to the observer acquires an additional incremental velocity which is perpendicular to the original velocity. In the observer's reference frame:

- The magnitude of the relative velocity will increase incrementally to produce an incremental increase in the relativistic mass of the object (particle).

- The energy ($E=M*C^2$) represented by the increase in relativistic mass equals the incremental work done in the observer's reference frame in producing the incremental transverse velocity change.

On casual evaluation, the revision of the Lorentz Transformation for Transverse Force would seem to lead to a different difficulty, as shown by the Compressed Spring Thought Experiment illustrated in Figure 7-1-1.

7.2 - Consider an object (particle) in which energy is stored anisotropically. Such an object might be a compressed spring as illustrated in Figure 7-1-1. In this thought Experiment, two Springs are compressed and tied to store energy. One spring is oriented so as to store its energy in a direction parallel to the velocity vector between reference frame A and B. The other spring is orientated so as to store its energy in a direction transverse to that velocity vector. For simplicity, it will be assumed that the energy storage and compressed distance of both springs are identical, as observed in reference frame B.

7.3 - As observed in reference frame B, the energy stored in the transverse spring is:

Eq. 7.3.1 $\quad e_T = d_T * f_T / 2$

And the energy, e_P, stored in the parallel spring is:

Eq. 7.3.2 $\quad e_P = d_P * f_P / 2$

As observed in reference frame A, the energy, E_T, stored in the transverse spring is:

Eq. 7.3.3 $\quad E_T = D_T * F_T / 2$

And the energy, E_P, stored in the parallel spring is:

Eq. 7.3.4 $E_P = D_P * F_P / 2$

Where:

- 'D' is the deflection of a spring as observed in reference frame A.

- 'd' is the deflection of the same spring as observed in reference frame B.

- 'E' is the energy stored in a spring as observed in reference frame A.

- 'e' is the energy stored in the same spring as observed in reference frame B.

All other symbols and subscripts are as previously defined. Since the symbols 'D' and 'd' refer to lengths and the symbols 'F' and 'f' refer to forces, the Lorentz Transformations for Parallel and Transverse Lengths and Forces apply.

Since it has been arbitrarily defined that, as observed in reference frame B, the stored energy and the compression distance of the parallel and transverse springs are equal, we may write:

Eq. 7.3.5 $e_T = e_P$

Eq. 7.3.6 $f_T = f_P$

Eq. 7.3.7 $d_T = d_P$

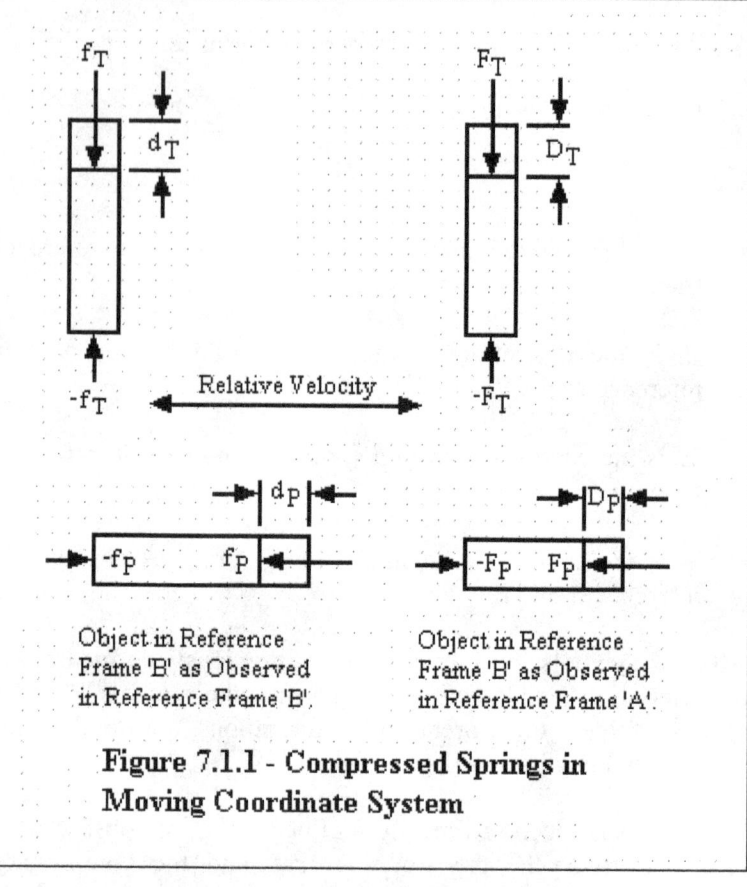

Figure 7.1.1 - Compressed Springs in Moving Coordinate System

Combining Equations 3.3.1, 3.3.2, 4.5.3, 4.8.3, 7.3.1, 7.3.2, 7.3.3, 7.3.4, 7.3.5, 7.3.6 and 7.3.7 provides:

Eq. 7.3.8 $E_P = E_T*(1-V^2/C^2)$

Eq. 7.3.9 $e_P = E_P/(1-V^2/C^2)^{0.5}$

Eq. 7.3.10 $e_T = E_T*(1-V^2/C^2)^{0.5}$

7.4 - In combination, Equations 7.3.5 and 7.3.8 appear to pose a dilemma. Special Relativity was derived based upon the principle that all velocity reference frames within the limits of +/-C are equally valid as a basis for all physical measurements (Principle of Relativity). Equation 7.3.8 shows that, as observed in reference frame A, the compressed spring that is parallel to the relative velocity vector has stored less energy than did the compressed spring which is transverse to the relative velocity vector. Equation 7.3.5 shows that, as observed in reference frame B, the energies stored in the two springs are equal.

7.5 - As observed in reference frame A, the transverse spring loses stored energy as it is rotated to the parallel direction, As observed in reference frame B it does not lose stored energy. One would expect that a loss of energy, as observed in reference frame A, stored in the spring as a result of the change in its orientation would result in a torque which would tend to rotate the spring towards the parallel orientation. The existence of such a torque in reference frame A and its non-existence in reference frame B would compromise the Principle of Relativity by allowing a 'preferred' velocity reference frame to be established.

7.6 - The effect suggested by the previous paragraph should not occur. In order for an object (particle) to acquire a velocity with respect to reference frame A, kinetic energy must be added to the object (particle) and the mass of that object (particle) is increased by the amount of the kinetic energy added in accordance with $E=M*C^2$. When the object (particle) is eventually brought to rest with respect to reference frame A, it gives up that kinetic energy and reverts to its original rest mass. One would conclude, therefore, that the energy is stored in the moving object (particle) and/or in the space in the immediate vicinity of the moving object (particle). One is then faced with the question as to where and how that kinetic energy is stored. To investigate that question, let us consider that the rest mass of an object (particle) results form the storage of energy within the object (particle) in a manner equivalent to the storage of energy in the parallel and transverse springs of Figure 7.1.1.

7.7 - Let us first consider the storage of kinetic energy in the spring which is transverse to the velocity vector, as revealed by Equation 7.3.10. Dividing this Equation by C^2 provides:

Eq.7.7.1 $m_T = M_T*(1-V^2/C^2)$

Since Equation 7.7.1 is the conventional Lorentz Transformation for Mass, it follows that, as observed in terms of reference frame A, the kinetic energy of energy stored along a transverse axis of the object (particle) in reference frame B is stored as an increase of the stored energy along that axis. This increase in stored energy results from an increase in stiffness of the 'spring' which stores the 'rest mass' energy. (The deflection of the 'spring' is the same in both reference frames while the force, in reference frame A, producing the deflection has increased. This is equivalent to an increase in the 'stiffness' of the 'spring'.)

7.8 - The storage of the kinetic energy of the compressional energy stored in the parallel spring is not so readily explained. Dividing Equation 7.3.9 by C^2 provides:

Eq.7.8.1 $m_P = M_P/(1-V^2/C^2)$

Since this Equation is the reciprocal of the Lorentz Transformation for Mass, it indicates that the relativistic mass of the compressional energy of the parallel spring decreases as observed in reference frame A as the velocity of reference frame B with respect to A is increased. The decrease in the relativistic mass of the compressed energy in the parallel spring results from a decrease in the deflection of the spring, as observed in reference frame A, while the force applied to the spring remains unchanged. Unlike the transverse spring, the parallel spring does not store the kinetic energy associated with its compressional energy within itself. In terms of reference frame A, its kinetic energy and part of its rest mass energy must be stored in the space (Figure 7-8-1) around the spring since that energy is transported with the spring but is not stored in the spring itself. (It should be noted that such a conclusion would also have been erroneously attributed to the transverse spring under the conventional Lorentz Transformation for Transverse Force, Equation 6.2.1)

7.10 - Since Equation 5.11 shows that, as observed in reference frame A, the storage of both rest mass and kinetic energy in the transverse spring, E_T, accounts for the relativistic mass of the energy stored in the transverse spring, it follows that:

Eq. 7.10.1 $E_{RT} = E_T$

Where:

- 'E_{RT}' is the energy equivalent of the transversely stored relativistic mass of the object (particle), as observed in reference frame A.

Similarly, the total energy stored in the parallel spring, E_P is given by:

Eq. 7.10.2 $E_P = E_{RP}*(1-V^2/C^2)$

In order for this to occur, the parallel spring must shed energy from within itself and carry that energy along with it in a disk shaped region of the space around it, with the disk oriented perpendicular to the velocity vector. The energy contained in that disk, as observed in reference frame A, must equal:

Eq. 7.10.3 $E_S = E_R*(V^2/C^2)/(1-V2/C^2)^{0.5}$

Where:

'E_S' is the energy, as observed in reference frame A, stored in the space in the vicinity of the object (particle) as a result of its velocity with respect to reference frame A.

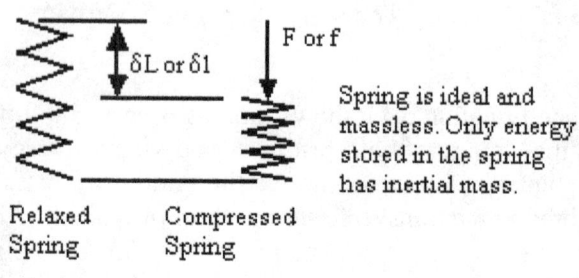

Spring is ideal and massless. Only energy stored in the spring has inertial mass.

Relaxed Spring   Compressed Spring

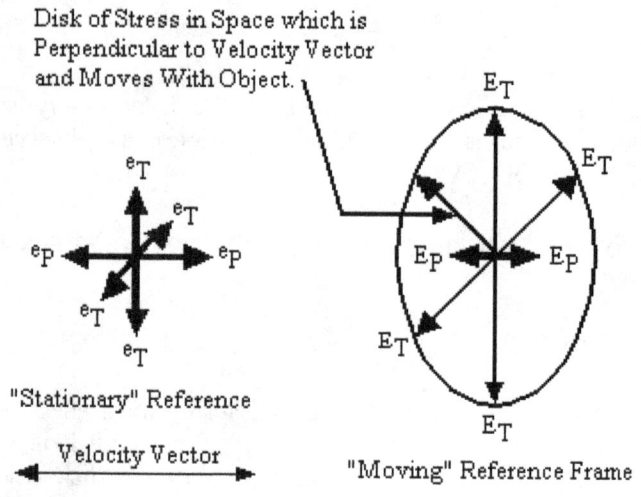

Object consisting of energy stored in three mutually perpendicular springs as observed in "moving" and "stationary" reference frames. Kinetic energy is stored in transverse springs. Kinetic energy plus stored energy leaves parallel spring and is transported in disk of "stressed space" in plane perpendicular to the velocity vector.

Figure 7.8.1 - The Location of Kinetic Energy

ABOUT THE AUTHOR

H. E. Retic (a penname chosen to protect his family from abuse by the academic community which postings to Newsgroups suggest would occur) is a retired Mechanical Engineer who graduated from Cornell University in 1948.

The author first worked for the world's largest manufacturer of public utility steam generators equipment and designed the oil firing equipment that is used unchanged today. In the mid 1950's, he switched to the military electronics field where, in a 9 month period, he designed and built the stable platform of the inertial guidance system used in the X-15 aircraft. Although his initial assignment was the mechanical design, he also designed its electronics when it became obvious that the electronic engineers assigned to the task did not have the versatility to design circuitry based upon the emerging power transistor technology.

In the late 1960's the author designed for sale to the Navy the first Low-Light Level Television System capable of operating from full daylight to the photon noise limit threshold without picture degradation. This was a goal that the Night Vision Laboratory at Ft. Belvoir had declared to be impossible after expending $150,000,000 with two major companies in the effort. The author achieved this result by the simple expedient of reading and understanding the data sheets for the components employed. Unfortunately for those involved, Low Light Level Television was made obsolete by infra-red technology and the achievement died. An investigation by Pentagon personnel as to how our company could succeed where the Night Vision Laboratories had failed led to the comment that "the explanation is obviously correct but we still can't believe that the solution is that simple."

The author holds over 25 patents and in one year was granted 10% of the patents issued to a major military contractor having over 5000 employees. His overall list of patents range from fuel firing equipment through military electronic equipment (including a means for making the use of highly accurate navigation systems of ballistic missile submarines unnecessary) to a practical gear assisted continuously variable transmission for automobiles which would significantly improve fuel economy and reduce the size of engines required to achieve a given level of acceleration.

To provide an intellectual outlet, the writer began to study gravitation in the mid 1960's and succeeded in deriving the material presented in the Appendix "Gravity" using a rigorous method based upon easily tested and accepted principles and which avoided the use of sophisticated mathematics and the obfuscation and opportunity for error that its usage incurs. The availability of word processors in the 1980's made the writing of "Gravity" practical and the availability of the Internet and the authors retirement made the publication of the "The Einstein Hoax" practical in the mid 1990's.

www.ingramcontent.com/pod-product-compliance
Lightning Source LLC
Chambersburg PA
CBHW021232100125
20189CB00011B/134